本书出版得到国家自然科学地区基金（31260634）、吉首大学重点学科建设项目、武陵山动物生理生态与动物产业科技创新团队项目、武陵译学丛书项目资助

武陵译学丛书

蛙病毒

——变温脊椎动物的致死性病原

Ranaviruses:
Lethal Pathogens of Ectothermic Vertebrates

马修·J.格雷　V.格雷戈里·钦察尔　编著

卓琦　译

彭小宁　蒋林　刘汝荣　审译

湖南大学出版社·长沙

著作合同登记号：图字：18-2018-163 号　　审图号：GS（2019）4565 号

图书在版编目（CIP）数据

蛙病毒：变温脊椎动物的致死性病原/（美）马修·J. 格雷（Matthew J. Gray），（美）V. 格雷戈里·钦察尔（V. Gregory Chinchar）编著；卓琦译. —长沙：湖南大学出版社，2020. 7

ISBN 978-7-5667-1545-6

Ⅰ. ①蛙…　Ⅱ. ①马… ②V… ③卓…　Ⅲ. ①蛙科—动物病毒病—研究　Ⅳ. ①S947. 2

中国版本图书馆 CIP 数据核字（2018）第 110647 号

蛙病毒——变温脊椎动物的致死性病原

WA BINGDU——BIANWEN JIZHUI DONGWU DE ZHISIXING BINGYUAN

编　　著：马修·J. 格雷　V. 格雷戈里·钦察尔

译　　者：卓　琦

责任编辑：邓素平

印　　装：湖南省众鑫印务有限公司

开　　本：710 mm×1000 mm　1/16　**印张**：20　**字数**：381 千

版　　次：2020 年 7 月第 1 版　**印次**：2020 年 7 月第 1 次印刷

书　　号：ISBN 978-7-5667-1545-6

定　　价：78. 00 元

出 版 人：李文邦

出版发行：湖南大学出版社

社　　址：湖南·长沙·岳麓山　**邮　　编**：410082

电　　话：0731-88822559(发行部),88822264(编辑室),88821006(出版部)

传　　真：0731-88649312(发行部),88822264(总编室)

网　　址：http://www. hnupress. com　**电子邮箱**：susie_press@ 126. com

“武陵译学丛书”总序

白晋湘

提起武陵，中国人大概都会想起陶渊明的《桃花源记》，因为那篇名文开首便说是武陵的一位渔人发现了桃花源这个世间美好的所在。在这里，谈论武陵渔人和桃花源不是为了寻找本区域的光荣史，而是因为桃花源与翻译这门学科密切相关。

桃花源里面的人“与外人间隔”，“不知有汉”。桃花源人与外界存在隔膜是因为桃花源人拒绝与外界沟通。他们告诉那位渔人，“不足为外人道也”。而我们之所以知道桃花源的存在，是因为那位渔人走出桃花源后，对外面的人“说如此”。我想，“说如此”就是沟通的开始，而沟通也是翻译的最高追求。武陵渔人最终没有完成沟通，他带人去寻找桃花源时，“不复得路”。武陵渔人没有找到去桃花源的路，而这正是今天的我们应该继续的事业，以翻译沟通世界。没有沟通的世界，就不会有人类的桃花源。

晚清的林纾先生翻译的《茶花女》，曾让当时的国人震惊于“外国人也有用情如此之专者”，同样告诉我们沟通的必要与紧迫。

20 世纪 60 年代，加拿大学者 M. 麦克卢汉创造了一个现在举世皆知的词语——地球村，来说明科技对人类的影响。无论是从共时性还是从历时性的角度来看，科技进步确实缩短了人与人之间的距离。但是，缩短绝对不等同于消除。只要距离还存在，《圣经》里所讲的“天下人都讲一样的语言”就只能是一个想象中的桃花源。

面对距离，人类需要沟通。人类要沟通，就需要翻译这门学科。从这个角度而言，翻译是一项人类永恒需要的事业；每一个从事翻译的人，都是通向人类能够顺畅沟通的美好世界的奠基人。

“异域文术新宗，自此始入华土”。1909 年，鲁迅为他和周作人翻译的《域外小说集》写的序言中的这句话发人深省。鲁迅等近现代中国第一代翻译人用这种大气魄为中国翻译事业树立了一个光辉的典范。

对于鲁迅这代翻译人来说，翻译事关民族的变革与发展，是一项让中国人了解世界，行走在整个世界“进化”道路上的大事业。确实如此，翻译将一种语言转换为另一种语言，不仅是教育的一个专业和学术的一个领域，还拥有着更远大的追求。也因为有着更远大的追求，翻译才成为一项伟大的事业。包括鲁迅在内的近现代中国最早的翻译者是中国的普罗米修斯，他们通过翻译将域外的火种带到中国，让我们这个古老的民族凤凰涅槃，实现浴火重生。

21 世纪的中国翻译学者们已接过先贤手里的火种，正着力推进中国翻译事业的发展。他们一方面将国外文明引入中国，另一方面将中国文明推介出去。作为武陵山区唯一的一所综合性大学，吉首大学有责任、有义务发展该区域的翻译事业，将沈从文等文学大师的作品译介到更多的民族和国家，同时也把武陵山区神奇的自然风光、悠久的历史文化、浓郁的民族风情推向全世界。

2017 年 11 月 21 日

于吉首大学凤凰楼

“武陵译学丛书”主编絮语

吉首大学是湖南省属综合性大学，是武陵山区规模最大、实力最强、层次最高、影响最广的综合性大学。2003 年获得硕士学位授予权，2012 年被确定为中西部高校基础能力建设单位，同年获得“服务国家特殊需求博士人才培养项目”，为实现转型发展战略，学校在“十三五”期间全面启动博士学位授权点的申报和建设工作。2016 年，学校再次被确定为中西部高校基础能力建设单位。

吉首大学外语本科办学历史悠久，人才培养质量高。从 1979 年开始招收英语本科学生，现有英语、翻译、商务英语和日语四个本科专业。每个专业师资力量雄厚，专业设置合理，教学质量有保障，人才培养有特色。近年来，吉首大学外语专业学生在各类各级专业技能竞赛中成绩优秀，名列前茅，在“中国大学生莎士比亚戏剧大赛”摘过银奖，在“湖南省大学英语演讲比赛”中夺过冠军。本科办学 39 年来，为社会各界培养了“下得去、留得住、用得上、干得好”的各类人才 5 000 余名，据不完全统计，其中 20 多位校友已经成为海内外具有影响力的行业引领人才。此外，吉首大学外语学科 2016 年申报的翻译硕士专业学位授权点成功获批，2017 年开始招收第一届翻译硕士专业学位研究生。这将为武陵山区的资源优势转化为经济优势储备人才，让武陵山区蕴含的特色旅游资源和民族文化资源等转化为国际竞争力，为武陵山区扶贫攻坚和打造湖南省旅游强省提供原动力。

吉首大学外语学科现有专任教师 114 人，其中高级职称 45 人，博士 18 人（含在读）。外语学科积淀深厚，学术水平稳步提升。多年积淀为外语学科夯实了基础、充实了内涵、提升了实力。近年来，通过主办“民族地区外语学科发展学术研讨会”“全国外语界面研究高层论坛”“湖南省翻译协会年会暨学术研讨会”等各种专业学术会议以及邀请国内相关领域知名专家学者来校讲学等方式，进一步活跃了学术氛围，助推了学科建设，催生了高水平的学术成果。2011 年至 2016 年，我校外语学科教师获得包括“中华学术外译”在内的国家社科基金项目6项；在《外语教学与研究》《中国翻译》等权威期刊上发表论文8篇，出版学术专著、译著和教材20余部。

为了把外语学科建设成为武陵山片区一流的、有影响力的学科，进一步丰富学科内涵，凝练学科特色，依据学校“立足大湘西，服务大武陵”的办学定位，充分考虑武陵山区经济、社会、文化发展的实际需求，鼓励外语学科教师致力于学术研究和翻译实践，产出一批高水平的研究成果和彰显本土文化特色的优秀翻译作品，白晋湘校长不仅亲自拟定“武陵译学丛书”名称，每年从校长专项经费中拨付出版经费，而且还就著作出版质量等事宜提出了很多切实可行的建议。在他的关心与支持下，我们一方面整合外语学科现有研究力量，聚焦学术前沿话题，成立科研团队，合力攻关具有重大研究意义的课题，推出一批高层次、高水平的学术专著；另一方面，我们以武陵山区丰富的旅游资源和灿烂的民族文化为依托，响应党的十七届六中全会提出的中国文化“走出去”伟大战略，将本土优秀的文化产品译介到国外，用翻译的方式对大湘西地区乃至整个武陵山区的民族文化和旅游资源进行保护、传承和弘扬。

我们相信，通过五至八年的建设，“武陵译学丛书”一定会助力外语学科结出累累硕果！

2017年11月22日

于吉首大学逸夫楼

译者序

蛙病毒是隶属虹彩病毒科蛙病毒属的一类感染鱼类、两栖类动物和爬行类动物，并在全球范围内新出现的病原。该类病毒既能感染人工饲养的动物，也能感染野生动物，并能在同一地区或同一地点感染多种动物，在变温脊椎动物不同物种间传播。目前已有报道显示，蛙病毒至少能感染52科175种变温脊椎动物。全世界除南极洲以外，其他六大洲都有蛙病毒的分布。蛙病毒的地理分布及宿主范围正逐步扩大，蛙病毒有可能引起种群数量下降和灭绝。因此，蛙病毒对地理隔离的宿主物种或低丰度物种存在重大威胁，并在国际贸易领域获得极大关注。

蛙病毒病作为新出现的传染病对全球生物多样性构成了严重威胁。同时，由于在历史上人们忽视了蛙病毒属的虹彩病毒类群是过去20年内全球野生和人工饲养的两栖类、爬行类和鱼类种群大规模死亡的元凶，为了共享当代蛙病毒研究的技术资料并确定今后的研究方向，世界各国的科学家针对蛙病毒—宿主相互关系的生态学、进化、潜在的贮藏宿主、传播动力学以及对蛙病毒感染的免疫学和组织病理学反应展开了研究，并取得了惊人的成就。

蛙病毒病与两栖类动物种群数量的下降有着密切的关系。蛙病毒属的一群病毒在美洲、欧洲和亚洲造成了两栖类动物大规模死亡。这些两栖类动物栖息地几乎遍布所有的纬度和海拔高度。蛙病毒引起的死亡，常常发生在物种丰度高且分布比较广泛的物种中，并且还会引起全球局部地区珍稀动物的死亡。蛙病毒也能造成许多爬行类物种和鱼类的发病死亡，并通过一年一度大规模死亡的反复发生造成种群数量下降甚至局部灭绝。蛙病毒呈地方性流行，在不同的病毒分离株之间有着显著的地理分布上的差异，甚至会在地理隔离的宿主种群中使病毒的致病性增强。研究进一步发现，从水产养殖场和饵料商店所得到的病毒分离株，相对于野外分离株来说毒力更强。因此，蛙病毒的毒性常常在变温脊椎动物长距离运输之后增强，有时也会通过病毒子污染物增强。人为的应激因子同样会诱发蛙病毒病的发生。蛙病毒病发生在一系列不同的变温脊椎动物中，并导致全球两栖类动物种群数量下降。研究表明，全球变温脊椎动物物种的商业贸易以及应激因子（或压力源）是蛙病毒病出现的原因。

蛙病毒病作为新出现的动物传染性疫病正席卷全球，不仅危害着野生变温脊椎动物的安全，给全球动物生态安全带来极大的威胁，还给鱼类、两栖类和爬行类等特种动物养殖业带来了灾难性打击，造成不可挽回的损失。近年来，蛙病毒属的病毒所引起的爬行类动物、两栖类动物和鱼类疾病已在美洲、欧洲、亚洲和大洋洲等地普遍流行，并引起全世界各国学者的广泛关注。据报道，蛙病毒病已成为制约中国水产养殖业发展的主要因素。1995 年，中国科学院水生生物研究所张奇亚课题组首次从患有传染性溃疡致死综合征的沼泽绿牛蛙（*Rana grylio*）中观察并分离到一株病原病毒，称之为 RGV（Rana grylio virus），即 RGV9506。该研究团队在对其做了超微观察和细胞感染的基础上，分析了病毒的形态发生及与宿主细胞间的相互作用，并确定为蛙病毒。1998 年，该研究团队又在湖南、湖北两地，从处在不同生长期、患有相似病症的蛙中分离到另外 2 株病原病毒，分别称为 RGV9807 和 RGV9808。经超微观察、血清学相关性检测等，明确了 3 个 RGV 分离株都属于蛙病毒。之后从濒危的中国大鲵（*Andrias davidianus*）中分离到了一种蛙病毒，该病毒是从人工饲养的大鲵种群中分离的，在所有中国大鲵的病例中，都出现了高发病率和高死亡率。系统发生分析显示，从中国大鲵人工养殖种群中分离的这种病毒与普通产婆蟾病毒（CMTV）最为接近。曾令兵博士对多个中国大鲵养殖场进行了研究，并对中国 11 个省的该病毒进行了记录。在中国，该蛙病毒的出现，对中国大鲵的保护构成了严重的威胁。以后，国内不少学者对不同种水生经济动物的蛙病毒进行了报道。

经过多年的努力，国内学者已在不同地域相继从中华鳖（*Trionyx sinensis*）、中国大鲵、鞍带石斑鱼（*Epinephelus lanceolatus*）、斜带石斑鱼（*Epinephelus coioides*）、巨石斑鱼（*Epinephelus tauvina*）、大口黑鲈（*Micropterus salmoides*）、大菱鲆（*Scophthalmus maximus*）、似鲶高原鳅（*Triplophysa siluroides*）、东北林蛙（*Rana dybowskii*）、沼泽绿牛蛙、美洲牛蛙（*Rana catesbiana*）、虎纹蛙（*Rana tigrina*）、黑斑蛙（*Rana nigromaculata*）、棘胸蛙（*Quasipaa spinosa*）、大鳄龟（*Macrochelys temminckii*）等动物中分离和鉴定了不同的蛙病毒株，并展开了蛙病毒分类学、生物学、生态学、病理学、免疫学、流行病学及分子生物学等方面的研究，为蛙病毒疾病的发生机制、预防和治疗提供了综合性的基础资料。

蛙病毒在我国正以不可阻挡之势威胁着变温脊椎动物养殖产业的发展，特别是水产养殖业，其中对石斑鱼产业、食用蛙养殖产业、中华鳖养殖产业和大鲵养殖产业影响最大。中国大鲵属两栖纲、有尾目、隐鳃鲵科，是现存最大的珍贵两栖动物，为我国特有珍稀物种，已被列为Ⅱ类国家重点保护野生动物。世界自然保护联盟物种生存委员会（IUCN/SSC）和《中国物种红色名录》均将大鲵的受威胁现状列为极危级（CR）。我国于20世纪70年代开始大鲵驯养繁殖试验，90年代后，随着水生野生动物特许利用政策的明确和大鲵人工繁殖的成功，在坚持“保护中开发，开发中促保护”的理念下，大鲵驯养、繁殖蓬勃发展，形成了“原生态保护、仿生态培育、人工繁殖、集约化养殖”的发展模式。在国内，不少学者将蛙病毒称为虹彩病毒。实际上，蛙病毒不是单一病毒种，而是虹彩病毒科蛙病毒属的一个病毒类群。国内先后有4个研究小组都证明了在国外引起广泛关注并导致鱼类、两栖类和爬行类动物感染原虹彩病毒科的蛙病毒在驯养大鲵中致病。该病毒种群对野生和驯养的中国大鲵具有生态学和经济学的意义，随着大鲵产业的发展，蛙病毒已造成中国大鲵的野生恢复减缓和驯养大鲵的经济损失，特别是2011年后，引起了驯养大鲵大量死亡，先后在陕西、四川、甘肃、湖南、重庆与贵州等地被发现。这种严重危害大鲵的传染性疾病，给各地的大鲵产业带来了毁灭性的打击。发病大鲵体表出现出血斑点、溃疡，头部与四肢肿胀，故这种病被称为“溃疡病”或“大脚病”等。该病一旦发生，常导致较高的死亡率，有的大鲵养殖场发病大鲵的死亡率甚至高达100%，给广大养殖户造成了严重的经济损失，故又被称为“大鲵癌症”。

张家界市是中国大鲵的主要原产地，也是我国大鲵国家级自然保护区所在地。依托良好的生态环境和优质的大鲵资源，并经过多年的技术攻关，张家界市取得了大鲵人工驯养繁殖技术、仿生态繁殖技术的成功。张家界市的大鲵产业也逐步成为全市农业提质升级的朝阳产业。2010年，国家质检总局批准对“张家界大鲵”实施国家地理标志产品保护。2013年，为证明特定品质，打造知名品牌，“张家界大鲵”经国家工商行政管理总局注册为地理标志证明商标。2016年，张家界市大鲵产业养殖规模持续扩大，市场销售快速增长，产品研发投资加大，产业融合不断深入，市场销售价格趋于稳定。目前，张家界市大鲵资源型产业给区域经济的发展带来了前所未有的机遇，同时也面临着极大的挑战，特别是面临着蛙病毒带来的重大威胁。国内有学者先后从张家界市不同地域的养殖场分离和鉴定了不同的蛙病毒株。这些蛙病毒给不少企业和养殖户带来了灭顶之灾。译者亲临其境，倍感伤叹，在进行综合防治技术指导的同时，也在查找相关的技术资料，偶遇国外学者编著的《蛙病毒——变温脊椎动物的致死性病原》一书。该书从蛙病毒的历史与未来、蛙病毒的分布与宿主范围、蛙病毒的分类学及种系发生、蛙病毒生态学和进化、蛙病毒复制、蛙病毒的宿主免疫与免疫逃避、蛙病毒的比较病理学及诊断技术、蛙病毒研究的设计与分析共八个方面对国内外蛙病毒研究的前沿进行了综述。译者对该书爱不释手，便将其翻译成中文，以飨读者，旨在为国内蛙病毒的研究和蛙病毒病的防治提供参考。为了方便读者联系，本译著脚注中的作者及通信地址，均未翻译。不当之处，请广大读者指正。

译者：卓琦

2017 年 10 月 18 日

“武陵译学丛书”专家委员会

蛙病毒是一群新出现的致病源，该病原导致了全球范围内人工饲养和野生两栖类、鱼类和爬行类动物种群的大规模死亡。全球变温脊椎动物物种的商业贸易以及应激因子（或压力源）有可能是这些蛙病毒病出现的原因。图片（从顶部左侧开始顺时针方向）由马修·阿蓝德（Matthew Allender）、纳撒尼尔·惠尔赖特（Nathaniel Wheelwright）、马修·娜米勒（Matthew Neimiller）、乔纳森·科尔比（Jonathan Kolby）、耿毅（Yi Geng）、耿毅、乔纳森·科尔比和罗兰多·马佐尼（Rolando Mazzoni）提供；中间插图由珍妮·琼斯（Jeanne Jones）提供。

目　次

撰稿作者

马修·C. 阿蓝德 比较生物科学系 兽医学院 伊利诺伊大学 乌尔班纳 伊利诺伊州 美国（Matthew C. Allender, Department of Comparative Biosciences, College of Veterinary Medicine, University of Illinois, Urbana, IL, USA）

弗朗西斯科·德·赫苏斯·安迪诺 微生物学与免疫学系 罗彻斯特大学医学中心 罗彻斯特市 纽约 美国（Francisco De Jesús Andino, Department of Microbiology and Immunology, University of Rochester Medical Center, Rochester, NY, USA）

埃伦·阿里尔 公共卫生与医学和兽医科学学院 詹姆斯库克大学 汤斯维尔 昆士兰州 澳大利亚（Ellen Ariel, College of Public Health, Medical and Veterinary Sciences, James Cook University, Townsville, QLD, Australia）

杰西·L. 布伦纳 生物科学学院 华盛顿州立大学 普尔曼市 华盛顿州 美国（Jesse L. Brunner, School of Biological Sciences, Washington State University, Pullman, WA, USA）

V. 格雷戈里·钦察尔 微生物学系 密西西比大学医学中心 杰克逊市 密西西比州 美国（V. Gregory Chinchar, Department of Microbiology, University of Mississippi Medical Center, Jackson, MS, USA）

阿曼达·L. J. 达弗斯 生物学系 戈登州立大学 巴恩斯维尔市 佐治亚州 美国（Amanda L. J. Duffus, Department of Biology, Gordon State College, Barnesville, GA, USA）

朱莉娅·E. 厄尔 国家数学与生物综合研究所 田纳西大学 诺克斯维尔市 田纳西州 美国（Julia E. Earl, National Institute for Mathematical and

Biological Synthesis, University of Tennessee, Knoxville, TN, USA)

伊娃-斯蒂娜·埃德霍尔姆 微生物学和免疫学系 罗彻斯特大学医学中心 罗彻斯特市 纽约 美国（Eva-Stina Edholm, Department of Microbiology and Immunology, University of Rochester Medical Center, Rochester, NY, USA）

迈克尔·戈特斯曼 传染病和病理学系 动物医学学院 佛罗里达大学 盖恩斯维尔市 佛罗里达州 美国（Michael Gotesman, Department of Infectious Diseases and Pathology, College of Veterinary Medicine, University of Florida, Gainesville, FL, USA）

马修·J. 格雷 野生动物健康中心 林业、野生动物和渔业系 田纳西大学 诺克斯维尔市 田纳西州 美国（Matthew J. Gray, Center for Wildlife Health, Department of Forestry, Wildlife and Fisheries, University of Tennessee, Knoxville, TN, USA）

利昂·格瑞福 微生物学和免疫学系 罗彻斯特大学医学中心 罗彻斯特市 纽约 美国（Leon Grayfer, Department of Microbiology and Immunology, University of Rochester Medical Center, Rochester, NY, USA）

保罗·希克 兽医学院 悉尼大学 悉尼 新南威尔士市 澳大利亚（Paul Hick, Faculty of Veterinary Science, University of Sydney, Sydney, NSW, Australia）

梅根·K. 海恩斯 兽医学院 威斯康星大学 麦迪逊市 威斯康星州 美国（Megan K. Hines, School of Veterinary Medicine, University of Wisconsin-Madison, Madison, WI, USA）

杰森·T. 霍韦尔曼 森林与自然资源系 普渡大学西拉法叶校区 印第安纳州 美国（Jason T. Hoverman, Department of Forestry and Natural Resources, Purdue University, West Lafayette, IN, USA）

詹姆斯·K. 贾柯维奇 生物科学系 加利福尼亚州立大学 圣马科斯 加利福尼亚州 美国（James K. Jancovich, Department of Biological Sciences, California State University, San Marcos, CA, USA）

雷切尔·E. 马尚 纳博科林动物临床检验实验室股份有限公司 巴特基

辛根　德国（Rachel E. Marschang，Laboklin GmbH & Co. KG，Bad Kissingen，Germany）

黛布拉·L. 米勒　野生动物健康中心及生物医学与诊断科学中心　田纳西大学　诺克斯维尔市　田纳西州　美国（Debra L. Miller，Center for Wildlife Health and Department of Biomedical and Diagnostic Sciences，University of Tennessee，Knoxville，TN，USA）

艾伦·P. 伯西尔　保护研究所　圣地亚哥动物园　圣地亚哥市　加利福尼亚州　美国（Allan P. Pessier，Institution for Conservation Research，San Diego Zoo，San Diego，CA，USA）

秦启伟　热带海洋生物资源与生态重点实验室　中国科学院南海海洋研究所　广州市　中国（Qiwei Qin，Key Laboratory of Tropical Marine Bio-Resources and Ecology，South China Sea Institute of Oceanology，Chinese Academy of Sciences，Guangzhou，China）

雅克·罗伯特　微生物学和免疫学系　罗彻斯特大学医学中心　罗彻斯特市　纽约　美国（Jacques Robert，Department of Microbiology and Immunology，University of Rochester Medical Center，Rochester，NY，USA）

纳塔利·K. 斯特克勒　传染病和病理学系　动物医学学院　佛罗里达大学　盖恩斯维尔市　佛罗里达州　美国（Natalie K. Steckler，Department of Infectious Diseases and Pathology，College of Veterinary Medicine，University of Florida，Gainesville，FL，USA）

安克·C. 索尔　环境与动物卫生研究所　霍恩海姆大学　斯图加特市　德国（Anke C. Stöhr，Institute of Environmental and Animal Hygiene，University of Hohenheim，Stuttgart，Germany）

安德鲁·斯托福　生物科学学院　华盛顿州立大学　普尔曼市　华盛顿州　美国（Andrew Storfer，School of Biological Sciences，Washington State University，Pullman，WA，USA）

托马斯·B. 华尔兹克　传染病和病理学系　动物医学学院　佛罗里达大学　盖恩斯维尔市　佛罗里达州　美国（Thomas B. Waltzek，Department of Infectious Diseases and Pathology，College of Veterinary Medicine，University of Florida，

Gainesville, FL, USA)

理查德 · J. 惠廷顿 兽医学院 悉尼大学 悉尼 新南威尔士市 澳大利亚（Richard J. Whittington, Faculty of Veterinary Science, University of Sydney, Sydney, NSW, Australia)

张奇亚 淡水生态与生物技术国家重点实验室 中国科学院水生生物研究所 武汉 湖北省 中国（Qiya Zhang, State Key Laboratory of Freshwater Ecology and Biotechnology, Institute of Hydrobiology, Chinese Academy of Sciences, Wuhan, Hubei, China)

前言　蛙病毒的历史与未来

马修·J. 格雷[1]和 V. 格雷戈里·钦察尔[2]

青蛙就像一只神奇的鸟儿，
它站着的时候几乎像坐着，
它跳跃的时候几乎像飞行，
它没有任何感觉，
它也没有尾巴，
当它坐着的时候，几乎就像没有坐在什么东西之上。

——匿名者

艾伦·格拉诺夫（Allan Granoff，1923—2012）在试图建立支持鲁克疱疹 1
病毒复制的细胞系时，第一次偶然分离到了蛙病毒（Granoff 等人，1966）。虽然艾伦分离到的蛙病毒 3（Frog virus 3，FV3）后来成为虹彩病毒科（Iridoviridae）和蛙病毒属（*Ranavirus*）最好鉴定的成员，但是该病毒暴发所产生的影响，在当时没有受到足够的重视。FV3 既不是第一种公认的低等脊椎动物致病源的虹彩病毒，也不是首次分离到的虹彩病毒。淋巴细胞囊肿病毒（Lymphocystis disease virus，LCDV）和无脊椎动物虹彩病毒 1（Invertebrate iridovirus 1，IIV1）是常见的致病因子。在鱼类中，LCDV 感染是常见的非致命性的疾病。除了损害外形外，该类疾病的主要特征是在皮肤及内部器官出现比较罕见的疣状物。而 IIV1 感染是大蚊幼虫（crane fly larvae）隐性感染和显性感染的病因。FV3 在研究中未处于主导地位。由于该病毒最初被认为与蛙腺癌有关，圣犹大医院为了完成研究使命，使蛙腺癌成为人类恶性肿瘤研究的模型，对 FV3 进行了研究。与 LCDV 和 IIV1 不同，FV3 在培养细胞中极易生长，

---

① M. J. Gray/Center for Wildlife Health, Department of Forestry, Wildlife and Fisheries, University of Tennessee, 274 Ellington Plant Sciences Building, Knoxville, TN 37996-4563, USA/e-mail: mgray11@ utk. edu.

② V. G. Chinchar/Department of Microbiology, University of Mississippi Medical Center, Jackson, MS 39216, USA/© The Author(s) 2015/M. J. Gray, V. G. Chinchar (eds.), Ranaviruses, DOI 10. 1007/978-3-319-13755-1_1.

因而容易进行详细的分子鉴定研究。虽然 FV3 在肿瘤发育中的作用很快得以证实是不正确的，但迄今为止，FV3 为弄清缺少研究的虹彩病毒科的复制策略提供了一个途径。此外，在后来的 20 年间，对 FV3 的研究不仅加深了对虹彩
2 病毒复制的重要认识，而且加深了对真核细胞生物学、病毒进化以及宿主病毒相互作用的重要认识。

为了阐明 FV3 复制的分子和细胞事件，艾伦与其同事以及在美国和欧洲地区的其他研究者，从 1965 年发现 FV3 开始直到 20 世纪 90 年代初，耗费了大量时间（Murti 等人，1985；Williams，1996）对 FV3 进行了相关研究。虽然在这些研究中获得了分子方面的重要认识，但是对 FV3 和虹彩病毒科的其他成员的研究，仍令人焦虑。人们最后弄清了无脊椎动物虹彩病毒如同杆状病毒一样不适合作为昆虫的生物控制原。FV3 和其他脊椎动物虹彩病毒开始被认为是致病轻微的病原体，因为几乎没有蛙病毒疾病暴发的报道，即使有报道，其对种群的影响也很小。此外，不像 LCDV，在生态学上和商业上重要的鱼类物种之间还缺乏 FV3 感染的证据。因此，即使是在最近强调转化研究（translational research）之前，虹彩病毒相对于医学上和商业上关注度高的痘病毒（Poxviruses）和疱疹病毒（Herpesviruses）来说，其研究地位相对靠后。

从 20 世纪 80 年代中叶开始，随着类似 FV3 但不完全等同于 FV3 的分离数量增加，这些分离到的病毒与生态学上和商业上重要的鱼类、爬行类以及两栖类的大规模死亡有着关联，“蛙病毒不会产生较大危害”这一乐观看法也慢慢地发生了改变（Chinchar 等人，2009）。到目前为止，蛙病毒感染和发病的案例在全球六大洲都有记载。据统计，蛙病毒至少存在于 175 种变温脊椎动物物种当中（Duffus 等人，2015）。目前还不清楚，全球出现的蛙病毒感染增多是由于病毒毒力的增强还是自然或与人类相关活动引起的传播增多，抑或是监督增加，以及诊断与检测手段改进。不管原因如何，蛙病毒现在被视为能感染所有变温脊椎动物类别（鱼类、爬行类和两栖类）的致病源，并且因特异性的病毒、宿主以及环境因子不同而引发显著的发病和死亡。

目前，虹彩病毒科含有 5 个属，其中 2 个属即虹彩病毒属（*Iridovirus*）和绿虹彩病毒属（*Chloriridovirus*）感染无脊椎动物，另外 3 个属即淋巴囊肿病毒属（*Lymphocystivirus*）、细胞肿大病毒属（*Megalocytivirus*）和蛙病毒属（*Ranavirus*）只感染变温脊椎动物（Jancovich 等人，2015a）。淋巴囊肿病毒和细胞肿大病毒只感染鱼类。然而，如上文所述，蛙病毒的靶宿主包括鱼类、两栖类和爬行类。比较高等的脊椎动物，如鸟类和哺乳类，目前未见感染的报道。感染受到阻碍有可能反映出温度限制，超过这一温度上限（接近 32 ℃），病毒不能复制。然而，比较高等的脊椎动物并不缺乏合适的细胞受体，当孵育

温度为 30 ℃时，蛙病毒也能在哺乳细胞系如仔仓鼠肾细胞中进行复制。蛙病毒即使通过热和辐射使病原失活，同样也能在哺乳动物中造成细胞凋亡（Grayfer 等人，2015）。因此，蛙病毒代表着一个类群的病原，该类病原具有广泛的宿主范围，并在全球具有感染不同脊椎动物种群的潜力。

我们常常会面临一个问题："蛙病毒对于野生动物来说，是一个重大的威
胁吗?"我们认为答案是"是"。该威胁的严重性取决于一系列不同的因素。布 3
伦纳（Brunner 等人，2015）运用流行病学理论及数学模拟的结果描述了蛙病毒是如何导致物种数量下降的。然而，由于缺乏种群因长期反复感染蛙病毒而大规模死亡的纵向研究，因此，解释种群数量下降所必需的数据也很少。最近的研究开始弥补这一不足，普赖斯（Price）及其同事报道，蛙病毒在西班牙北部几个不同地点的两栖类 3 个物种中导致了物种数量的下降（Price 等人，2014）。蒂彻（Teacher）及其同事分析了在英国 11 年所累积的数据，发现在蛙病毒引起大规模死亡的地点，林蛙的丰度下降了大约 80%（Teacher 等人，2010）。彼得兰卡（Petranka）及其他几位生态学家曾观察到，在蛙病毒引起大规模死亡的地点，连续数年宿主种群没有得到恢复（Petranka 等人，2003；Wheelwright 等人，2014）。朱莉娅・厄尔（Julia Earl）的研究显示，封闭的林蛙种群，再次暴发蛙病毒会在 5 年之内导致种群的灭绝（Earl 和 Gray，2014）。这些研究提示，蛙病毒引起局部物种的灭绝存在数种因素，如在几个宿主间存在高易感性物种，同时还可能存在不依赖种群密度的传播方式，进而导致物种数量的下降。然而，到目前为止，还没有文献报道蛙病毒导致的宿主物种的灭绝。这种不确定性不得不强调需要对蛙病毒和宿主种群的监测做更深入的研究，这方面格雷等人（Gray 等人，2015）已作出了相应的总结。更为重要的是，在未找出因蛙病毒引起物种灭绝的证据之前，我们不应坐视不理。显而易见，目前的研究提示，蛙病毒存在着引起物种灭绝的潜在威胁。特别是许多稀有物种，它们是蛙病毒的宿主。例如，高度濒危的中国大鲵（*Andrias davidianus*）（Geng 等人，2010）、穴居沙龟（*Gopherus polyphemus*）（Westhouse 等人，1996）、隙穴蛙多斑亚种（*Lithobates sevosus*）（Sutton 等人，2014）、北方蟾蜍（*Anaxyrus boreas boreas*）（J. Chaney，M. Gray 和 D. Miller，田纳西大学，未发表资料）对蛙病毒的易感性非常高。需要进一步研究来证实其他稀有物种的高易感性（Gray 等人，2015）。在人工饲养条件下，常常会观察到宿主 100% 的死亡率，这有可能是宿主病毒丰度的增加、蛙病毒有效的传播及与这些环境相关联的应激因子引起的（Waltzek 等人，2014）。少数几个经济物种，例如牛蛙（Mazzoni 等人，2009）、石斑鱼（Qin 等人，2001），以及受保护的物种，例如白鲟（Waltzek 等人，2014）、中国大鲵（Geng 等人，2010；Cunningham

等人，2015)，在人工饲养条件下，蛙病毒的感染已造成了灾难性的损失。综合蛙病毒对高易感性宿主影响的初步资料，我们认为，将蛙病毒这一病原看作变温脊椎动物生物多样性的严重威胁是合理的。

另外一个常常提及的问题就是“蛙病毒是新出现的吗?”，换句话说，“蛙病毒感染在分布、流行或宿主范围正不断增长吗?”对这一问题的回答具有挑战性。但是有资料提示，该问题的答案是“是”。斯托弗（Storfer）依据病毒与宿主之间缺乏共进化史，提供的证据显示，虎纹钝口螈病毒（Ambystoma tigrinum virus，ATV）在一些地方也出现了感染（Storfer 等人，2007）。该学者
4 的研究提示，ATV 的出现是由作为鱼饵的幼蝾螈贸易，以及跨越长地理距离转运亚致死性感染蝾螈引起的（Storfer 等人，2007；Picco 和 Collins，2008）。目前，佛罗里达州大学的研究者华尔兹克（Waltzek）对取自世界各地的几十个蛙病毒进行了全基因组的测序，这一研究工作使得他能够考察蛙病毒系统进化的地理格局，并确定了最近引入蛙病毒的区域。总而言之，种群中蛙病毒致病源的出现既可能是由于它们是新型病原，也可能是使宿主免疫功能降低的环境应激因子增加引起的。如上所述，第一种假设存在的支持证据是病原的污染，也就是人们长距离转运已感染动物的结果或者转运病原污染物的结果（Cunningham 等人，2003）。此外，尽管研究受限，但是有证据显示，杀虫剂、除草剂及利用湿地养牛都可以作为应激因子（stressors），从而增加蛙病毒出现的机会（Forson 和 Storfer，2006；Gray 等人，2007；Kerby 等人，2011）。在过去的 4 年中，已有报道，蛙病毒感染的病例高达 90% 以上（Duffus 等人，2015）。对蛙病毒防范意识增强的同时，增加采样的频度有可能导致检测阳性结果的增加，因此，蛙病毒感染的增加不可能是这些因素单独起作用。

本书从蛙病毒分类学（Jancovich 等人，2015a）、蛙病毒的分布（Duffus 等人，2015）、宿主病原生态学与进化（Brunner 等人，2015）、蛙病毒复制策略（Jancovich 等人，2015b）、宿主抗病毒免疫及病毒对策（Grayfer 等人，2015）、蛙病毒比较病理学和诊断（Miller 等人，2015）、蛙病毒研究设计与分析的建议（Gray 等人，2015）等方面对蛙病毒进行了综述。从整体上讲，本书对蛙病毒及其对宿主生物的影响提供了一个全新的整体认识，反映了具有不同研究技能的科学家的贡献（这些科学家包括分子病毒学家、免疫学家、生态学家、兽医学家及种群生物学家）并将他们对蛙病毒及其所引起的疾病的研究方向进行了整合。

此外，本书还提供了全球专业人员共同探讨和学习蛙病毒的机会。比如说，为了编写本书，先后召开了两次针对蛙病毒的国际专题学术讨论会（明尼苏达州明尼阿波里斯市，2011 年；田纳西州诺克斯维尔市，2013 年）。这两

次会议让试图弄清楚蛙病毒及其致病潜力的科学家能集中在一起进行学术交流（Lesbarrères 等人，2012；Duffus 等人，2014）。第三次专题学术讨论会 2015 年在美国的佛罗里达州的盖恩斯维尔市召开。在 2011 年与 2013 年专题学术讨论会之间，对蛙病毒感兴趣的研究者发起并建立了全球蛙病毒联盟（Global Ranavirus Consortium，GRC）。建立 GRC 的目的在于促进蛙病毒研究和蛙病毒疾病诊断的科学家和兽医工作者之间的交流与合作。具体来讲，在于：（1）增进蛙病毒生物学和与疾病有关的所有领域的知识；（2）促进多学科之间的科学合作；（3）传播有关蛙病毒的知识和信息；（4）提供专家指导与培训的机会。GRC 通过主办两年一次的专题学术讲座、组织区域性研讨会和讨论组，维持具有各种资源的网站的运行，其中包括有关蛙病毒出版物列表和检测蛙病毒病原的实验室的运行，来达到这些目的。他们也做出了不懈努力来创建全球蛙病毒报告系统（Global Ranavirus Reporting System），该系统将成为在线数据管理系统，允许用户进行蛙病毒感染与疾病病例的上传、修改和下载。GRC 在 5
2015 年发布了会员章程。

蛙病毒研究的未来将是一个什么情形呢？我们刚刚对这些病原与它们不同宿主之间的复杂关系有了肤浅认识，需要掌握更多有关蛙病毒的基础分子生物学、宿主的免疫反应和引起的病理变化的资料，可参阅贾柯维奇（Jancovich）等人（2015b）、格瑞福（Grayfer）等人（2015）和米勒（Miller）等人（2015）进行的总结。这些是认识蛙病毒宿主相互作用的基本机制的最基础资料，还需要进行更多的研究，以弄清楚蛙病毒为什么会出现在特定的区域。蛙病毒的众多致病因素与基本流行病学原理（例如不依赖种群密度的传播）、自然应激因子（例如繁殖）、人为应激因子（例如杀虫剂）或最近病原的引入有关吗？为了回答这些问题，实验室的实验需要与野外研究和免疫学家及病理学家所进行的宿主健康评估结合起来。我们还不清楚气候变化对蛙病毒分布和致病性的潜在影响。考虑到许多蛙病毒在更温暖的条件下复制得更快（Ariel 等人，2009），因而存在着大气变暖导致蛙病毒病出现的可能。我们还知道一些快速干燥繁殖地点由于天气变化在增加，对于两栖类动物来说，这种快速干燥会使两栖类幼体产生应激反应，从而导致蛙病毒疾病的进一步加剧。另外一个因素就是在人工饲养设施中，与大规模死亡相关的蛙病毒的致病性明显增强，那些与水产养殖场或蛙场相关的蛙病毒感染（Brunner 等人，2015）就是蛙病毒致病性明显增强的结果。假如这一假设是正确的，那么变温脊椎动物的贸易会使高毒力的蛙病毒株流向世界各地，从而需要实施世界动物卫生组织（英语：World Organisation for Animal Health；法语：Office international desépizooties，OIE，也称国际兽疫局）所推荐的对动物进行预检验的相关规定

（Schloegel 等人，2010）。当我们想到商业的时候，国际贸易是普遍关心的问题。然而，斯托弗的研究显示，短距离（几百公里）转运动物足以导致蛙病毒病的发生（Storfer 等人，2007）。有趣的是，我们最近完成了对照性实验，该实验结果显示，运输距离即使少到 100 公里也足以导致蛙病毒与宿主之间共进化史的差别，从而导致死亡率的增加（P. Reilly，M. Gray，D. Miller，田纳西大学，未发表资料）。总体来讲，这些资料提示，蛙病毒的出现很可能反映出人类诱导的传播、环境应激的增加、宿主免疫力的衰退以及蛙病毒毒力的提高等的综合作用。

自从发现 FV3 以后，在接近 50 年的时间内，蛙病毒已经走过了从研究对象（即具有分子方面兴趣而没有商业或医学重要性的一个病毒科）到成为一个属的历程，这个属的病毒成员无论是潜在地还是实际地都对动物的健康产生了深远的影响。而且对两栖类动物感染蛙病毒的免疫应答反应的研究已经超出了我们对低等脊椎动物抗蛙病毒免疫的认识，并且提出了蛙病毒疫苗开发的途径。本著作已经验证了 30 年前对非普通生物研究后提出的观点，绝不仅仅因为这些生物是奇特的，还因为它们为深入认识所有生物共有的基础生物学过程提供了证据。

致谢

6 开放存取出版物的发行得到了田纳西大学（林业、野生动物与渔业系，研究与交流办公室，农业研究所）、华盛顿州立大学图书馆、戈登州立大学（学术事务办公室）、两栖爬行动物兽医协会（the Association of Reptilian and Amphibian Veterinarians）以及两栖爬行动物保护协会（Amphibian and Reptile Conservancy）的资金赞助。

开放存取

参考文献

Ariel E，Nicolajsen N，Christophersen MB，Holopainen R，Tapiovaara H，Jensen BB（2009）. Propagation and isolation of ranaviruses in cell culture. Aquaculture 294：159-164.

Brunner JL，Storfer A，Gray MJ，Hoverman JT（2015）. Ranavirus ecology and evolution: from epidemiology to extinction. In: Gray MJ, Chinchar VG (eds) Ranaviruses: lethal pathogens of ectothermic vertebrates. Springer, New York.

Chinchar VG, Hyatt A, Miyazaki T, Williams T (2009). Family *Iridoviridae*: poor viral relations no longer. Curr Top Microbiol Immunol 328: 123-170.

Cunningham AA, Daszak P, Rodriguez JP (2003). Pathogen pollution: defining a parasitological threat to biodiversity conservation. J Parasitol 89 (suppl): S78-S83.

Cunningham AA, Turvey ST, Zhou F (2015). Development of the Chinese giant salamander *Andrias davidianus* farming industry in Shaanxi Province, China: conservation threats and opportunities. Fauna & Flora International, *Oryx* doi: 10. 1017/S0030605314000842. http: // journals. cambridge. org/orx/ salamander china.

Duffus ALJ, Gray MJ, Miller DL, Brunner JL (2014). Second international symposium on ranaviruses: a North American herpetological perspective. J North Am Herpetol 2014: 105-107.

Duffus ALJ, Waltzek TB, Stöhr AC, Allender MC, Gotesman M, Whittington RJ, Hick P, Hines MK, Marschang RE (2015). Distribution and host range of ranaviruses. In: Gray MJ, Chinchar VG (eds) Ranaviruses: lethal pathogens of ectothermic vertebrates. Springer, New York.

Earl JE, Gray MJ (2014). Introduction of ranavirus to isolated wood frog populations could cause local extinction. EcoHealth 11: 581-592.

Forson DD, Storfer A (2006). Atrazine increases ranavirus susceptibility in the tiger salamander, *Ambystoma tigrinum*. Ecol Appl 16: 2325-2332.

Geng Y, Wang KY, Zhou ZY, Li CW, Wang J, He M, Yin ZQ, Lai WM (2010). First report of a ranavirus associated with morbidity and mortality in farmed Chinese giant salamanders (Andrias davidianus). J Comp Pathol. doi: 10. 1016/j. jcpa. 2010. 11. 012.

Granoff A, Came PE, Breeze DC (1966). Viruses and renal carcinoma of *Rana pipiens*: I. The isolation and properties of virus from normal and tumor tissues. Virology 29: 133-148.

Gray MJ, Miller DL, Schmutzer AC, Baldwin CA (2007). Frog virus 3 prevalence in tadpole populations inhabiting cattle-access and non-access wetlands in Tennessee, USA. Dis Aquat Organ 77: 97-103.

Gray MJ, Brunner JL, Earl JE, Ariel E (2015). Design and analysis of ranavirus studies: surveillance and assessing risk. In: Gray MJ, Chinchar VG (eds) Ranaviruses: lethal pathogens of ectothermic vertebrates. Springer, New York.

Grayfer L, Edholm E-S, De Jesús Andino F, Chinchar VG, Robert J (2015). Ranavirus host immunity and immune evasion. In: Gray MJ, Chinchar VG (eds) Ranaviruses: lethal pathogens of ectothermic vertebrates. Springer, New York.

Jancovich JK, Steckler N, Waltzek TB (2015a). Ranavirus taxonomy and phylogeny. In: Gray MJ, Chinchar VG (eds) Ranaviruses: lethal pathogens of ectothermic vertebrates. Springer, New York.

Jancovich JK, Qin Q, Zhang Q-Y, Chinchar VG (2015b). Ranavirus replication: molecular, cellular, and immunological events. In: Gray MJ, Chinchar VG (eds) Ranaviruses: lethal pathogens of ectothermic vertebrates. Springer, New York.

Kerby JL, Hart AJ, Storfer A (2011). Combined effects of virus, pesticide, and predator cue on the larval tiger salamander (*Ambystoma tigrinum*). Ecohealth 8: 46–54.

Lesbarrères D, Balseiro A, Brunner J, Chinchar VG, Duffus A, Kerby J, Miller DL, Robert J, Schock DM, Waltzek T, Gray MJ (2012). Ranavirus: past, present and future. Biol Lett 8: 481–483.

Mazzoni R, de Mesquita AJ, Fleury LFF, de Brito W et al. (2009). Mass mortality associated with a frog virus 3-like ranavirus infection in farmed tadpoles *Rana catesbeiana* from Brazil. Dis Aquat Organ 86: 181–191.

Miller DL, Pessier AP, Hick P, Whittington RJ (2015). Comparative pathology of ranaviruses and diagnostic techniques. In: Gray MJ, Chinchar VG (eds) Ranaviruses: lethal pathogens of ectothermic vertebrates. Springer, New York.

Murti KG, Goorha R, Granoff A (1985). An unusual replication strategy of an animal virus. Adv Virus Res 30: 1–19.

Price, SJ, Garner TWJ, Nichols RA, et al. (2014). Collapse of amphibian communities due to an introduced *Ranavirus*. Curr Biol 24: 2586–2591. http://www. cell. com/current-biology/pdfExtended/ S0960-9822(14)01149-X.

Petranka JW, Murray SS, Kennedy CA (2003). Responses of amphibians to restoration of a southern Appalachian wetland: perturbations confound post-restoration assessment. Wetlands 23: 278–290.

Pico AM, Collins JP (2008). Amphibian commerce as a likely source of pathogen pollution. Conserv Biol 22：1582–1589.

Qin QW，Lam TJ，Sin YM，Shen H，Chang SF，Ngoh GH，Chen CL（2001）. Electron microscopic observations of a marine fish iridovirus isolated from brown-spotted grouper，*Epinephеous tauvina*. J Virol Methods 98：17–24.

Schloegel LM，Daszak P，Cunningham AA，Speare R，Hill B（2010）. Two amphibian diseases，chytridiomycosis and ranaviral disease，are now globally notifiable to the world organization for animal health（OIE）：an assessment. Dis Aquat Organ 92：101–108.

Storfer A，Alfaro ME，Ridenhour BJ，Jancovich JK，Mech SG，Parris MJ，Collins JP（2007）. Phylogenetic concordance analysis shows an emerging pathogen is novel and endemic. Ecol Lett 10：1075–1083.

Sutton WB，Gray MJ，Hardman RH，Wilkes RP，Kouba A，Miller DL（2014）. High susceptibility of the endangered dusky gopher frog to ranavirus. Dis Aquat Organ 112：9–16.

Teacher AGF，Cunningham AA，Garner TWJ（2010）. Assessing the long-term impact of ranavirus infection in wild common frog populations. Anim Conserv 13：514–522.

Waltzek TB，Miller DB，Gray MJ，Drecktrah B，Briggler JT，MacConnel B，Hudson C，Hopper L，Friary J，Yun SC，Maim KV，Weber ES，Hedrick RP（2014）. New disease records for hatchery-reared sturgeon I. Expansion of frog virus 3 host range into *Scaphirhynchus albus*. Dis Aquat Org 111：219–227.

Westhouse RA，Jacobson ER，Harris RK et al.（1996）. Respiratory and pharyngoesophageal iridovirus infection in a gopher tortoise（*Gopherus polyphemus*）. J Wild Dis 32：682–686.

Wheelwright NT，Gray MJ，Hill RD，Miller DL（2014）. Sudden mass die-off of a large population of wood frog（*Lithobates sylvaticus*）tadpoles in Maine，USA，likely due to ranavirus. Herpetol Rev 45：240–242.

Williams T（1996）. The Iridoviruses. Adv Virus Res 46：345–411.

Williams T，Barbosa-Solomieu V，Chinchar VG（2005）. A decade of advances in iridovirus research. Adv Virus Res 65：173–248.

Wissenberg R（1965）. Fifty years of research on the lymphocystis virus disease of fishes（1914–1964）. Ann N Y Acad Sci 126：362–374.

Xeros N（1954）. A second virus disease of the leather jacket，*Tipula paludosa*. Nature（London）174：562–563.

第一章　蛙病毒的分布与宿主范围

阿曼达·L. J. 达弗斯[①]，托马斯·B. 华尔兹克[②]，
安克·C. 索尔[③]，马修·C. 阿蓝德[④]，迈克尔·戈特斯曼，
理查德·J. 惠廷顿，保罗·希克[⑤]，梅根·K. 海恩斯[⑥]和雷切尔·E. 马尚[⑦]

1　引言

9 蛙病毒是一群感染两栖类、鱼类和爬行类动物的病原，并且还是全球范围
内新出现的一类病原。该病原既感染人工饲养的动物，也感染野生动物。而且
蛙病毒还能在一个地点同时感染多个物种（Mao 等人，1999a；Duffus 等人，
10 2008），并能在不同变温脊椎动物类别之间进行传播（Brenes 等人，2014a，b；
Brunner 等人，2015）。目前，已知蛙病毒至少能够感染 52 科 175 种变温脊椎
动物。除南极洲以外，其他六大洲都有蛙病毒的发现（见表 1；图 1 与图 2）。

---

① A. L. J. Duffus/Department of Biology, Gordon State College, Barnesville, GA, USA/e-mail: aduffus@gordonstate. edu.

② T. B. Waltzek · M. Gotesman/Department of Infectious Diseases and Pathology, College of Veterinary Medicine, University of Florida, Gainesville, FL, USA/e-mail: tbwaltzek@ufl. edu; mgotesman@ufl. edu.

③ A. C. Stöhr/Institute of Environmental and Animal Hygiene, University of Hohenheim, /Stuttgart, Germany/e-mail: anke. stoehr@freenet. de.

④ M. C. Allender/Department of Comparative Biosciences, College of Veterinary Medicine, /University of Illinois, Urbana, IL, USA/e-mail: mcallend@illinois. edu.

⑤ R. J. Whittington · P. Hick/Faculty of Veterinary Science, University of Sydney, Sydney, NSW, Australia/e-mail: richard. whittington@sydney. edu. au; paul. hick@sydney. edu. au.

⑥ M. K. Hines/University of Wisconsin-Madison, School of Veterinary Medicine, Madison, WI, USA/e-mail: mkhines@wisc. edu.

⑦ R. E. Marschang/Laboklin GmbH & Co. KG, Bad Kissingen, Germany/e-mail: rachel. marschang@gmail. com-© The Author (s) 2015-M. J. Gray, V. G. Chinchar (eds.), Ranaviruses, DOI 10. 1007/978-3-319-13755-1_2.

表 1　蛙病毒在两栖类、鱼类和爬行类宿主中感染病例的分布

类群	科名	感染的物种数量
两栖类	产婆蟾科	1
	钝口螈科	8[a]
	蟾蜍科	8
	绿骨蛙科	1
	丛蛙科	3
	隐鳃鲵科	2
	箭毒蛙科	5
	雨蛙科	15
	小鲵科	1
	细趾蛙科	2[a]
	角蟾科	1
	龟蟾科	2
	负子蟾科	1
	无肺螈科	21
	蛙科	22[a]
	树蛙科	1
	蝾螈科	8
	北美锄足蟾科	1
鱼类	鲟科	3
	鳗鲡科	1
	太阳鱼科	9
	蛇头鱼科	1
	胭脂鱼科	1
	鲤科	2
	塘鳢科	1
	狗鱼科	2
	鳕科	1
	刺鱼科	1
	北美鲶科	2
	隆头鱼科	1
	尖嘴鲈科	1
	笛鲷科	1
	梦鲈科	3
	鲈科	2
	花鲋科	1

续表

类群	科名	感染的物种数量
鱼类	鲑科	1
	石首鱼科	1
	菱鲆科	1
	鮨科	4
	鲶科	1
爬行类	飞蜥科	2
	蛇蜥科	1
	蟒蛇科	1
	安乐蜥科	2
	泽龟科	4
	壁虎科	1
	鬣蜥科	1
	草蜥科	2
	蟒科	4
	陆龟科	8
	鳖科	1
	巨蜥科	1

注：a 表示数据只包括属。

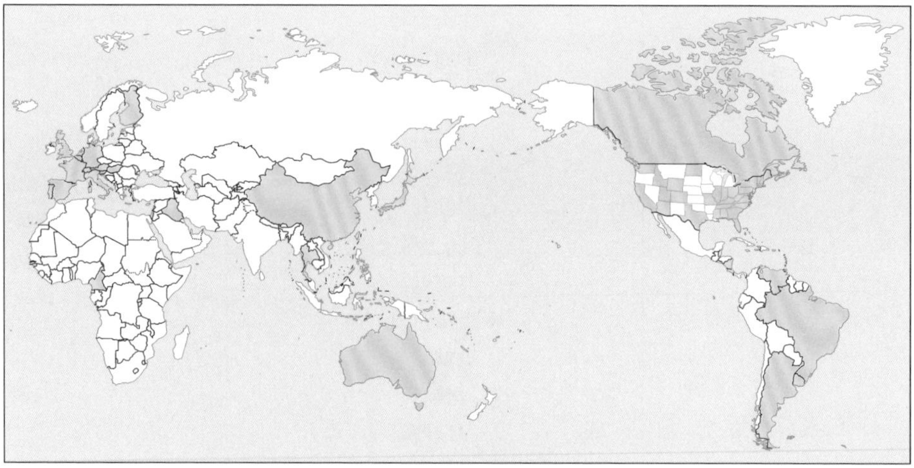

图 1 全球蛙病毒病例的分布

大多数有关蛙病毒的流行病学、地理学和宿主范围的资料来自对大规模死 11
亡的研究、一个或两个时间点对小种群的零散监测研究以及少量较大规模的监测研究，并集中在少数具有经济重要性或保护价值的宿主物种（Grizzle 和 Brunner，2003；Gray 等人，2009b；Whittington 等人，2010；Miller 等人，2011；Gray 等人，2015）。由于被感染动物的大体症状不明显并难以发现，因而蛙病毒作为重要病原的意识缺乏，并且偶尔会造成误诊。因此，现存蛙病毒的地理分布和宿主范围很可能被低估。

蛙病毒之所以归类于新出现的病原，是因为蛙病毒的地理分布及宿主范围 12
正逐步扩大（Daszak 等人，1999）。越来越多的证据显示，动物的区域交易或国际贸易常常会引入蛙病毒。例如，横带虎斑钝口螈（*Ambystoma mavortium*）幼体在美国西南部作为鱼饵销售，使得蛙病毒和虎纹钝口螈病毒（Ambystoma tigrinum virus，ATV）感染率高达 100%（Picco 和 Collins，2008；Brunner 等人，2015）。基于一系列不同的原因，已发现两栖类蛙病毒存在于边境贸易的动物当中，其中包括人类消费的动物和贸易宠物（Schloegel 等人，2009；Kolby 等人，2014）。施勒格尔（Schloegel）等人（2009）发现从三个主要港口城市进
口到美国的两栖类动物中，有 8.5%感染了蛙病毒。类似的，科尔比（Kolby） 13
等人（2014）发现经过香港国际机场出口的两栖类动物中超过 50%感染了蛙病毒。同时，还发现感染蛙病毒的爬行类动物存在于国际贸易的动物当中（Hyatt 等人，2002；Stöhr 等人，2013b，2015）。

最后要提出的是，在国际贸易的观赏鱼类中，也显示有蛙病毒的感染（Hedrick 和 McDowell，1995）。此外，蛙病毒感染的结果会在不同宿主和蛙病毒株之间存在差异。目前已弄清楚，蛙病毒能引起种群数量下降和灭绝（Teacher 等人，2010；Price 等人，2014；Earl 和 Gray，2014）。因此，蛙病毒对地理隔离的宿主物种或低丰度的物种存在着重大威胁（Heard 等人，2013；Price 等人，2014；Earl 和 Gray，2014），一些高易感性的稀有物种也存在着巨大风险（Earl 和 Gray，2014）。但是，普通物种同样也会被感染。例如，在蛙病毒反复引起大规模死亡的英国，林蛙（*Rana temporaria*）种群数量平均下降了 80%（Teacher 等人，2010）。因此，弄清楚蛙病毒的地理分布、宿主范围以及这些新出现病原的系统发生关系显得尤为重要（Jancovich 等人，2015）。

2　感染两栖类的蛙病毒

1960 年，在美国中西部，有学者从北部豹蛙（*Lithobates pipiens*）中首次

图 2 蛙病毒在三类变温脊椎动物中造成疾病。(a) 荷兰死亡的成年食用蛙（*Rana esculenta*）（照片由 Jeiger Herder 提供）；（b）美国患病的平原旱掘蟾（*Spea bombifrons*）（照片由 Drew Davis 提供）；（c）中国感染的史氏鲟（*Acipenser schrenckii*）（照片由曾令兵提供）；（d）患病的中国大鲵幼鲵（照片由耿毅提供）；（e）美国死亡的东部箱龟（*Terrapene Carolina carolina*）（照片由 Matthew Allender 提供）；（f）德国患有蛙病毒感染性皮炎的丽纹龙蜥（*Japalura splendida*）（照片由 Helge Behncke 提供）。

分离到了蛙病毒（Granoff 等人，1965；Clark 等人，1968）。该病毒是从一只患有腺癌的蛙中分离到的，并被命名为蛙病毒 3（Frog virus 3，FV3）。（Granoff 等人，1965）后来，蛙病毒 3 成为蛙病毒属（*Ranavirus*）的一个典型种类，在以后的几十年里，FV3 在病毒学的很多方面得到很好的鉴定（Chinchar，

2002）。蛙病毒病与两栖类动物有关联的报道很少，因而该病原并没有引起注意。蛙病毒属中第二个被发现的病毒就是博乐虹彩病毒（Bohle iridovirus，BIV），该病毒直到20世纪初叶才从两栖动物中分离到。博乐虹彩病毒是从澳大利亚人工饲养的动物中检测到的（Speare 和 Smith，1992）。几乎同时，动物流行病引发的大规模死亡在美国和英国也连续不断地被报道（Collins 等人，1988；Cunningham 等人，1993）。但是，直到20世纪中叶，大规模死亡的原因才确定为蛙病毒感染（Drury 等人，1995；Cunningham 等人，1996；Jancovich 等人，1997；见图2）。在两栖类动物当中，与蛙病毒相关的死亡和感染的报道呈指数增长，其中，超过90%的报道是在2010年以后。虽然研究者对蛙病毒是病原的意识增强和对蛙病毒的监测增加，而且这一趋势也在不断加强，但是蛙病毒病例的增加不可能单单是采样的技术问题。现在，我们认识到，两栖类蛙病毒的分布是全球性的，因为在发生大规模死亡的地方，原先并没有检测出蛙病毒。

蛙病毒呈全球性分布（见图1和表2），并已被确定为两栖类种群的重要威胁（Duffus 和 Cunningham，2010；Teacher 等人，2010；Miller 等人，2011）。目前的报道显示，两栖类蛙病毒至少存在于25个国家，18科105种两栖类动物当中（见图3和表2）。这些数字有可能被低估，因为许多两栖类动物在自然条件下呈隐性感染，或蛙病毒感染极少有症状，即使有一些症状但不明显，23
而且常常与其他致病因子感染的症状相混淆。由于蛙病毒感染的病程进展很快，再加上宿主死亡后的快速分解，因而，死亡事件不容易被观察到（Brunner 等人，2015；Miller 等人，2015）。令人惊奇的是蚓螈，该类动物喜欢掘土且人们对其缺少研究，蚓螈科是至今两栖类动物中唯一没有被报道蛙病毒感染的一科。据我们所知，目前还没有人做蚓螈对蛙病毒的易感性实验，也没有人对其野生种群进行监测。

已知有三大公认感染两栖类动物的蛙病毒种，即FV3、ATV以及BIV。除此之外，研究者还从两栖类动物中分离到了其他几种蛙病毒，例如普通产婆蟾病毒（Common midwife toad virus，CMTV）等（Balseiro 等人，2009）。但是，到目前为止，普通产婆蟾病毒作为蛙病毒种还没有得到公认。如同贾柯维奇（Jancovich）等人（2015）在讨论中所提出的一样，宣布一个特定的分离病毒作为一个独特的病毒种是复杂的，要求在序列上高度保守，常常基于蛙病毒分离株之间的氨基酸序列高度保守，同源性高于95%等事实的综合判断。未来研究的一个挑战就是要鉴定病毒的遗传序列，只有这样，才允许在进化背景条件下对蛙病毒进行独特的鉴定。下面我们将讨论已知的有关蛙病毒感染两栖类动物的内容。

表 2 野外和人工饲养两栖类蛙病毒感染的分布，包括进口动物原产地的地理位置；该表引自 Miller 等人（2011）

大陆	国家	原产地[a]	州或省[a]	科名	拉丁名	I, M[b]	W, C[c]	参考文献
非洲	喀麦隆			负子蟾科	欧口湖爪蟾	I	W	Docherty-Bone et al.（2013）
亚洲	中国	—	—	隐鳃鲵科	中国大鲵	M	C	Geng et al.（2011），Zhou et al.（2013），Ma et al.（2014）
		—	—	蛙科	东北林蛙或海参威蛙	I	W	Xu et al.（2010）
		—	—		美国青蛙	M	C	Zhang et al.（1996，2001）
		—	—		虎纹蛙 曾用名 *Rana tigrina*	M	C	Weng et al.（2002）
	日本	—	—	小鲵科	云斑小鲵	M	C	Une et al.（2009a）
		—	—	蛙科	美国牛蛙 曾用名 *Rana catesbieana*	M	W	Une et al.（2009b）
	泰国	—	—	蛙科	虎纹蛙	M	W	Kanchanakhan（1998）
大洋洲	澳大利亚	—	—	雨蛙科	澳大利亚绿色雨蛙	M	W，C	Cullen and Owens（2002），Weir et al.（2012）
		—	—		大雨蛙	M	C	Weir et al.（2012）
		—	—	龟蟾科	华丽穴居蛙	M	C	Speare and Smith（1992）
		—	—		红背小蟾	M	C	Cullen and Owens（2002）
欧洲	比利时	—	—	蟾蜍科	大蟾蜍	I	W	Martel et al.（2012）
		—	—	蛙科	美国牛蛙	I	W	Sharifi an-Fard et al.（2011）
		CN[d]	—	蝾螈科	贵州疣螈	M	C	Pasmans et al.（2008）
	克罗地亚	—	—	蛙科	虎皮蛙	M	W	Fijan et al.（1991）
	丹麦	—	—	蛙科	虎皮蛙	M	W	Ariel et al.（2009）

续表

大陆	国家	原产地[a]	州或省[a]	科名	拉丁名	I, M[b]	W, C[c]	参考文献
欧洲	法国	—	—	蛙科	中国林蛙	M	W	C. Miaud（个人通讯资料）
	德国	CH, CZ, PL, SE, SK	—	蛙科	虎皮蛙	M	C	Stöhr et al.（2013a）
		Iraq	—	蝾螈科	桔乌尔米螈	M	C/W	Stöhr et al.（2013c）
	以色列	—	—	蟾蜍科	绿蟾蜍	I	W	Miller et al.（2011）
	意大利	—	—	蛙科	虎皮蛙	—	—	Ariel et al.（2010）
	荷兰	—	—	箭毒蛙科	迷彩箭毒蛙	M	C	Kik et al.（2012）
		—	—		两色叶毒蛙	M	C	Kik et al.（2012）
		—	—		红画眉箭毒蛙	M	C	Kik et al.（2012）
		—	—	蛙科	侧褶蛙属未定种	M	W	Kik et al.（2011）
		—	—	蝾螈科	欧洲滑螈	M	W	Kik et al.（2011）
	葡萄牙	—	—	雨蛙科	产婆蟾		W	R. Marschang（个人通讯资料）
		—	—	蝾螈科	利比亚蝾螈		W	R. Marschang（个人通讯资料）
		—	—		斑纹蝾螈	M	W	Alves de Matos et al.（2008）
		—	—		博斯欧螈	M	W	Alves de Matos et al.（2008）
	西班牙	—	—	雨蛙科	产婆蟾	M	W	Balseiro et al.（2009）
		—	—	蝾螈科	高山蝾螈	M	W	Balseiro et al.（2010）
	瑞士	—	—	蛙科	湖侧褶蛙	M	C	Stöhr et al.（2013a）
	英国	—	—	雨蛙科	产婆蟾	M	W	Duffus et al.（2014）
		—	—	蟾蜍科	大蟾蜍	M	W	Hyatt et al.（2000），Duffus et al.（2014）
		—	—	蛙科	中国林蛙	M	W	Cunningham et al.（1993），Drury et al.（1995），Cunningham et al.（1996），Teacher et al.（2010），Duffus et al.（2013）
		—	—	蝾螈科	欧洲滑螈	I	W	Duffus et al.（2014）

续表

大陆	国家	原产地[a]	州或省[a]	科名	拉丁名	I，M[b]	W，C[c]	参考文献
北美洲	加拿大	—	AB，SK，MB	钝口螈科	横带虎斑钝口螈	M	W	Bollinger et al.（1999），Schock et al.（2008），Miller et al.（2011）
		—	ON		钝口螈未定种	I	W	Duffus et al.（2008）
		—	ON	雨蛙科	灰树蛙	I	W	Duffus et al.（2008）
		—	—		春雨蛙	M	W	Miller et al.（2011）
		—	ON		拟蝗蛙未定种	I	W	Duffus et al.（2008）
		—	ON，PEI	蛙科	青铜蛙	I，M	W	St Amour et al.（2008），Miller et al.（2011），Forzán and Wood（2013）
		—	SK，ON，QU		北美豹蛙	I，M	W，C	Greer et al.（2005）Schock et al.（2008），Echaubard et al.（2010），Paetow et al.（2011）
		—	AB，NU，NWT，ON，OC，SK		木材蛙	I，M	W	Greer et al.（2005）Duffus et al.（2008）Schock et al.（2008），Miller et al.（2011），Schock et al.（2010）；D. Lesbarreres（personal communication），D. Schock（个人通讯资料）
		—	BC		俄勒冈斑点青蛙	M	C	D. Schock et al.（2008）
		—	ON	蝾螈科	绿红东美螈	I	W	Duffus et al.（2008）
	哥斯达黎加	—	—	蟾蜍科	洪都拉斯蟾蜍	I	W	Whitfi eld et al.（2013）
		—	—		海蟾蜍	I	W，C	Speare et al.（1991），Whitfi eld et al.（2013）
		—	—	雨蛙科	橄榄鼻树蛙	I	W	Whitfi eld et al.（2013）
		—	—		包迪树蛙	I	W	Whitfi eld et al.（2013）
		—	—	绿骨蛙科	矮玻璃蛙	I	W	Whitfi eld et al.（2013）

续表

大陆	国家	原产地[a]	州或省[a]	科名	拉丁名	I, M[b]	W, C[c]	参考文献
北美洲	哥斯达黎加	—	—	丛蛙科	布氏从蛙	I	W	Whitfield et al.（2013）
		—	—		佛氏丛蛙	I	W	Whitfield et al.（2013）
		—	—		大头丛蛙	I	W	Whitfield et al.（2013）
		—	—	箭毒蛙科	草莓箭毒蛙	I	W	Whitfield et al.（2013）
	尼加拉瓜	—	—	雨蛙科	红眼树蛙	M	W	Stark et al.（2014）
		—	—	蛙科	弗氏林蛙	M	W	Stark et al.（2014）
	美国	—	—	钝口螈科	杰斐逊钝口螈	M	W	Miller et al.（2011）
		—	—		长趾钝口螈	M	W	Miller et al.（2011）
		—	MA, ME, NC, NY, OH, TN		斑点钝口螈	I, M	W	Green et al.（2002）, Petranka et al.（2003）, Gahl and Calhoun（2010）, Todd-Thompson（2010）, Brunner et al.（2011）, O'Bryan et al.（2012）, Homan et al.（2013）
		—	AZ, ND		横带虎斑钝口螈	I, M	W	Jancovich et al.（1997, 2005）, Docherty et al.（2003）, Picco and Collins（2008）, Greer et al.（2009）
		—	TN		暗斑钝口蛙	M	W	Todd-Thompson（2010）
		—	TN		鼹钝口螈	I	W	O'Bryan et al.（2012）
		—	CO, ID, ND, UT, WY		虎纹钝口螈	I, M	W, C	Green et al.（2002）, Hoverman et al.（2012a）
		—	—	蟾蜍科	美国蟾蜍	I	W	Hoverman et al.（2012a）
		—	—		北方蟾蜍	I, M	C	Cheng et al.（2014）
		—	NJ		北美弗氏蟾蜍	M	W	Monson-Collar et al.（2013）
		—	—		大黄蜂步行蟾	I	C	Cheng et al.（2014）

续表

大陆	国家	原产地[a]	州或省[a]	科名	拉丁名	I，M[b]	W，C[c]	参考文献
北美洲	美国	—	TN	隐鳃鲵科	美国隐鳃鲵	I	W	Souza et al.（2012）
		—	—	箭毒蛙科	迷彩箭毒蛙	I	C	Miller et al.（2008）
		—	—		金色箭毒蛙	I	C	Miller et al.（2008）
		—	—	雨蛙科	北蝗蛙	I	W	Hoverman et al.（2012b）
		—	—		可普灰树蛙	I，M	W，C	Miller et al.（2011），Driskell et al.（2009）
		—	TN		可普灰树蛙或树蛙复合种	I	W	O'Bryan et al.（2012）
		—	—		美国树蛙	M	W	Green and Converse（2005）
		—	TX		斑纹合唱蛙	M	W	Torrence et al.（2010）
		—	TN，MI		春雨蛙	M	W	Green et al.（2002），Gahl and Calhoun（2010），Todd-Thompson（2010）
		—	TN		山地合唱蛙	I，M	W	Todd-Thompson（2010），Hoverman et al.（2012a）
		—	—		太平洋树蛙	M	W	Miller et al.（2011）
		—	ID		内华达树蛙	M	W	Russell et al.（2011）
		—	—	角蟾科	三角枯叶蛙	I	C	Cheng et al.（2014）
		—	TN	无肺螈科	微暗蝾螈	I	W	Gray et al.（2009a）
		—	GA，NC		黑腹蝾螈	I	W	Rothermel et al.（2013）
		—	VA		暗棕脊口螈	I	W	Davidson and Chambers（2011），Hamed et al.（2013）
		—	TN		横仿脊口螈	I	W	Gray et al.（2009a，b）

续表

大陆	国家	原产地[a]	州或省[a]	科名	拉丁名	I, M[b]	W, C[c]	参考文献
北美洲	美国	—	GA, NC	无肺螈科	铲鼻蝾螈	I	W	Rothermel et al. (2013)
		—	GA, NC, TN, VA		山脊口螈	I	W	Gray et al. (2009a, b), Davidson and Chambers (2011), Hamed et al. (2013), Rothermel et al. (2013)
		—	GA, NC, TN		奥科伊蝾螈	I	W	Gray et al. (2009a, b) Rothermel et al. (2013)
		—	VA		蓝脊灰暗蝾螈	I	W	Hamed et al. (2013)
		—	VA		北方小蝾螈	I	W	Hamed et al. (2013)
		—	GA, NC, TN, VA		黑腹脊口螈	I	W	Gray et al. (2009a, b), Davidson and Chambers (2011), Hamed et al.(2013), Rothermel et al. (2013)
		—	TN		桑提拉灰暗蝾螈	I	W	Gray et al. (2009a, b)
		—	TN		侏儒蝾螈	I	W	Gray et al. (2009a, b)
		—	VA		南方双带河溪螈	I	W	Davidson and Chambers (2011)
		—	VA		金丝蝾螈	I	W	Davidson and Chambers (2011)
		—	VA		斑尾洞螈	I	W	Davidson and Chambers (2011)
		—	TN		蓝脊双线蝾螈	I	W	Gray et al. (2009a, b)
		—	TN		辣椒蝾螈	I	W	Gray et al. (2009a, b)
		—	VA		白斑蝾螈复合种	I	W	Davidson and Chambers (2011)
		—	TN		乔氏无肺螈	I	W	Gray et al. (2009a, b)
		—	VA		北灰颊蝾螈	I	W	Hamed et al. (2013)
		—	VA		威勒蝾螈	I	W	Hamed et al. (2013)
		—	—	蛙科	平原豹蛙	M	W	Miller et al. (2011)

续表

大陆	国家	原产地[a]	州或省[a]	科名	拉丁名	I，M[b]	W，C[c]	参考文献
北美洲	美国	—	FL，OH，MA，ME，NC，TN，VA	蛙科	北美牛蛙	I，M	W，C	Wolf et al.（1969）Green et al.（2002），Gray et al.（2007），Majji et al.（2006），Miller et al.（2007；2009），Gahl and Calhoun（2010），Davidson and Chambers（2011），Homan et al.（2013），Landsberg et al.（2013）
		—	MA，ME，NJ，OH，TN		青铜蛙	I，M	W	Green et al.（2002），Gahl and Calhoun（2010），Gray et al.（2007），Johnson et al.（2007），Miller et al.（2009），Homan et al.（2013），Monson-Collar et al.（2013）
		—	MA，NH，TN，VA		北美沼泽蛙	I，M	W	Green et al.（2002），Hoverman et al.（2012b），Davidson and Chambers（2011）
		—	MA，MN，VT		北部豹蛙	I，M	W	Granoff et al.（1965），Clark et al.（1968），Green et al.（2002），Uyehara et al.（2010）
		—	MA，MN		貂青蛙	M	W	Green et al.（2002）
		—	NJ，TN，FL		南方豹蛙	I，M	W	Johnson et al.（2007），Miller et al.（2011），Hoverman et al.（2012a），O'Bryan et al.（2012），Landsberg et al.（2013），Monson-Collar et al.（2013）
		—	CT，MA，ME，MN，NC，ND，NY，TN		林蛙	I，M	W	Green et al.（2002），Petranka et al.（2003），Harp and Petranka（2006），Gahl and Calhoun（2010），Todd-Thompson（2010），Uyehara et al.（2010），Brunner et al.（2011），T. Rittenhouse（个人通讯资料）
		—	—		非洲牛箱头蛙	M	C	Miller et al.（2007）
		—	CA		红腿林蛙	M	W	Mao et al.（1999a）

续表

大陆	国家	原产地[a]	州或省[a]	科名	拉丁名	I，M[b]	W，C[c]	参考文献
北美洲	美国	—	—	蛙科	加洲红腿林蛙	M	W	Miller et al.（2011）
		—	—		河蛙	M	W	Miller et al.（2011）
		—	ID		哥伦比亚斑林蛙	M	W	Converse and Green（2005），Green and Converse（2005），Russell et al.(2011)
		—	—		黄腿山蛙	M	W	Converse and Green（2005）
		—	—	树蛙科	大树蛙	M	C	Miller et al.（2008）
		—	KY	蝾螈科	北美东蝾螈	I，M	W	Granoff et al.（1965），Green et al.（2002），Glenney et al.（2010），Richter et al.（2013）
		—	—	北美锄足蟾科	东方锄足蟾	I，M	W	Miller et al.（2011）
南美洲	阿根廷	—	—	细趾蛙科	巴塔哥尼亚蛙	M	W	Fox et al.（2006）
	巴西	—	—	蛙科	美国牛蛙	M	C	Mazzoni et al.（2009）
	乌拉圭	—	—	蛙科	美国牛蛙	I	C	Galli et al.（2006）
	委内瑞拉	—	—	蟾蜍科	海蟾蜍	I	W	Zupanovic et al.（1998a）
		—	—	细趾蛙科	细趾蛙未定种	I	W	Zupanovic et al.（1998b）

[a] 表示当有信息的时候才提供。

[b] I 表示没有蛙病毒疾病症状的感染；M 表示由蛙病毒疾病引起的死亡。

[c] W 表示野生种群；C 表示人工饲养种群，包括动物园和养蛙场饲养的种群，但不包括用来对照的病毒接种研究的种群。

[d] 这些动物被认为是从中国进口的。

（表中“原产地[a]”列）CN：中国；CH：瑞士；CZ：捷克；PL：波兰；SE：瑞典；SK：斯洛伐克；Iraq：伊拉克；（表中“州或省[a]”列）AB：阿尔伯塔省；SK：萨斯喀彻温省；MB：曼尼托巴省；ON：安大略省；PEI：爱德华王子岛；QU：；NU：努纳武特省；NWT：加拿大西北地区；QC：魁北克省；BC：不列颠哥伦比亚省；MA：马萨诸塞州；ME：缅因州；NC：北卡罗来纳州；NY：纽约州；OH：俄亥俄州；TN：田纳西州；AZ：亚利桑那州；ND：北达科他州；CO：科罗拉多州；ID：爱达荷州；UT：犹他州；WY：怀俄明州；NJ：新泽西州；TX：得克萨斯州；GA：佐治亚州；VA：弗吉尼亚州；FL：佛罗里达州；NH：新罕布什尔州；MN：明尼苏达州；VT：佛蒙特州；CT：康涅狄格州；CA：加利福尼亚州；KY：肯塔基州

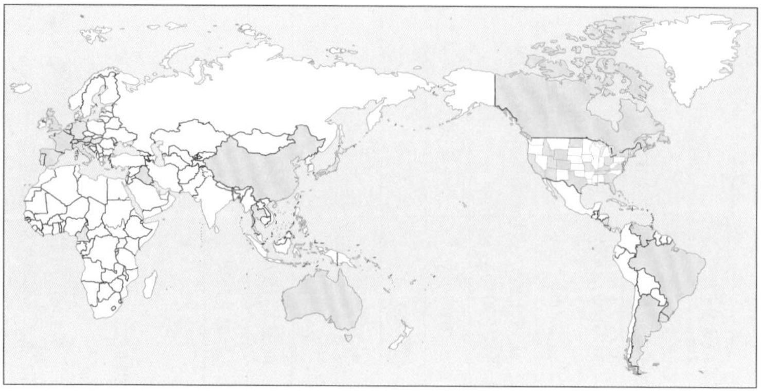

图 3　两栖类感染蛙病毒病例的分布（注：此病毒分布没有覆盖台湾省）

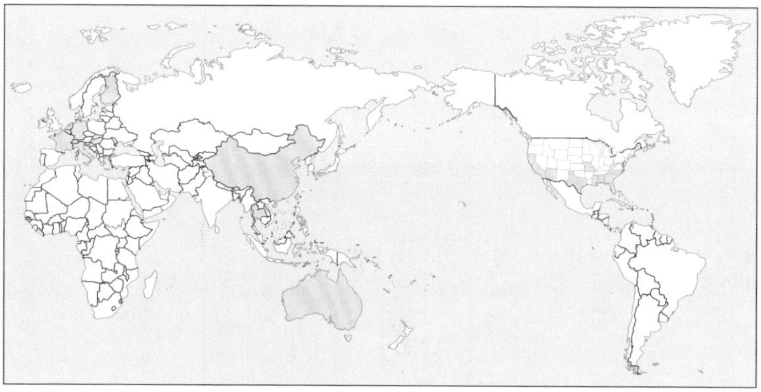

图 4　鱼类感染蛙病毒病例的分布

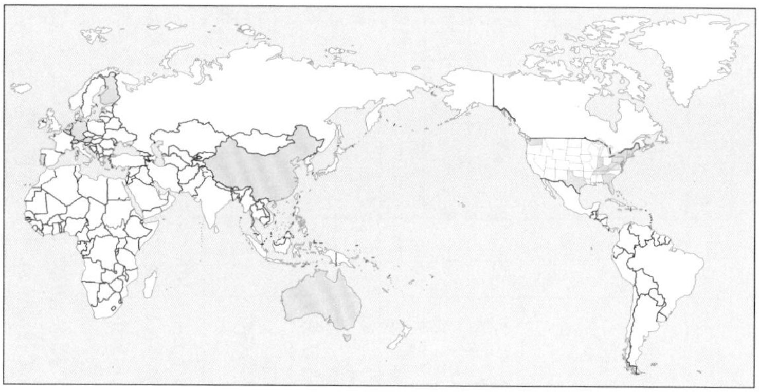

图 5　爬行类动物感染蛙病毒病例的分布（注：此病毒分布没有覆盖台湾省）

2.1　蛙病毒 3

蛙病毒最初是从豹蚊蛙（leopard frog）中分离到的。该病毒自分离以来，两栖类确认感染 FV3 和 FV3 样病毒的病例的物种数量在不断增长，鱼类和爬行类感染蛙病毒的物种数量也是如此，详见第 3 节和第 4 节的内容。研究发现，感染的动物包括那些可见到明显发病或死亡的病例以及明显健康的个体。 24
在两栖类动物中，已发现 FV3 和 FV3 样病毒病在北美洲的大部分地区暴发，其中主要发生于北美东部无尾两栖类（anurans）和有尾两栖类（urodeles）动物当中。无论是在美国还是在加拿大，由 FV3 或 FV3 样病毒造成的感染数量目前还不清楚，因为许多研究者还没有报道在他们的研究中所检测出的蛙病毒株。在墨西哥，还没有蛙病毒感染的报道，这很可能是缺乏研究的结果。在中美洲，至少在两栖类 10 个物种中检测到了一种 FV3 样病毒。特别是在哥斯达黎加，至少有 8 个物种发生了 FV3 样病毒感染（Whitfield 等人，2013）。此外，尼加拉瓜已在 2 个物种中检测出了 FV3 样蛙病毒（Stark 等人，2014）。在南美洲，FV3 感染和发病的病例主要与美国牛蛙（*L. catesbeianus*）养殖场有关（Mazzoni 等人，2009）。但是，也有学者对野生巴塔哥尼亚蛙（*Atelognathus patagonicus*）作了单一感染病例的报道（Fox 等人，2006）。

在欧洲，FV3 样病毒感染的第一次大暴发发生在英国东南部的林蛙中（Cunningham 等人，1993，1996；Drury 等人，1995）。这些病毒在林蛙中出现以后，很快大蟾蜍（*Bufo bufo*）也发生了感染（Hyatt 等人，2000；Cunningham 等人，2007）。自此以后，有文记载在普通蝾螈（*Lissotriton vulgaris*）和普通产婆蟾（*Alytes obstetricans*）中也发现了 FV3 样病毒感染（Duffus 等人，2014）。此外，FV3 样病毒已在欧洲大陆的两栖类动物中被检测出来（Ariel 等人，2009；Stöhr 等人，2013c）。总的来说，在欧洲，已有文献记载，FV3 样蛙病毒至少存在于两栖类 5 个物种中。

在亚洲，已有数篇文献报道，FV3 样病毒既存在于野生两栖类动物种群当中，也存在于人工饲养的两栖类动物种群当中。在中国，已发现整个黑龙江省有 5.7% 的成年东北林蛙和 42.5% 的东北林蛙幼体感染了 FV3 样病毒（Xu 等人，2010）。在日本，有文献报道，FV3 样病毒是引进物种美国牛蛙蝌蚪大规模死亡的病因（Une 等人，2009a）。在这个报道当中，记载了在牛蛙蝌蚪大规模死亡的同时，同域池塘当中的成年牛蛙和鱼（*Gnathopogon spp.* 颌须鮈属，未鉴定种）没有发生死亡或濒临死亡的情况。有记录还显示，蛙病毒存在于鱼的肝脏中（Une 等人，2009b），这一结果提示，FV3 样病毒存在着不同动物类别间的传播（interclass transmission）。在非洲，多彻蒂-博恩（Docherty-Bone）等人（2013）报道了蛙病毒感染的唯一病例，其病原很可能是FV3样病毒。所使用的引物（primers）是针对FV3主衣壳蛋白（major capsid protein，MCP）所设计的，但扩增出来的PCR产物质量很差，且不能用来测序（Docherty-Bone等人，2013）。总的来说，在亚洲和非洲对蛙病毒的监测数据已落后于北美洲和欧洲的数据。这些数据对于弄清楚蛙病毒的分布、宿主范围及其威胁是必不可少的。

正如所提到的一样，FV3和FV3样病毒已影响到水产养殖。在日本，一种与FV3主衣壳蛋白有99%同源的蛙病毒已从养殖的日本云斑小鲵（*Hynobius nebulosus*）中得到了分离株（Une等人，2009a）。在美洲，也同样有文献记载FV3样病毒与水产养殖中的大规模死亡事件有关。在美国，FV3样病毒感染是养殖场中蝌蚪和刚变态牛蛙大规模死亡的原因（Majji等人，2006；Miller等人，
25 2007）。在巴西，已经确认FV3样病毒是好几个美国牛蛙养殖场发生大规模死亡事件的元凶（Majji等人，2006；Miller等人，2007）。有证据显示，存在于水产养殖场的FV3样病毒，其毒力比在自然种群中发现的病毒株的毒力高得多（Majji等人，2006；Hoverman等人，2010，2011）。

2.2 虎纹钝口螈病毒

虎纹钝口螈病毒（Ambystoma tigrinum virus，ATV）是1995年在美国亚利桑那州于圣拉斐尔谷（San Rafael Valley）的索诺拉虎纹钝口螈（*Ambystoma tigrinum stebbinsi*）的幼体中首次采集到的（Jancovich等人，1997）。该病毒在健康的和有发病症状的幼螈中同时得到了分离（Jancovich等人，1997）。后来在实验室，分离到的该病毒株通过水体以及采食感染动物身体组织传播到了健康的个体（Jancovich等人，1997）。通过科赫法则（Koch's Postulates），ATV感染被确定是虎纹钝口螈幼体发病的原因，同时也是1985年曾描述过的、反复发生的动物流行病的原因（Collins等人，1988）。

野外的虎纹钝口螈病毒似乎仅限于北美西部（Jancovich等人，2005；Ridenhour和Storfer，2008）。ATV病毒株的系统地理学研究提示，局域范围的扩大和远距离的病毒定殖事件，要归结为人为的传播（Jancovich等人，2005）。有研究者发现，ATV存在于商业上作为鱼饵销售的虎纹钝口螈的幼体当中，这为ATV分布范围的扩大提供了一个人为的解释。

由于存在着引入ATV的潜在危险，因而ATV被认为是当地受保护的有尾两栖类动物的重大威胁。在实验室，濒危的加州虎纹钝口螈（*Ambystoma californiense*）对 ATV 非常易感，并经历了与感染相关的死亡（Picco 等人，2007）。到目前为止，还没有 ATV 感染野生加州虎纹钝口螈或与之相关的死亡报道，部分原因可能是禁止进口钝口螈科（Ambystomatidae）动物到这个州。

相对于不同年份的感染来说，同一年份不同池塘之间 ATV 感染的模式类似（Greer 等人，2009）。在美国亚利桑那州北部凯巴布国家森林（Kaibab National Forest）的凯巴布高原虎纹钝口螈的自然种群中，ATV 在各个池塘的暴发几乎是同时发生的（Greer 等人，2009）。有趣的是，虽然进行了四年的观察研究，但是没有看到明显的疾病症状，即使是感染率高于 50% 的池塘也是如此（Greer 等人，2009）。有学者认为，在这些种群中没有观察到发病和死亡，是由于在 ATV 与宿主之间存在共进化（Greer 等人，2009）。这有可能是事实，因为有证据表明，从美国西部分离到的 ATV 病毒株具有区域适应性（Ridenhour 和 Storfer，2008）。

2.3　博乐虹彩病毒

对博乐虹彩病毒（Bohle iridovirus，BIV）的首次描述是在 20 世纪初叶， 26
该病毒是从澳大利亚刚变态的华丽穴居蛙（*Limnodynastes ornatus*）中分离到的，这种蛙当时被人工饲养并突然发生死亡（Speare 和 Smith，1992）。运用 BIV 的主衣壳蛋白序列进行分析，BIV 被确定与动物流行性造血器官坏死病毒（Epizootic hematopoietic necrosis virus，EHNV）最接近。EHNV 是澳大利亚已经分离到的、存在于鱼类中的蛙病毒（Hyatt 等人，2000）。之后的实验显示，BIV 对澳大利亚无尾目的其他动物具有致病性，并与人工饲养与野生环境中动物的死亡事件相关（Cullen 等人，1995；Cullen 和 Owens，2002）。

已知 BIV 和 BIV 样病毒仅存在于澳大利亚的无尾目动物中。直到最近，也就是 2010 年，饲养在美国艾奥瓦州的水族馆中的北方蟾蜍（*Anaxyrus boreas boreas*）发生了大规模死亡。一种 BIV 样病毒从北方蟾蜍中得到了分离，该病毒当时被临时命名为动物园蛙病毒（Zoo ranavirus，ZRV），研究发现，该病毒与 BIV 具有高度的序列同源性（Cheng 等人，2014）。这些北方蟾蜍与其他多个物种饲养在一起，其中一些物种是从东南亚野外收集的，因而该病毒可能存在从国外引入的传播途径。感染动物园蛙病毒并在疾病暴发中不发生死亡的其他物种包括三角枯叶蛙（*Megophrys nasut*a）和大黄蜂蟾蜍（*Melanophryniscus stelzneri*）（Cheng 等人，2014）。目前还不清楚动物园蛙病毒是否是 BIV 的一种新颖的北美蛙病毒分离株，BIV 是否来自人工饲养的其他动物。

2.4 两栖类中的其他蛙病毒

普通产婆蟾病毒（CMTV）首先是在 2007 年，于西班牙北部大规模死亡的普通产婆蟾蝌蚪中分离到的（Balseiro 等人，2009）。2008 年，在西班牙发生普通产婆蟾蝌蚪大规模死亡的地区，CMTV 引起的第二次大规模死亡不仅影响到了普通产婆蟾蝌蚪，还影响到了快成年的高山蝾螈（*Mesotriton alpestris*）（Balseiro 等人，2010）。有文献报道，在西班牙，研究者经过长期研究发现，在几个不同地点，由 CMTV 引起的大规模死亡，至少发生在两栖类 6 个物种中，其中还记录 CMTV 引起了 3 个物种种群数量的下降（Price 等人，2014）。在比利时，CMTV 感染不仅存在于美国牛蛙这一外来入侵的种群当中（Sharifian-Fard 等人，2011），还存在于荷兰人工饲养及野生动物当中（Kik 等人，2011，2012）。在欧洲，CMTV 似乎是最常见的蛙病毒，而在其他地方还未发现 CMTV 的存在。研究发现，与其他蛙病毒相比，该病毒在进化上显得比较独特（Mavian 等人，2012）。因此，对这个病毒种命名很有必要。

另外一个比较独特的蛙病毒是最近从濒危的中国大鲵分离到的（Geng 等人，2011；Chen 等人，2013；Ma 等人，2014）。该病毒是从人工饲养的种群中分离到的，在所有的病例中，都引起了高发病率和高死亡率（Geng 等人，
27 2011；Chen 等人，2013；Ma 等人，2014）。系统发生分析显示，从中国大鲵人工养殖种群中分离到的这种病毒与 CMTV 最为接近（Chen 等人，2013）。曾令兵博士对好几个中国大鲵养殖场进行了研究，并对遍布中国 11 个省的该病毒的分布进行了记录（Zeng 和 Ma，长江水产研究所，未发表资料）。在中国，该蛙病毒的出现和流行，对中国大鲵的保护构成了严重的威胁（Cunningham 等人，2015）。

还有其他几种蛙病毒已从两栖类动物中分离出来，并被命名，但是这几种病毒还没有被确认为单独的病毒种。蛙虹彩病毒（Rana grylio virus，RGV）与 FV3 十分接近，是 20 世纪中叶在中国分离到的一种病毒（Zhang 等人，1996；Lei 等人，2012）。虎纹蛙病毒（Tiger frog virus，TFV）是于 2000 年在中国蛙养殖场发生大规模死亡后分离到的一种病毒（Weng 等人，2002），这两种病毒对亚洲水产养殖场构成了威胁。

3 感染鱼类的蛙病毒

无论是在海洋环境中，还是在淡水环境中，蛙病毒都可以造成有鳍鱼类严

重的全身性疾病（Whittington 等人，2010）。动物流行性造血器官坏死病毒是第一个与鱼类大规模死亡相关的蛙病毒，于 1985 年在澳大利亚首次得到分离（Langdon 等人，1986b）。不久以后，在欧洲检测到了与蛙病毒在遗传存在显著不同但紧密关联的欧洲鲶鱼病毒（European catfish virus，ECV）（Ahne 等人，1989）。桑堤-库珀蛙病毒（Santee-Cooper ranavirus，SCRV）曾非正式地被称为大嘴鲈病毒（Largemouth bass virus，LMBV）。目前，桑堤-库珀蛙病毒已被国际病毒分类委员会（International Committee on the Taxonomy of Viruses，ICTV）归类为蛙病毒，该病毒在美国与野生鱼类的动物流行病有着密切的联系（Plumb 等人，1996）。在两栖类和爬行类动物中，FV3 与典型的发病有着密切的联系。FV3 在同域的北方红腿蛙（*Rana aurora*）发生动物流行病期间，于发病后濒临死亡的三刺鱼（*Gasterosteus aculeatus*）中得到了分离（Mao 等人，1999a）。目前，三种主要感染鱼类的蛙病毒得到了国际病毒分类委员会的公认：EHNV 在自然状态下，只存在于澳大利亚；ECV 在分布上局限于欧洲；SCRV 主要在北美洲的鱼类中被检测到。

EHNV 和 ECV 两种病毒均对水产养殖产生影响（Whittington 等人，2010），而 BIV 似乎仅于澳大利亚孵化后饲养的尼罗罗非鱼鱼苗（*Oreochromis niloticus*）中暴发一次（Ariel 和 Owens，1997）。最近，在北美洲和亚洲孵化场饲养的淡水鱼中反复检测到了 SCRV 和 FV3（Woodland 等人，2002b；Prasankok 等人，2005；Deng 等人，2011；George 等人，2014；Chinchar 和 Waltzek，2014；Waltzek 等人，2014）。自 20 世纪 90 年代以后，两种在遗传上不同但密切关联的蛙病毒即新加坡石斑鱼虹彩病毒（Singapore grouper iridovirus，SGIV）和石斑鱼虹彩病毒（Grouper iridovirus，GIV）在亚洲对石斑鱼海洋养殖业产生了负面影响（Chua 等人，1994；Murali 等人，2002；Qin 等人，2003）。蛙病毒作为病原出现在自然环境中和人工饲养的有鳍鱼类种群中，其原因目前还不清楚。但是，在全球范围内，同种有鳍鱼类的蛙病毒，例如 SCRV 被反复检测出来，则提示活 28
动物及其产品的跨国境转运对于其动物流行病学起着很重要的作用（Hedrick 和 McDowell，1995；Plumb 和 Zilberg，1999a；Grant 等人，2005；Schramm 和 Davis，2006；Deng 等人，2011；George 等人，2014）。

3.1 动物流行性造血器官坏死病毒

在脊椎动物中，与全身性感染及大规模死亡有关联的第一个蛙病毒就是动物流行性造血器官坏死病毒（EHNV）。1985 年，该病毒感染被鉴定为发生在澳大利亚红鳍鲈（*Perca fluviatilis*）和虹鳟（*Oncorhynchus mykiss*）中的流行性死亡的原因（Langdon 等人，1986b，1988；Langdon 和 Humphrey，1987），但并没有确定疾病的暴发源。1981 年至 1984 年，在对澳大利亚鲑科鱼类进行的病毒感染检测调查中，并没有鉴定出任何病毒（Langdon 等人，1986a）。因此，动物流行性造血器官坏死病毒在首次发生大规模死亡事件之前，不存在于虹鳟当中。虽然对红鳍鲈（redfin perch）种群没有进行调查，但以前并没有大规模死亡的记载（Whittington 等人，1996）。到目前为止，除红鳍鲈以外，在野外，还没有动物流行性造血器官坏死病毒致死的案例。尽管如此，还是存在这样一个事实，就是按照实验室的感染实验，已知其他 14 个物种对该蛙病毒易感（Whittington 等人，2010；Becker 等人，2013）。观察野外有鳍鱼类的困难，再加上没有可信的临床疾病的报道，是澳大利亚目前还没有检出 EHNV 的原因。

在澳大利亚东南部，EHNV 对水产养殖的影响还只局限于渔场养殖的虹鳟中，而在澳大利亚的塔斯马尼亚及澳大利亚西部的鲑科鱼类，未遭到 EHNV 感染的侵扰。除了塔斯马尼亚，在整个澳大利亚的东南部，EHNV 感染是野生红鳍鲈的一种流行病。红鳍鲈对 EHNV 高度易感，而虹鳟则具有较强的抵抗力（Whittington 和 Reddacliff，1995）。

在受感染的虹鳟养殖场，EHNV 只引起小部分鱼发病（Whittington 等人，1994，1999）。通常，在整个同龄鱼类中，总死亡率≤4%。同时，一旦少数鱼被感染，那么感染个体的死亡率就会很高（Whittington 等人，1994，1999）。相反，EHNV 在红鳍鲈中会引起严重的疾病，在疾病的流行地区，感染比例高的种群是水花和幼鱼以及引入到其他区域的成鱼（Langdon 等人，1986b；Langdon 和 Humphrey，1987；Whittington 等人，1996）。大量的证据显示，感染过 EHNV 的红鳍鲈种群，会在随后几年得以恢复。依据病毒分离和血清学研究，有证据显示，无论是红鳍鲈还是虹鳟，都能在 EHNV 亚临床感染的条件下生存。因此，它们可能是病原的贮藏宿主（reservoirs）（Whittington 等人，2010）。

EHNV 会逐渐从红鳍鲈传播到河流系统中，这可能是由自然条件下鱼的洄游、鱼的放流（fish releases）以及鸟类（avifauna）引起的（Whittington 等人，1996）。经水传染和吞食被感染的鱼类是 EHNV 在一个种群内部易感宿主之间传播的途径。但远距离传播有可能是人类活动的结果，特别是在水产养殖中被
29 感染的虹鳟水花的远距离运输（Langdon 等人，1988；Whittington 等人，1994，1999）。在鳟鱼养殖场，一年一度的 EHNV 暴发可能是由于 EHNV 在局部环境中持续存在或从野生红鳍鲈中再次感染。

红鳍鲈 EHNV 引起的自然流行病，通常在夏天极易发生。有证据显示，EHNV 的致病性与水温呈正相关关系。在低于 10 ℃ 的条件下，红鳍鲈对

EHNV 不易感，而在比较高的温度条件下，该病毒的潜伏期（incubation periods）会大大缩短（Whittington 和 Reddacliff，1995）。在虹鳟鱼中，EHNV 的暴发常常发生在 11 ℃ ~20 ℃（Whittington 和 Reddacliff，1995；Whittington 等人，1994，1999）。这种依赖温度的致病性可能与病毒复制速率有关（Ariel 等人，2009）。

兰登（Langdon，1989）率先进行了 EHNV 的传播研究。该学者确定了易感宿主范围，这是世界动物卫生组织将该病毒列入清单的重要原因。虽然人们公认，该病毒缺乏实验室感染模型来确定宿主鱼对自然条件下分离到的病毒株的易感性，但是，有预测表明，EHNV 感染的宿主范围在不断增加。在三个独立的人工感染实验研究中，黑鮰（*Ameiurus melas*）、白斑狗鱼（*Esox Lucius*）和白梭吻鲈（*Sander lucioperca*）在进行 EHNV 水浴接种之后经历了显著的死亡（Bang-Jensen 等人，2009，2011a；Gobbo 等人，2010）。另外，金鱼（*Carassius auratus*）、鲤鱼（*Cyprinus carpio*）和欧洲鲶（*Silurus glanis*）在进行 EHNV 水浴接种之后，没有发生显著的死亡（Bang-Jensen 等人，2011b；Leimbach 等人，2014）。如同其他蛙病毒一样，EHNV 感染的结果可能取决于不同的病毒、宿主和环境因素，其中包括病毒的浓度和传播途径（route of delivery）、病毒株、宿主遗传学、宿主密度和年龄以及水温（Brunner 等人，2015）。例如，当对欧洲红鳍鲈鱼群接种 EHNV 时，观察到其死亡率较低。但是，正是这些被接种的个体造成了病原的传播，并造成了与之共同饲养的澳大利亚红鳍鲈的大量死亡（Ariel 和 Bang-Jensen，2009）。

3.2　欧洲鲶鱼病毒

欧洲鲶鱼病毒是欧洲引起鱼发病最重要的蛙病毒。该病毒在正式归类之前被称为欧洲六须鲶病毒（European sheatfish virus，ESV）。该病毒在德国养殖的六须鲶（Ahne 等人，1989，1991）以及法国与意大利的野生黑鮰（black bullheads）中触发了广泛的流行（Pozet 等人，1992；Bovo 等人，1993；Bigarré 等人，2008）。很明显的是，欧洲鲶鱼病毒在某些地方造成了疾病的流行，例如在法国的布尔歇湖和阿普勒蒙湖（Bigarré 等人，2008）。在意大利，该疾病既发生在人工养殖的黑鮰中，也发生在野生黑鮰以及渔场养殖的云斑鮰（*A. nebulosus*）中，并影响这些鱼类的养殖产量（Ariel 等人，2010）。此外，2008 年，在匈牙利云斑鮰中也检测出了 ECV 的暴发（Juhász 等人，2013）。

在欧洲，欧洲鲶鱼病毒病毒株的宿主范围、地理分布和多样性还不完全清 30
楚。该病毒很容易通过一系列接种方法，包括水浴接种、共养和肌肉内注射传播给鲶鱼（Ahne 等人，1990；Pozet 等人，1992），并导致接种鲶鱼大量死亡，而仅仅只有一小部分接种过病毒的鲶鱼存活下来（Pozet 等人，1992）。有趣的是，戈博（Gobbo）等人（2010）发现，依据与蛙病毒宿主的亲缘关系，其易感性有不同的方式，黑鮰对 ECV 易感，但对用于本研究的 ESV 分离株不易感。最近的实验显示，ECV 不同的分离株呈现不同的致病性，并且水温对疾病的结果产生强烈的影响，在接种病毒的六须鲶中，其死亡率为 8%~10%。在三个独立的人工感染实验研究中，黑鮰、白斑狗鱼和六须鲶通过水浴接种 ECV 病毒株以后，出现了显著的死亡现象（Bang-Jensen 等人，2009；Gobbo 等人，2010；Leimbach 等人，2014）。但是，金鱼、鲤鱼和白梭吻鲈通过水浴接种 ECV 病毒株以后，并没有出现显著的死亡（Bang-Jensen 等人，2011a，b）。

3.3 桑提-库珀蛙病毒

1995 年曾报道，在美国南卡罗来纳州桑堤河-库珀水库的大嘴鲈中发现虹彩病毒暴发的流行病（Plumb 等人，1996）。作者依据宿主的名称将病原命名为大嘴鲈病毒（LMBV）。后来，经过遗传学分析，确立 LMBV 是蛙病毒属一个独特的成员（Mao 等人，1997，1999b），并且与医生鱼病毒（doctor fish virus，DFV）和孔雀鱼病毒 6（guppy virus 6，GV6）几乎完全相同，此外，该病毒曾从原产于东南亚进口的观赏鱼中得到了分离（Hedrick 和 McDowell，1995）。依据该病毒分离的地点，LMBV 后来被命名为桑堤-库珀蛙病毒（Santee-Cooper ranavirus，SCRV）。然而，格里兹尔（Grizzle）等人（2002）发表异议，引用名的改变应尊重事实，就是 LMBV 最先（1991）是从佛罗里达州韦尔湖的大嘴鲈中分离到的（Francis-Floyd，1992）。前面所提及的研究和最近的种系发生分析支持 LMBV、GV6 和 DFV 是同一病毒种的不同病毒株，该病毒种被国际病毒分类委员会（ICTV）正式命名为 SCRV（Holopainen 等人，2009）。更为重要的是，依据遗传序列分析以及流行病学和病理学特征分析，有学者认为，SCRV 与蛙病毒属的病毒差异太大，以至于不能归于蛙病毒属（Hyatt 等人，2000；Whittington 等人，2010；Jancovich 等人，2015）。

曾有多次报道，北美洲大嘴鲈野生种群多次发生 SCRV 感染（Grizzle 和 Brunner，2003；Plumb 和 Hanson，2011）。SCRV 的毒力似乎在自然环境和实验环境条件下存在差异，在成鱼中，典型的疾病发生在夏天。成鱼发病时，能在水面观察到病鱼的浮力和平衡出现问题。到目前为止，引起 SCRV 流行病的各种因素仍不清楚，可能存在病例的特异性。然而，大嘴鲈的遗传背景、种群中接触 SCRV 的病史、SCRV 病毒株以及环境因素，例如与水温升高相关的溶解氧浓度的降低，被认为是引起 SCRV 病毒病的潜在因素（Grizzle 和 Brunner，2003；Plumb 和 Hanson，2011）。在大嘴鲈和条纹鲈（*Morone saxatilis*）中，实验感染条纹鲈显示，无论是浸泡还是注射接种病毒，病毒滴度与死亡率之间都有着直接的关联。然而，条纹鲈整个累计死亡率较低（Plumb 和 Zilberg，1999b；Zilberg 等人，2000）。在大嘴鲈口服 SCRV 的传播实验中，大嘴鲈出现了皮肤和内脏器官的感染（例如鱼鳔的感染），但未发生死亡（Woodland 等人，2002a）。 31

在美国的 31 个州中，南至佛罗里达州，西至亚利桑那州，北至威斯康星州、密歇根州、纽约州、佛蒙特州以及特拉华州，在许多不同种的野生无症状淡水鱼类包括 6 科 17 种中分别分离到了 SCRV（Goldberg，2002；Woodland 等人，2002b；Groocock 等人，2008；USFWS，2011；Iwanowicz 等人，2013；见表 3）。最近，有学者还从刚引进到切萨皮克湾水域的外来物种乌鳢（*Channa argus*，又称北方蛇头鱼）中分离到了 SCRV 病毒株（Iwanowicz 等人，2013）。有实验资料显示，SCRV（DFV/GV6）病毒株在虹鳟和大鳞大麻哈鱼（*Oncorhynchus tshawytscha*）中诱发了感染，并产生了低的死亡率，但在斑点叉尾鮰（*Ictalurus punctatus*）中则并非如此（Hedrick 和 McDowell，1995）。

在中国，从濒临死亡的、孵化后饲养的大嘴鲈中分离到了一个 SCRV 病毒株。研究发现，该病毒株对大嘴鲈具有很高的致死性（Deng 等人，2011），然而，其他 7 种实验鱼包括鲤鱼（*Cyprinus carpio*；见表 3）则很少发生死亡，或没有发生死亡。相反，最近印度南部养殖场的鲤鱼发生了大规模的死亡，虽然文献作者对病毒诱导的病理变化没有提供详细的描述，但其元凶就是一种 SCRV 病毒株（George 等人，2014）。

SCRV 有可能通过活鱼及与观赏鱼养殖有关的产品（Hedrick 和 McDowell，1995；Deng 等人，2011；George 等人，2014）、饵料（Plumb 和 Zilberg，1999a）以及垂钓业（Grant 等人，2005；Schramm 和 Davis，2006）等在整个美国乃至全球范围内进行传播。例如，在美国举办的大嘴鲈垂钓锦标赛中，SCRV 就会通过感染的鱼和未感染的鱼近距离接触而传播给本地鱼类。然而，因垂钓引起的应激还没有显示出会大大增加与 SCRV 相关的死亡率（Grant 等人，2005；Schramm 和 Davis，2006）。考虑到 SCRV 在冰冻宿主组织中仍保持感染状态，进出口冰冻鱼组织有可能是病毒传播的另外一种机制（Plumb 和 Zilberg，1999a）。在未来，需要我们通过共同努力进行监测，以确定在全球范围内由 SCRV 传播给上述产业所带来的风险。

表 3 野生鱼类和人工饲养鱼类蛙病毒感染或死亡分布(包括进口动物原产地的地理位置)

大陆	国家和地区	原产地[a]	州或省[a]	科名	拉丁名	I, M[b]	W, C[c]	参考文献
亚洲	中国	—	广东	太阳鱼科	加州鲈	M	C	Deng et al.(2011)
	印度	—	—	鲤科	鲤鱼	M	C	George et al.(2014)
	新加坡	—	—	鮨科	巨石斑鱼	M	C	Chua et al.(1994), Qin
		—	—	太阳鱼科	加州鲈	—	C	Huang et al.(2011)
	中国台湾	—	—	尖吻鲈科	尖吻鲈	—	C	Huang et al.(2011)
		—	—	笛鲷科	红鳍笛鲷	—	C	Huang et al.(2011)
		—	—	鮨科	青石斑鱼	M	C	Murali et al.(2002), Huang et al.(2011)
		—	—		点带石斑鱼	—	C	Huang et al.(2011)
		—	—		鞍带石斑鱼	—	C	Huang et al.(2011)
	泰国	—	佛统	塘鳢科	云斑尖塘鳢	M	C	Prasankok et al.(2005)
大洋洲	澳大利亚	—	NSW, SA, Vic	鲈科	赤鲈	M	W	Whittington et al.(2010)
		—	NSW, SA, Vic	鲑科	虹鳟	M	C	Whittington et al.(2010)
欧洲	丹麦	—	—	鳕科	大西洋鳕	I	W	Ariel et al.(2010)
		—	—	菱鲆科	大菱鲆	M	C	Ariel et al.(2010)
	芬兰	—	—	鲈科	白梭吻鲈	I	W	Tapiovaara et al.(1998)
	法国	—	—	鮰科	黑鮰	M	W	Pozet et al.(1992), Bigarré et al.(2008)
	德国	—	—	鲶科	欧洲巨鲶	M	C	Ahne et al.(1989, 1991)
	匈牙利	—	—	鮰科	褐首鲶	M	W	Juhász et al.(2013)
	意大利	—	—	鮰科	黑鮰	M	W	Bovo et al.(1993)
		新西兰	—	鳗鲡科	澳洲鳗鲡	—	W	Bovo et al.(1999)

续表

大陆	国家和地区	原产地[a]	州或省[a]	科名	拉丁名	I, M[b]	W, C[c]	参考文献
北美洲	美国	亚洲	CA	隆头鱼科	裂唇鱼	—	—	Hedrick and McDowell(1995)
		亚洲	VA	蛇头鱼科	乌鳢	I	W	Iwanowicz et al.（2013）
		—	MN	太阳鱼科	岩钝鲈	—	—	USFWS（2011）
		—	MD，TN	太阳鱼科	红胸太阳鱼	—	—	USFWS（2011）
		—	GA，IL，MN，OH	太阳鱼科	蓝鳃太阳鱼	—	—	Woodland et al.（2002a），USFWS（2011）
		—	AR	太阳鱼科	长耳太阳鱼	—	—	USFWS（2011）
		—	AZ，KY，MI，MN，NY，OH，OK，PA，WI，WV	太阳鱼科	小口黑鲈	M	W	Groocock et al.（2008），USFWS（2011）
		—	FL	太阳鱼科	南方黑鲈	—	—	USFWS（2011）
		—	AL，AR，GA，KY，LA，MI，OK，TN	太阳鱼科	大口黑鲈	—	—	USFWS（2011）
		—	AL，AR，AZ，CT，DE，FL，GA，IA，IL，KS，KY，LA，MD，MI，MN，MO，MS，NC，NJ，NY，OH，OK，PA，SC，TN，TX，VA，VT，WI	太阳鱼科	加州鲈	I，M	W，C	Grant et al.（2005），Groocock et al.（2008），Grizzle et al.（2002），Hanson et al.（2001），Mao et al.（1999b），Neal et al.（2009），Plumb et al.（1996，1999），Southard et al.（2009），USFWS（2011），Woodland et al.（2002b）
		—	WI	太阳鱼科	黑斑刺盖太阳鱼	—	—	USFWS（2011）
		—	KY	胭脂鱼科	小孔亚口鱼	—	—	USFWS（2011）
		—	NC	鲤科	小头美鲅	—	—	USFWS（2011）

续表

大陆	国家和地区	原产地[a]	州或省[a]	科名	拉丁名	I, M[b]	W, C[c]	参考文献
北美洲	美国	—	IL	狗鱼科	北美狗鱼	—	—	USFWS（2011）
		—	—	狗鱼科	暗色狗鱼	—	—	Goldberg（2002）
		—	SC	梦鲈科	美洲狼鲈	—	—	USFWS（2011）
		—	KS	梦鲈科	金眼狼鲈	—	—	USFWS（2011）
		—	AZ	梦鲈科	密西西比狼鲈	—	—	USFWS（2011）
		—	MN，OH	石首鱼科	淡水石首鱼	—	—	USFWS（2011）
		亚洲	CA	胎鳉科	孔雀花鳉	—	—	Hedrick and McDowell（1995）
		—	CA	刺鱼科	三刺鱼	M	W	Mao et al.（1999a）
		—	GA	鲟科	俄罗斯鲟	M	C	Waltzek et al.（2014）
		—	CA	鲟科	白鲟	M	C	Waltzek et al.（2014）
		—	MO	鲟科	密苏里铲鲟	M	C	Waltzek et al.（2014）

[a] 表示当有的时候才提供相应信息。

[b] I 表示没有蛙病毒疾病症状的感染，M 表示由蛙病毒疾病引起的死亡。

[c] W 表示野生种群，C 表示人工饲养种群，包括动物园和鱼类养殖场的种群，不包括用来对照的病毒接种研究的种群。

NSW：新南威尔士州；SA：南澳大利亚州；Vic：维多利亚州；CA：加利福尼亚州；VA：弗吉尼亚州；MN：明尼苏达州；MD：马里兰州；TN：田纳西州；GA：佐治亚州；IL：伊利诺伊州；OH：俄亥俄州；AR：阿肯色州；AZ：亚利桑那州；KY：肯塔基州；MI：密歇根州；NY：纽约州；OK：俄克拉何马州；PA：宾夕法尼亚州；WI：威斯康星州；WV：西弗吉尼亚州；FL：佛罗里达州；AL：亚拉巴马州 LA：路易斯安那州；CT：康涅狄格州；DE：特拉华州；IA：艾奥瓦州；KS：堪萨斯州；MO：密苏里州；MS：密西西比州；NC：北卡罗来纳州；NJ：新泽西州；SC：南卡罗来纳州；TX：得克萨斯州；VT：佛蒙特州

3.4　蛙病毒3

虽然在野生鱼类中，有关 FV3 的感染只有唯一一例文献报道（Mao 等人，1999a），但是在人工养殖或渔场养殖（captive/cultured）的鱼类当中，有关鱼类感染 FV3 已有无数病例报道。在野生鱼类的感染当中，FV3 样病毒出现在一个濒临死亡的三刺鱼（threespine stickleback）体中，在发病的鱼体中还发现 35
了黏孢子虫或称黏体动物寄生虫（myxozoan parasites）共感染，这就使得病毒在发病中的作用显得模糊（Mao 等人，1999a）。然而，FV3 样病毒已多次从人工饲养的鱼类中得到分离。据报道，在泰国养殖的云斑尖塘鳢（*Oxyeleotris marmoratus*）当中分离到一种蛙病毒，其主衣壳蛋白基因的片段显示出与 FV3 的核甘酸序列有 98%～99% 的同源性（Prasankok 等人，2005）。此外，在美国密苏里河流域，FV3 的暴发极大地妨碍了重要濒危物种密苏里铲鲟（*Scaphirhynchus albus*）的种群恢复（Waltzek 等人，2014）。在美国密苏里河斯威特斯普林斯的 Blind Pony 孵化场，分别于 2001 年、2009 年以及 2013 年在当年所产的密苏里幼铲鲟中发生高死亡率的流行病（Chinchar 和 Waltzek，2014；Waltzek 等人，2014）。研究者利用 2009 年的病毒分离株进行传播实验，结果使本地的稚密苏里铲鲟（*Juvenile pallid sturgeon*）在水浴感染接种后，复制了同样高死亡率的疾病（Waltzek 等人，2014）。此外，从濒临死亡的、孵卵饲养的俄罗斯鲟（*Acipenser gueldenstaedtii*）中分离到了一株 FV3 病毒株，结果发现，该病毒无论是对俄罗斯鲟，还是对湖鲟（*A. fluvescens*），腹腔注射接种病毒后，都引起了死亡（Waltzek 等人，2014）。最后，FV3 样病毒还于 1998 年在美国加利福尼亚养殖场中的稚美国白鲟（*A. transmontanus*）中得到了分离（Waltzek 等人，2014）。

运用一系列不同类别变温脊椎动物中分离到的 FV3 样病毒进行传播实验，结果显示，FV3 样病毒能感染黑鮰（*Ameiurus melas*）、白斑狗鱼、梭鲈、食蚊鱼（*Gambusia affinis*）以及蓝鳃太阳鱼（*Lepomis macrochirus*）。然而，在这些被感染的品种中，没有观察到死亡或很少发生死亡（Gobbo 等人，2010；Bang-Jensen 等人，2009，2011a，b；Brenes 等人，2014a）。类似的情况还有，最近北美洲鱼类健康调查结果显示，从外观显得健康的黑头呆鱼（*Pimephales promelas*）、玻璃梭鲈（*Sander vitreus*）以及白斑狗鱼中都分离到了 FV3 病毒（Waltzek 等人，2014）。尽管这些只是研究的初步资料，但这些资料提示，濒危的鲟可能对 FV3 样病毒具有易感性，而其他鱼类有可能只是简单地作为病毒的携带者或终末宿主（dead-end hosts）。未来的研究需要进一步探讨 FV3 样病毒在整个野生和人工饲养鱼类中的重要性，以及这些重要病原在全球传播对水产养殖的潜在影响。

3.5 博乐虹彩病毒

正如上文所指出的一样，博乐虹彩病毒（BIV）首次是从澳大利亚患病的华丽穴居蛙蝌蚪中分离到的。尽管该病毒被国际病毒分类委员会命名为一个不同的种，但是其主衣壳蛋白基因与 EHNV 有 97.8% 的同源性，而 EHNV 是澳大利亚的一种流行病病原（Marsh 等人，2002）。如同 FV3 一样，实验室人工感染接种研究显示，BIV 对鱼类同样具有致病性，这种致病性也见于尖吻鲈（*Lates calcarifer*）。在澳大利亚，该鱼是一种受普遍欢迎的体育垂钓性鱼类（Moody 和 Owens，1994）。一次偶然的机会，在澳大利亚发现 BIV 与孵化场饲养的尼罗罗非鱼鱼苗大量死亡有关（Ariel 和 Owens，1997）。虽然作者没有从遗传上鉴定该虹彩病毒，但是用濒临死亡的尼罗罗非鱼鱼苗饲养尖吻鲈幼鱼，
36 复制了尖吻鲈在接种 BIV 后所报道的发病情况（Moody 和 Owens，1994）。

3.6 感染鱼类而在分类学上没有被命名的蛙病毒

虽然感染淡水鱼类的蛙病毒比较好鉴定，但是对感染海洋养殖鱼类和野生种群的蛙病毒则知之甚少。SGIV 和 GIV 这两种蛙病毒除外。这两种病毒在亚洲给石斑鱼海洋养殖业带来了严重影响（Chua 等人，1994；Murali 等人，2002；Qin 等人，2003）。依据 26 个保守虹彩病毒基因的系统发生分析，结果显示，GIV 和 SGIV 互相是最近的亲缘种。然而，这些病毒在遗传上是有别于其他蛙病毒的一个分支（Eaton 等人，2007）。因此，需要考虑将 CIV 和 SGIV 划归为虹彩病毒科的一个新属（Jancovich 等人，2015）。

SGIV 是 1994 年在新加坡网箱养殖的褐斑石斑鱼（*Epinephelus tauvina*）发生大规模死亡之后首次被分离鉴定的（Chua 等人，1994）。当时，作者将这一流行病称为石斑鱼沉睡病（sleepy grouper disease），遗憾的是，他们没有分离到该病毒并进行遗传学上的鉴定。1998 年，在新加坡的养鱼场，从中国台湾省进口的褐斑石斑鱼鱼苗中再次观察到了这一相同的疾病（Qin 等人，2003）。这些研究者通过进行遗传学上的鉴定，认为该病毒是一种新型的蛙病毒，并取名为 SGIV（Qin 等人，2003）。类似的是，GIV 在中国台湾省也给青石斑鱼（*Epinephelus awoara*）的生产带来了不良影响（Murali 等人，2002）。在实验感染过程中，通过注射将病毒接种给青石斑鱼，结果引起了 100% 的死亡（Murali 等人，2002）。最近有报道，在中国台湾省，从养殖的石斑鱼和非石斑鱼中分离到了 SGIV 和 GIV 病毒株（Huang 等人，2011，表 3）。这一研究说明，

SGIV 和 GIV 包括第一次从淡水鱼大嘴鲈和下海产卵的尖吻鲈中分离到的这些病毒，其宿主范围明显扩大。

其他从野生海洋鱼类中分离并作了部分鉴定的蛙病毒包括从丹麦大西洋鳕鱼（*Gadus morhua*）中分离到的鳕鱼蛙病毒（Cod ranavirus，CoIV）（Ariel 等人，2010）以及从新西兰沿海短鳍鳗（*Anguilla australis*）中分离到的短鳍鳗蛙病毒（Short-finned eel ranavirus，SERV）（Bang-Jensen 等人，2009）。从外观上看起来健康的淡水和海水养殖鱼类中分离到的蛙病毒包括从芬兰梭鲈小鱼（Finnish pike-perch fingerlings）中分离到的梭鲈虹彩病毒（Pike-perch iridovirus，PPIV）（Tapiovaara 等人，1998）以及从丹麦大菱鲆（*Scophthalmus maximus*）鱼苗中分离到的大菱鲆蛙病毒（Ranavirus maxima，Rmax）（Ariel 等人，2010）。对这些鱼类病毒种系发生的初步分析显示，它们代表着以前未知的病毒，并要求对这些病毒的生物学以及这些病毒对养殖的和野生的鱼类种群的影响做更进一步的综合研究（Holopainen 等人，2009；Jancovich 等人，2015）。

4　感染爬行类动物的蛙病毒

尽管在爬行类动物中有无数个蛙病毒感染的病例描述，但是，由于缺乏认 37
识，缺乏长期的研究，以及在生物学研究上缺乏对该疾病的监测，爬行类动物的蛙病毒病极有可能被少报（Daszak 等人，1999；Johnson 等人，2010；Allender，2012）。值得注意的是，爬行类动物蛙病毒的感染在过去的十年间已明显加速。在龟类动物中，这些病毒作为重要传染原的病例报道不断增多，从而，人们对这些病毒的认识也在不断加深。而且，从总体上来讲，在爬行类动物当中，人们对这些病毒作为病原的意识也不断增强（Shaver，2012）。据推测，两栖类及爬行类动物在全球的贸易以及蛙病毒具有广泛的宿主范围是蛙病毒病发生的主要原因（Stöhr 等人，2013a），这一点对野生的和人工养殖的爬行类、两栖类和鱼类具有重要的意义。

爬行类动物感染蛙病毒的信息量在不断增大。到目前为止，大多数在爬行类动物中检测出来的蛙病毒一直是 FV3 样病毒（Huang 等人，2009；Allender 等人，2011）。除此之外，ECV 样病毒、BIV 样病毒以及 CMTV 样病毒已在人工饲养的好几种爬行动物中被检测出来（Marschang 等人，2013；Stöhr 等人，2015）。对这些病毒的鉴定大多是依据主衣壳蛋白（MCP）基因的序列片段。但是，其他的序列数据也有助于我们弄清在爬行类动物中所发现的蛙病毒之间的关系。在美国，目前还只在爬行类动物中检测出 FV3 样病毒，该病毒是无尾类动物中报道最多的蛙病毒。在欧洲，已有文献进行过描述，蛙病毒类型广泛，其中既包括 FV3 样病毒以及 ECV、BIV，也包括 CMTV 的病毒代表。在亚洲，存在于龟鳖类动物中的中华鳖虹彩病毒（Soft-shelled turtle iridovirus，STIV）为完全测序的蛙病毒，并且显示出与 FV3 的亲缘关系极近（Huang 等人，2009）。

4.1 爬行动物蛙病毒感染的病史

20 世纪 80 年代，在瑞士，有学者对乌龟感染虹彩病毒的两个病例在文献中作了描述（Heldstab 和 Bestetti，1982；Müller 等人，1988）。根据所描述的临床症状、病理学变化以及电子显微镜的观察发现，这些动物被认为感染了蛙病毒，这也是对爬行类动物最早进行蛙病毒感染和发病的病例记录。20 世纪 90 年代中叶，在人工饲养的和野生的龟鳖类动物中，有学者记录了为数不多，并证实是蛙病毒感染爬行类动物的病例（见表 4）。蛙病毒感染爬行动物的第一批病例为北美洲的箱龟（*Terrapene c. carolina*）和四爪陆龟（*Testudo horsfieldii*），现在已有该病毒基因组的数据。依据主衣壳蛋白基因序列片段和限制性内切酶分析，这两例蛙病毒感染似乎都是由 FV3 样病毒引起的（Mao 等人，1997）。关于蛙病毒感染这两种龟，到目前为止，还没有相关的临床资
40 料发表。最近，在世界范围内，蛙病毒感染龟鳖类的报道和病例数量在不断增加，在美国感染箱龟的情况尤其如此（De Voe 等人，2004；Allender 等人，2006；Johnson 等人，2008；Allender，2012，见表 4）。虽然在龟鳖类动物中检测出蛙病毒的报道最为常见，但是在蜥蜴和蛇类动物中，蛙病毒的检出率也在不断增加，并且大部分来自人工饲养的种群（Stöhr 等人，2013b；Behncke 等人，2013；Marschang 等人，2013）。在野生的和人工饲养的爬行类动物中，蛙病毒检出率升高的原因，有可能是认识的提高、监测的增加、检测方法的改进或真正病原的出现。

4.2 龟鳖类蛙病毒的暴发

成年龟鳖发生 FV3 样病毒感染的报道比稚龟鳖更为常见（Johnson，2006）。然而，最近对东方箱龟（eastern box turtles）的监测显示，稚龟 FV3 阳性率很高（Allender，2012）。因此，其中的原因有可能是龟鳖类与两栖类动物类似，在不同的发育阶段对蛙病毒的敏感性不同（Haislip 等人，2011）。有些箱龟蛙病毒的暴发，涉及许多个体挤压在一起进行长距离转运，从而导致很高

表 4　野生和人工饲养爬行类动物蛙病毒感染或死亡分布（包括进口动物原产地的地理位置）

大陆	国家	原产地[a]	州或省[a]	科名	拉丁名	I，M[b]	W，C[c]	参考文献
亚洲	中国	—	—	鳖科	中华鳖	M	C	Chen et al.（1999）
大洋洲	澳大利亚	印度尼西亚	—	蟒科	绿树蟒	M	C/W	Hyatt et al.（2002）
欧洲	奥地利	埃塞俄比亚	—	陆龟科	豹纹陆龟	M	C	Benetka et al.（2007）
	德国	经佛罗里达到亚洲	—	鬣蜥科	丽纹攀蜥	M	C/W	Behncke et al.（2013）
		—	—		鬃狮蜥	I，M	C	Stöhr et al.（2013b），Marschang et al.（2013）
		亚洲	—	蛇蜥科	亚洲玻璃蜥蜴	M	C/W	Stöhr et al.（2013a）
		—	—	蚺科	水蚺属未定种	I	C	Marschang et al.（2013）
		美国	FL	安乐蜥科	沙氏变色蜥	M	C/W	Stöhr et al.（2013b）
		美国	FL		绿安禄蜥	M	C/W	Stöhr et al.（2013b）
		—	—	水龟科	欧洲泽龟	I	C	Stöhr et al.（2013d）
		—	—		红耳龟	I	C	Marschang et al.（2013）
		—	—	壁虎科	巨型叶尾壁虎	M	C	Marschang et al.（2005）
		—	—	美洲鬣蜥科	绿鬣蜥	M	C	Stöhr et al.（2013b）
		—	—	陆龟科	欧洲陆龟	I	C	Marschang et al.（2013）
		—	—		赫曼陆龟	M	C	Blahak and Uhlenbrok（2010）
		—	—		四爪陆龟	I	C	Marschang et al.（2013）
		—	—		埃及陆龟	M	C	Blahak and Uhlenbrok（2010）
		—	—		缘翘陆龟	M	C	Blahak and Uhlenbrok（2010）
		—	—		豹纹陆龟	I	C	Marschang et al.（2013）

续表

大陆	国家	原产地[a]	州或省[a]	科名	拉丁名	I, M[b]	W, C[c]	参考文献
欧洲	德国	印度尼西亚	—	蟒科	血蟒	M	C/W	Stöhr et al.（2015）
		—	—		球蟒	I	C	Marschang et al.（2013）
		—	—		缅甸蟒	I	C	Marschang et al.（2013）
		—	—	巨蜥科	蓝树巨蜥	I	C	Marschang et al.（2013）
	葡萄牙	—	—	草蜥科	伊比利岩蜥	I	W	Alves de Matos et al.（2011）
	瑞士	南斯拉夫	—	陆龟科	赫曼陆龟	M	C/W	Müller et al.（1988）
		—	—		赫曼陆龟	M	C	Heldstab and Bestetti（1982）, Marschang et al.（1999）
	英国	—	—	草蜥科	捷蜥蜴	I	W	Marschang et al.（2013）
		—	—	陆龟科	赫曼陆龟	M	C	Marschang et al.（2013）
北美洲	美国	—	RI, VA, WA	水龟科	锦龟	M	W	USGS（2005, 2008）, Goodman et al.（2013）
		—	FL		佛罗里达箱龟	M	C, W	Johnson et al.（2008）
		—	IL, IN, KY, MD, NC, NY, PA, TN, TX, VA, WV		东部箱龟	I, M	C, W	Mao et al.（1997）, De Voe et al.（2004）, Allender et al.（2006）, Johnson et al.（2008）, Ruder et al.（2010）, Farnsworth（2012）, Sim et al.（2012）, Allender（personal communication）, Currylow et al.(2014), Kimble et al.（2014）
		—	—		红耳龟	I	W	Allender（personal communication）
		—	—	陆龟科	缅甸星龟	M	C	Johnson et al.（2008）
		—	—		哥法地鼠龟	M	W	Westhouse et al.（1996）, Johnson et al.（2008）
		—	—		四爪陆龟	—	C	Mao et al.（1997）

[a] 表示当有的时候才提供相应信息。

[b] I 表示没有蛙病毒症状的感染，M 表示由蛙病毒病引起的死亡。

[c] W 表示野生种群；C 表示人工饲养种群，包括动物园和鱼类养殖场的种群，不包括用来对照的病毒接种研究的种群。

FL：佛罗里达州；RI：罗得岛州；VA：弗吉尼亚州；WA：华盛顿州；IL：伊利诺伊州；IN：印第安纳州；KY：肯塔基州；MD：马里兰州；NC：北卡罗来纳州；NY：纽约州；PA：宾夕法尼亚州；TN：田纳西州；TX：得克萨斯州；WV：西弗吉尼亚州

的流行性感染与死亡率（Belzer 和 Seibert，2011；Farnsworth 和 Seigel，2013；Kimble 等人，2014）。在美国，对箱龟多年的调查研究发现，蛙病毒的流行性感染，在一个种群中还没有超出 5% 的报道，并且也没有发生异常的死亡事件（Allender 等人，2013）。

在爬行动物中，单一病例和蛙病毒病暴发的报道，到目前为止，主要发生在北美的箱龟中（De Voe 等人，2004；Allender 等人，2006；Johnson 等人，2008，2010；Ruder 等人，2010；Allender，2012；Kimble 等人，2014）。虽然东方箱龟主要生活在陆地上，但它们会花费大量的时间待在临时性的池塘里（Donaldson 和 Echternacht，2005），这样就有可能通过水体或在同一池塘共生的两栖类动物接触蛙病毒（Belzer 和 Seibert，2011；Currylow 等人，2014）。有些研究表明，箱龟中蛙病毒感染引起死亡率的增加，有可能与箱龟不断接触感染的且在同域生存的两栖类动物有关，并极有可能通过捕食感染的两栖类、接触由于两栖类排毒形成的含有蛙病毒的水体或通过食血昆虫而感染（Belzer 和 Seibert，2011；Kimble 等人，2014）。有学者报道，在美国弗吉尼亚州，对一种水生种类的野生东方彩龟研究表明，不同的池塘，东方彩龟蛙病毒感染流行率为 4. 8% ~31. 6%，且没有明显的发病表现（Goodman 等人，2013）。

科赫法则（Koch postulates）已成功地用于龟鳖类动物（chelonians），如箱龟的蛙病毒病验证。实验接种从缅甸星龟（*Geochelone platynota*）中分离到的 FV3 样病毒株和从东方箱龟中分离到的 FV3 样病毒株，结果都导致了巴西红耳龟（red-eared sliders）的大量死亡（Johnson 等人，2007；Allender，2012），特征性的临床症状可见流鼻液（nasal discharge）和口腔菌斑（oral plaques）。暴露在 22 ℃条件下的巴西红耳龟与暴露在 28 ℃条件下的巴西红耳龟相比，其死亡率要高得多，所观察到的临床症状也明显得多。龟在较低温度条件下，病毒的拷贝数量增加以及平均生存时间（median survival time）缩短（Allender 等人，2013）。

除上述箱龟病例外，其他 FV3 样病毒感染的报道还见于人工养殖的龟鳖 41
类动物（Marschang 等人，1999；De Voe 等人，2004；Benetka 等人，2007；Johnson 等人，2008；Blahak 和 Uhlenbrok，2010，见表 4）。在德国，有学者运用 PCR 技术检测出 2 个稚赫曼陆龟（*Testudo hermanni*）发生了蛙病毒感染。感染组的 7 个动物全部死亡，并具有相同的症状（Marschang 等人，1999）。相关联的病毒最先被描述为 FV3 样病毒，但之后，有研究显示，感染赫曼陆龟的蛙病毒在亲缘关系上更接近 CMTV（Stöhr 等人，2015）。PCR 证实，伴有鼻分泌物、胃炎和昏睡症状的豹纹陆龟（*Stigmochelys pardalis pardalis*）并发了蛙病毒和疱疹病毒的感染（Benetka 等人，2007）。在德国，蛙病毒与人工饲养的赫曼陆龟、埃及陆龟（*T. kleinmanni*）以及缘翘陆龟（*T. marginata*）死亡事件有着密切的联系。在有些病例当中，被感染的动物发生口腔炎、脾脏坏死、肠炎、肝炎、胰腺炎以及皮炎（Blahak 和 Uhlenbrok，2010）。与这些疾病暴发相关的病毒即龟病毒 1 和龟病毒 2（缩写为 ToRV-1 和 ToRV-2），其基因组分析显示，在序列分析聚类上更接近 FV3，同时，这些病毒在基因组的排列上更像 CMTV（Stöhr 等人，2015）。在中国，人工养殖的中华鳖发生了红脖子病，该疾病也与蛙病毒有关（Chen 等人，1999）。中华鳖虹彩病毒是第一个完整测序的爬行类动物蛙病毒，并且该病毒就是 FV3 样病毒（Huang 等人，2009）。蛙病毒感染龟鳖类动物的比较研究显示，无论是 FV3 样病毒，还是 CMTV 样病毒，都引起了发病，且表现出来自不同暴发地点的病毒株所引起的感染彼此也有差别，相对来自爬行类动物的其他蛙病毒来说，这些病毒株与先前描述的两栖类蛙病毒亲缘关系更近（Stöhr 等人，2015）。

蛙病毒对龟鳖类动物种群的影响是短期的还是长期的，目前还不清楚。针对蛙病毒暴发给种群稳定性带来的影响，一直存在着争论，特别是针对易感的东方箱龟所产生的争论更为激烈。由于这些龟鳖类动物繁殖率较低，且性成熟的时间也较长，所以由这一疾病所引起的成年雌性动物损失，随着时间的推移，将有可能导致种群数量的显著下降（Farnsworth 和 Seigel，2013）。

4.3 有鳞类爬行动物蛙病毒暴发

到目前为止，有鳞类爬行动物（包括蛇类与蜥蜴类）蛙病毒暴发的相关报道还很少。这些动物当中的第一个蛙病毒感染的报道是发生在 10 个稚绿树蟒群（*Morelia viridis*）中，这些稚蟒是从巴布亚新几内亚进口到澳大利亚的，且有口腔和肝脏损伤。从池中死亡尸体组织中分离到的蛙病毒，与 FV3 有 97%的同源性（Hyatt 等人，2002）。一种 FV3 样病毒从一条红血蟒（*Python brongersmai*）中具有相同病理学变化的几个器官中得到了分离，这条红血蟒是从德国进口到印度尼西亚的。所分离到的蛙病毒与虎纹蛙病毒（TFV）亲缘关系最为接近，中国最早对该病毒进行描述（Stöhr 等人，2015）。一只叶尾壁虎（*Uroplatus fimbriatus*）意外死亡，并诊断出具有 BIV 样病毒感染（Marschang
42 等人，2005；Stöhr 等人，2015）。在葡萄牙，从一只野外捕捉的伊比利亚山蜥蜴（*Lacerta monticola*）中分离到了一种蛙病毒，该蜥蜴没有显现出任何发病的临床症状。分离到的这一病毒株在亲缘关系上与 FV3 最近，同时还发现与红细胞坏死病毒有合并感染（Alves de Matos 等人，2011）。根据一份对蜥蜴样品进行病毒学筛选研究的描述（Stöhr 等人，2013b），在沙氏变色蜥（*Anolis sagrei*）、亚洲玻璃蜥蜴（*Dopasia gracilis*）、安乐蜥（*Anolis carolinensis*）、绿鬣蜥（*Iguana iguana*）以及鬃狮蜥（*Pogona vitticeps*）5 个蜥蜴物种中检测出了蛙病毒感染，所有感染的蜥蜴都可见到皮肤病变。病毒主衣壳蛋白（MCP）基因的片段测序显示，5 种被检测的病毒之间存在差异，且与 FV3 基因组相对应的部分有 98.4% ~ 100% 的同源性。研究还发现，从绿鬣蜥中检测到的蛙病毒与 ECV 有 100% 的同源性，在德国鬃狮蜥（bearded dragon）中发现的蛙病毒与乌龟中检测出的病毒（ToRV-1）完全同源，而 ToRV-1 与 FV3 亲缘关系最近（Stöhr 等人，2015）。对所分离到的病毒作做一步分析显示，从变色龙中分离的病毒基因组与 FV3 亲缘关系较近，而从亚洲玻璃蜥蜴中分离到的病毒，在系统发生的聚类分析上则与 TFV 亲缘关系较近（Stöhr 等人，2015）。从中国西南进口，经过美国佛罗里达州，然后进入到德国的丽纹龙蜥（*Japalura splendida*），在发生大规模死亡事件的过程中，分离到了一种蛙病毒，这种蛙病毒似乎与 FV3 的亲缘关系最近（Behncke 等人，2013）。在德国进行的一项研究中，研究者利用病毒学检测对爬行类不同动物样本进行蛙病毒筛选，结果发现，从 2010 年到 2013 年，被检测出来的感染数量不断增加，被感染的物种包括各种龟鳖类动物和有鳞类爬行动物（见表 4）。有些被检测出来的病毒与 ECV 亲缘关系最近，而其他病毒则与先前在欧洲两栖类或爬行类动物中所检测出来的蛙病毒聚类在一起，其中一种病毒就是 FV3 样病毒（A Stöhr，未发表资料）。在爬行动物中，蛙病毒检出率的升高既反映出这些病毒感染爬行动物是真正的新发现，也反映出监测的增加。蛙病毒研究中所发现的基因组差别表明，不是单一蛙病毒株已适应爬行动物作为宿主，而是发生了蛙病毒从两栖类和鱼类到爬行动物之间的多重传播（Jancovich 等人，2010），并且这种传播还将持续。很有趣的是，不少研究显示，在合法的爬行动物国际贸易中，通过欧洲进行贸易的爬行动物数量最大，这些动物既包括人工养殖的，也包括野外捕捉的（Bush 等人，2013）。在欧洲，有文献报道，爬行动物感染遗传上不同的蛙病毒，大多数与宠物贸易有关（Stöhr 等人，2013b；Stöhr 等人，2015）。在爬行动物中，有关外来爬行类宠物的非法贸易在蛙病毒感染的流行病学中的作用，还没有进行研究。然而，也有一些迹象表明，爬行类宠物的非法贸易在爬行类宠物蛙病毒的暴发中起着一定的作用（S Blahak，CVUA-OWL，个人交流资料）。在全球贸易中，野外捕捉的和人工养殖的爬行动物常常与其他动物有接触，如与其他爬行类和两栖类动物接触，并且没有定期进行蛙病毒感染的检测。在爬行类动物中，蛙病毒感染监测和报道增加的另外一个原因，就是蛙病毒与其他病原合并感染的发现在增加，使得蛙病毒病在有些情况之下的诊断，变得尤为困难。

5 不同动物种类之间蛙病毒的传播

43 如上所述，在整个三大类变温脊椎动物即两栖类、爬行类和硬骨鱼类中，蛙病毒作为病原能感染一系列广泛的宿主。学者们一直怀疑这群病毒可能在不同动物种类之间进行传播。但是，直到最近，有学者在控制实验条件的情况下，才揭示蛙病毒在不同动物种类之间传播的存在（Brenes 等人，2014a）。目前，现存蛙病毒在不同动物种类之间传播的证据显示，至少 BIV、ATV 以及 FV3 可以在不同动物种类的野生种群之间进行传播。

蛙病毒在不同动物种类之间传播的第一证据是由穆迪（Moody）和欧文斯（Owens）于 1994 年提供的。研究者将一种从两栖类中分离到的 BIV 通过水浴接种或注射接种尖吻鲈（*Lates calcarifer*），结果被接种的鱼类全部发病，并且 100% 死亡。同时，BIV 还成功地传染给了短颈龟（*Emydura macquarii krefftii*）和锯齿侧颈龟（*Myuchelys latisternum*）这两种龟的稚龟，但是对这两种龟的成年龟以及稚鳄鱼（*Crocodylus johnstoni*）的接种感染没有成功。此外，棕树蛇（*Boiga irregularis*）、普通绿树蛇（*Dendrelaphis punctulatus*）以及棱背蛇（*Tropidonophis mairii*）三种蛇的病毒传播实验表明，任何一种蛇都没有发病。但是，棕树蛇在接种 BIV 四周之后，从其体内分离到了 BIV（Ariel，1997）。BIV 在开始接种感染以后，如此长的时间没有表现出临床症状，却分离到了病毒，因而这一物种可能是病毒存活的贮藏宿主（Ariel，1997）。

虽然研究者最初认为 ATV 的感染严格限于有尾两栖类动物（Jancovich 等人，2001），但后来有研究显示，ATV 对无尾两栖类动物也有致病性（Schock 等人，2008）。实验感染大嘴鲈已获得成功，但被接种的动物没有发生死亡或发病（Picco 等人，2010）。还有证据显示，在池塘中存在着 FV3 样病毒株和 ATV 样病毒株的循环，并且既能感染有尾两栖类动物，又能感染无尾两栖类动物（Schock 等人，2008）。到目前为止，在爬行类动物中，还没有关于 ATV 实验感染的报道。

由贝利（Bayley）等人（2013）所做的另外一个研究显示，最初从鱼类中分离到的一种蛙病毒（PPIV）能够引起林蛙蝌蚪的死亡。将林蛙蝌蚪与一组 6 条带有不同蛙病毒分离株的带毒鱼进行水浴接种病毒，结果发现，在 6 个病毒分离株中，仅仅只有 1 个病毒株即 PPIV 造成了感染动物的死亡。这是接种从鱼类中分离的蛙病毒株引起两栖类动物死亡的第一例。后来，布雷内斯（Brenes）等人（2014a）所做的实验显示，从鱼类中分离到的一种 FV3 样病毒可以传染给两栖类。

已发现，FV3 样病毒存在于鱼类、两栖类动物和爬行类动物中，同时，还有来自野生种群、人工饲养环境以及来自传播实验研究的证据显示，这些病毒存在着非常广的宿主范围。毛静荷（Jinghe Mao）等人（1999a）研究发现，在同一地区，从三刺鱼和北方红腿蛙（*Rana aurora*）蝌蚪中分离到的蛙病毒株完全相同。在泰国，有学者从人工养殖并濒临死亡的云斑尖塘鳢中分离到了 FV3 样病毒（Prasankok 等人，2005）。同时，在美国，于传染病流行期间，从孵化场饲养的好几种鲟鱼中也分离到了 FV3 样病毒（Waltzek 等人，2014）。并且有资料显示，FV3 样病毒的作用，在白鲟（pallid sturgeon）中是研究得比较彻底并得到确认的。

布雷内斯等人（2014a）展示了从濒临死亡的白鲟中分离到的 FV3 样病 44
毒，通过水浴传染给了两栖类动物中的可普灰树蛙（*Hyla chrysoscelis*）、鱼类中的食蚊鱼（mosquito fish）以及爬行类动物中的巴西红耳龟。在一个类似的实验中，对佛罗里达甲鱼（*Apalone ferox*）、甜甜圈龟（*Pseudemys concinna*）和密西西比地图龟（*G. pseudogeographica kohnii*）三种龟鳖类动物进行了传染实验，以评估龟鳖类动物对相同 FV3 样病毒株、东方箱龟中分离到的 FV3 样病毒株以及美国牛蛙中分离到的 FV3 样病毒株的敏感性（Brenes 等人，2014b）。然而，在这些实验当中，没有观察到动物发病或发生死亡。有文献记载，接种鱼类蛙病毒分离株以及接种龟蛙病毒分离株的中华鳖发生了感染（Brenes 等人，2014b）。此外，密西西比地图龟在接种龟蛙病毒分离株时，观察到了感染的发生（Brenes 等人，2014b）。这些结果显示，爬行类动物、鱼类以及两栖类动物可能是其他类动物 FV3 样病毒的贮藏宿主。

不同类别的宿主在蛙病毒流行病学中的作用仍在研究之中。一系列野外研究表明，蛙病毒在一个宿主类群中，既可感染健康的，也可感染幸存的同地域变温脊椎动物。但是，作为病毒长期携带者的作用以及传播的流行病动态学，目前还不清楚。不少研究显示，不同的病毒对不同宿主的影响迥然不同，这一事实对于评估野生动物和人工饲养动物，特别是交易动物，在临床上显得健康的动物的感染状态显得尤为重要。显得健康而又感染了蛙病毒，并且没有受到怀疑的动物，可以通过直接接触或环境污染形成其他类动物中易感种类的一个传染源。所有这些发现都强调，需要我们对蛙病毒作为多物种致病源的认识重新作出评估，不仅仅限于蛙病毒作为特定动物类群的致病源。

6 结论

两栖类动物、鱼类以及爬行类动物蛙病毒的感染非常普遍，并在这些动物类别当中感染一系列不同的物种（见表 1 和图 6）。被感染的物种不仅包括经济上比较重要的一些动物，例如虹鳟、中华鳖以及牛蛙，还包括其他受保护的动物，例如中国大鲵、地鼠龟、隙穴蛙多斑亚种、白鲟。这些多物种致病源在经济上和保护上的风险取决于许多因素，其中包括宿主物种的特征。有一件事已得到肯定，就是有些宿主物种对蛙病毒高度易感，这些高易感物种极有可能在蛙病毒暴发期间受到感染。因此，认识宿主对不同蛙病毒的易感性是定量评估风险的关键。蛙病毒与两栖类动物之间的宿主—病原关系是最好鉴定的，在这方面的研究需要继续进行，同时，也需要密切关注爬行类动物和鱼类在蛙病毒流行病事件中的作用。

45 全世界的不同地区，蛙病毒感染的所有三类脊椎动物宿主都是高密度养殖的。人工养殖场的条件常常使遗传上相似的个体保持高密度，这样很容易造成蛙病毒病的重复暴发（Pearman 和 Garner，2005；见图 2）。另外，按照毒力权衡假说（virulence trade-off hypothesis），适合病毒传播的环境可以导致病毒毒力的增加（Alizon 等人，2009）。因此，反复暴发蛙病毒病的人工养殖场，有可能促进形成蛙病毒的进化型，这种蛙病毒进化型的毒力要比野生型的毒力更强（Brunner 等人，2015）。

46 蛙病毒宿主的商业贸易以及不同宿主类别间病原的传播有可能是促使蛙病毒在全球分布的重要因素。实验证据提示，一旦蛙病毒的新病毒株被引入本地种群当中，很可能会造成毁灭性的灾难（Pearman 等人，2004；Storfer 等人，2007; Hoverman 等人，2010）。随着动物国际贸易的全球化，蛙病毒伴随着这种全球化向世界各地传播（Schloegel 等人，2009；Kolby 等人，2014）。因此，搞清楚蛙病毒是如何被转运以及它们在哪里结束显得非常重要。

随着世界动物卫生组织（World Health Organization for Animals）宣布两栖类动物感染 EHNV 及蛙病毒是“野生动物必须报告的传染病”（国际兽疫局，OIE，2008）。这一规定需要多个已经同意国际兽疫局（OIE）政策的国家筛选出一个蛙病毒宿主样本，从而能够跨国境进行蛙病毒的检测（Schloegel 等人，2010）。然而，目前还没有哪个国家采取具体措施来实施重要的进口政策，即要求必须申报无蛙病毒动物。此外，因蛙病毒（EHNV 除外）感染的鱼类和爬行动物（无论是哪种蛙病毒感染）不包括在国际兽疫局规定条款之内。

图 6　蛙病毒暴发的地点，包括一些没有受到人为干扰的地方。例如：(a) 美国的缅因州（照片由 Nathaniel Wheelwright 提供）；(b) 荷兰的德温海尔德（照片由 Jeiger Herder 提供）；(c) 城市环境中所建造的池塘（照片由 Jeiger Herder 提供）；(d) 水产养殖场（照片由 Rolando Mazzoni 提供）。高密度的宿主和环境应激因子有可能导致蛙病毒的暴发，特别是在人工养殖场。例如：(e) 巴西的牛蛙养殖场（照片由 Rolando Mazzoni 提供）；(f) 一些高度濒危的物种也发生感染（如中国大鲵）（照片由耿毅提供）。

我们对蛙病毒分布和宿主范围的认识才刚刚开始，对野生的和人工养殖的种群以及商业贸易动物进行持续的监测，结合对蛙病毒的不同病毒株进行鉴定，对于全面认识蛙病毒分布及宿主的多样性是很有必要的。因此，当进行蛙病毒研究时，一旦检测到蛙病毒，至少应对蛙病毒的部分鉴定给予足够的资金支持。这就需要跨学科团队研究者共同进行攻关研究。

致谢

本研究得到了提供给 ALJD 的校长教师 A 型发展基金的支持。作者还要感谢杰西·布伦纳（Jesse Brunner，华盛顿州立大学）和两位匿名的评审人员对本手稿之前版本的评论。

开放存取出版物的发行得到了田纳西大学（林业、野生动物与渔业，系研究与交流办公室，农业研究所）、华盛顿州立大学图书馆、戈登州立大学（学术事务办公室）、两栖爬行动物兽医协会以及两栖爬行动物保护协会的资金赞助。

开放存取

参考文献

Ahne W，Schlotfeldt HJ，Thomsen I（1989）. Fish viruses：isolation of an icosahedral cytoplasmic deoxyribovirus from sheatfish（*Silurus glanis*）. J Vet Med B 36：333-336.

Ahne W，Ogawa M，Schlotfeldt HJ（1990）. Fish viruses：transmission and pathogenicity of an icosahedral cytoplasmic deoxyribovirus isolated from sheatfish（*Silurus glanis*）. J Vet Med B 37：187-190.

Ahne W，Schlotfeldt HJ，Ogawa M（1991）. Iridovirus infection of adult sheatfish（*Silurus glanis*）. Bull Eur Assoc Fish Pathol 11：97-98.

Alizon S，Hurford A，Mideo N et al.（2009）. Virulence evolution and the trade-off hypothesis：history，current state of affairs and the future. J Evol Biol 22：245-259.

Allender MC（2012）. Characterizing the epidemiology of ranavirus in North American chelonians：diagnosis，surveillance，pathogenesis，and treatment. Ph. D. Thesis，Department of Veterinary Clinical Medicine，University of Illinois at Urbana-Champaign，Urbana，p 219.

Allender MC, Fry MM, Irizarry AR et al. (2006). Intracytoplasmic inclusions in circulating leukocytes from an eastern box turtle (*Terrapene carolina carolina*) with iridoviral infection. J Wildl Dis 42: 677-684.

Allender MC, Abd-Eldaim M, Schumacher J et al. (2011). PCR prevalence of ranavirus in free ranging eastern box turtles (*Terrapene carolina carolina*) at rehabilitation centers in three southeastern US states. J Wildl Dis 47: 759-764.

Allender MC, Mitchell MA, McRuer D et al. (2013). Prevalence, clinical signs, and natural history characteristics of frog virus 3-like infections in eastern box turtles (*Terrapene carolina carolina*). Herpetol Conserv Biol 8: 308-320.

Alves de Matos AP, Caeiro MF, Marschang RE et al. (2008). Adaptation of ranaviruses from Peneda-Gerês National Park (Portugal) to cell cultures and their characterizations. Microsc Microanal 14: 139-140.

Alves de Matos AP, Caeiro MF, Papp R et al. (2011). New viruses from Lacerta monticola (Serra da Estrela, Portugal): further evidence for a new group of nucleo-cytoplasmic large deoxyriboviruses. Microsc Microanal 17: 101-108.

Ariel E (1997). Pathology and serological aspects of Bohle iridovirus infections in six selected water-associated reptiles in North Queensland. Ph. D. Thesis, Department of Microbiology and Immunology, James Cook University, North Queensland, p 168.

Ariel, E. & Bang-Jensen, B. (2009). Challenge studies of European stocks of redfin perch, *Perca fluviatilis* L., and rainbow trout, *Oncorhynchus mykiss* (Walbaum), with epizootic haematopoietic necrosis virus. *Journal of Fish Diseases*, 32, 1017-1025.

Ariel E, Owens L (1997). Epizootic mortalities in tilapia *Oreochromis mossambicus*. Dis Aquat Organ 29: 1-6.

Ariel E, Kielgas J, Svart HE et al. (2009). Ranavirus in wild edible frogs, *Pelophylax* kl. *esculentus* in Denmark. Dis Aquat Organ 85: 7-14.

Ariel E, Holopainen R, Olesen NJ et al. (2010). Comparative study of ranavirus isolates from cod (*Gadus morhua*) and turbot (*Psetta maxima*) with reference to other ranaviruses. Arch Virol 155: 1261-1271.

Balseiro A, Dalton KP, Cerrol A et al. (2009). Pathology, isolation and molecular characterization of a ranavirus from the common midwife toad *Alytes obstetricans* on the Iberian Peninsula. Dis Aquat Organ 84: 95-104.

Balseiro A, Dalton KP, Cerro A et al. (2010). Outbreak of common midwife toad virus in alpine newts (*Mesotriton alpestris cyreni*) and common midwife toads (*Alytes obstetricans*) in northern Spain: comparative pathological study of an emerging ranavirus. Vet J 186: 256–258.

Bang-Jensen B, Ersboll AK, Ariel E (2009). Susceptibility of pike *Esox lucius* to a panel of Ranavirus isolates. Dis Aquat Organ 83: 169–179.

Bang-Jensen B, Holopainen R, Tapiovaara H et al. (2011a). Susceptibility of pike-perch *Sander lucioperca* to a panel of ranavirus isolates. Aquaculture 313: 24–30.

Bang-Jensen B, Reschova S, Cinkova K et al. (2011b). Common carp (*Cyprinus carpio*) and gold fish (*Carassius auratus*) were not susceptible to challenge with ranavirus under certain challenge conditions. Bull Eur Assoc Fish Pathol 31: 112–118.

Bayley AE, Hill BJ, Feist SW (2013). Susceptibility of the European common frog *Rana temporaria* to a panel of ranavirus isolates from fish and amphibian hosts. Dis Aquat Organ 103: 171–183.

Becker JA, Tweedie A, Gilligan D et al. (2013). Experimental infection of Australian freshwater fish with epizootic haematopoietic necrosis virus (EHNV). J Aquat Anim Health 25: 66–76.

Behncke H, Stöhr AC, Heckers KO et al. (2013). Mass mortality in green striped tree dragons (*Japalura splendida*) associated with multiple viral infections. Vet Rec 173: 248.

Belzer WR, Seibert S (2011). A natural history of Ranavirus in an eastern box turtle population. Turtle Tortoise Newsl 15: 18–25.

Benetka V, Grabensteiner E, Gumpenberger M et al. (2007). First report of an iridovirus (Genus *Ranavirus*) infection in a Leopard tortoise (*Geochelone pardalis pardalis*). Wien Tierarztl Monatsschr 94: 243–248.

Bigarré L, Cabon J, Baud M et al. (2008). Ranaviruses associated with high mortalities in catfish in France. Bull Eur Assoc Fish Pathol 28: 163–168.

Blahak S, Uhlenbrok C (2010). Ranavirus infections in European terrestrial tortoises in Germany. In: Öfner S, Weinzierl F (eds) Proceedings of the 1st international conference on reptile and amphibian medicine, Munich, Germany, 4–7 March 2010, pp 17–23.

Bollinger RK, Mao J, Schock DM et al. (1999). Pathology, isolation, and preliminary molecular characterization of a novel iridovirus from tiger salamanders in Saskatchewan. J Wildl Dis 35: 413-429.

Bovo G, Comuzzi M, De Mas S et al. (1993). Isolation of an irido-like viral agent from breeding catfish (*Ictalurus melas*). Boll Soc It Patol Ittica 11: 3-10.

Bovo G, Giacometti P, Montesi F et al. (1999). Isolation of an irido-like agent from New Zealand eel. In: Proceedings of the European Association of Fish Pathologists, Rhodes, Greece, p 53.

Brenes R, Gray MJ, Waltzek TB et al. (2014a). Transmission of ranavirus between ectothermic vertebrate hosts. PLoS One 9: e92476.

Brenes R, Miller DL, Waltzek TB et al. (2014b). Susceptibility of fish and turtles to three ranaviruses isolated from different ectothermic vertebrate classes. J Aquat Anim Health 26: 118-126.

Brunner JL, Barnet KE, Gosier CJ et al. (2011). Ranavirus infection in die-offs of vernal pool amphibians in New York, USA. Herpetol Rev 42: 76-79.

Brunner JL, Storfer A, Gray MJ, Hoverman JT (2015). Ranavirus ecology and evolution: from epidemiology to extinction. In: Gray MJ, Chinchar VG (eds) Ranaviruses: lethal pathogens of ectothermic vertebrates. Springer, New York.

Bush ER, Baker SE, MacDonald DW (2013). Global trade in exotic pets 2006-2012. Conserv Biol 28: 663-676.

Chen ZX, Zheng JC, Jiang YL (1999). A new iridovirus isolated from soft-shelled turtle. Virus Res 63: 147-151.

Chen Z, Gui X, Gao C et al. (2013). Genome architecture changes and major gene variations of *Andrias davidianus ranavirus* (ADRV). Vet Res 44: 101.

Cheng K, Jones MEB, Jancovich JK et al. (2014). Isolation of a Bohle iridovirus-like agent from boreal toads housed within a cosmopolitan aquarium collection. Dis Aquat Organ 111(2).: 139-152.

Chinchar VG (2002). Ranaviruses (family Iridoviridae): emerging cold-blooded killers. Arch Virol 147: 447-470.

Chinchar VG, Waltzek TB (2014). Ranaviruses: not just for frogs. PLoS Pathog 10: e1003850.

Chua FHC, Ng ML, Ng KL et al. (1994). Investigation of outbreaks of a novel disease, 'Sleepy Grouper Disease', affecting the brown-spotted grouper, *Epinephelus tauvina* Forskal. J Fish Dis 17: 417-427.

Clark HF, Brennan JC, Zeigel RF et al. (1968). Isolation and characterization of viruses from the kidneys of *Rana pipiens* with renal adenocarcinoma before and after passage in the red eft (*Triturus viridescens*). J Virol 2: 629–640.

Collins JP, Jones TR, Berna HJ (1988). Conserving genetically distinct populations: the case of the Huachuca tiger salamander (*Ambystoma tigrinum stebbinsi Lowe*). In: Proceedings of the symposium on management of amphibians, reptiles and small mammals in North America, Flagstaff, Arizona, 19–21 July 1988.

Converse KA, Green DE (2005). Diseases of tadpoles. In: Majumadar SK, Huffman JE, Brenner FJ et al. (eds) Wildlife diseases: landscape epidemiology, spatial distribution and utilization of remote sensing techniques. The Pennsylvania Academy of Science, Easton, pp 72–88.

Cullen BR, Owens L (2002). Experimental challenge and clinical cases of Bohle iridovirus (BIV) in native Australian anurans. Dis Aquat Organ 49: 83–92.

Cullen BR, Owens L, Whittington RJ (1995). Experimental infection of Australian anurans (*Limnodynastes terraereginae* and *Litoria latopalmata*) with Bohle iridovirus. Dis Aquat Organ 23: 83–92.

Cunningham AA, Langton TES, Bennett PM et al. (1993). Unusual mortality associated with poxvirus-like particles in frogs (*Rana temporaria*). Vet Rec 133: 141–142.

Cunningham AA, Langton TES, Bennet PM et al. (1996). Pathological and microbial findings from incidents of unusual mortality on the common frog (*Rana temporaria*). Philos Trans R Soc Lond B Biol Sci 315: 1539–1557.

Cunningham AA, Hyatt AD, Russell P et al. (2007). Experimental transmission of a ranavirus disease of common toads (*Bufo bufo*) to common frogs (*Rana temporaria*). Epidemiol Infect 135: 1213–1216.

Cunningham AA, Turvey ST, Zhou F (2015). Development of the Chinese giant salamander *Andrias davidianus* farming industry in Shaanxi Province, China: conservation threats and opportunities. Oryx doi: 10.1017/S0030605314000842.

Currylow AF, Johnson AJ, Williams RN (2014). Evidence of ranavirus infections among sympatric larval amphibians and box turtles. J Herpetol 48: 117–121.

Daszak P, Berger L, Cunningham AA et al. (1999). Emerging infectious diseases and amphibian population declines. Emerg Infect Dis 6: 735–748.

Davidson SRA, Chambers DL (2011). Ranavirus prevalence in amphibian populations of Wise County, Virginia, USA. Herpetol Rev 42: 214–215.

De Voe RK, Geissler K, Elmore S et al. (2004). Ranavirus-associated morbidity and mortality in a group of captive eastern box turtles (*Terrapene carolina carolina*). J Zoo Wildl Med 35: 534-543.

Deng G, Li S, Xie J, Bai J et al. (2011). Characterization of a ranavirus isolated from cultured large mouth bass (*Micropterus salmoides*) in China. Aquaculture 312: 198-204.

Docherty DE, Meteyer CU, Wang J et al. (2003). Diagnostic and molecular evaluation of three iridovirus associated salamander mortality events. J Wildl Dis 39: 556-566.

Docherty-Bone TM, Ndifon RK, Nyingchia ON et al. (2013). Morbidity and mortality of the critically endangered Lake Oku clawed frog (*Xenopus longipes*). Endanger Species Res 21: 115-128.

Donaldson BM, Echternacht AC (2005). Aquatic habitat use relative to home range and seasonal movement of eastern box turtles (*Terrapene carolina carolina*: Emydidae) in Eastern Tennessee. J Herpetol 39: 278-284.

Driskell EA, Miller DL, Swist SL et al. (2009). PCR detection of ranavirus in adult anurans from the Louisville Zoological Gardens. J Zoo Wildl Med 40: 559-563.

Drury SEN, Gough RE, Cunningham AA et al. (1995). Isolation of an iridovirus-like agent from common frogs (*Rana temporaria*). Vet Rec 137: 72-73.

Duffus ALJ, Cunningham AA (2010). Major disease threats to European amphibians. Herpetol J 20: 117-127.

Duffus ALJ, Pauli PD, Wozney K et al. (2008). Frog virus 3-like infections in aquatic amphibian communities. J Wildl Dis 44: 109-120.

Duffus ALJ, Nichols RA, Garner TWJ (2013). Investigations into the life history stages of the common frog (*Rana temporaria*) affected by an amphibian ranavirus in the United Kingdom. Herpetol Rev 44: 460-463.

Duffus ALJ, Nichols RA, Garner TWJ, (2014). Detection off a frog virus 3-like ranavirus in native and introduced amphibians in the United Kingdom in 2007 and 2008. Herpetol Rev 45: 608-610.

Earl JE, Gray MJ (2014). Introduction of ranavirus to isolated wood frog populations could cause local extinction. Ecohealth. doi: 10. 1007/s10393-014-0950-y.

Eaton HE, Metcalf J, Penny E, Tcherepanov V, Upton C, Brunetti CR (2007). Comparative genomic analysis of the family *Iridoviridae*: re-annotating and defining the core set of iridovirus genes. Virol J 4: 11-28.

Echaubard P, Little K, Pauli B, Lesbarrères D (2010). Context-dependent effects of ranaviral infection on northern leopard frog life history traits. PLoS One 5: e13723.

Farnsworth SD (2012). On-site and off-site translocations of eastern box turtles (*Terrapene carolina carolina*): Comparisons of survival and overwintering ecology. MSc Thesis, Department of Biological Sciences, Towson University, Towson, MD, USA. http://www.academia.edu/5781810/ON-SITE_AND_OFF-SITE_TRANSLOCATIONS_OF_EASTERN_BOX_TURTLES_TERRAPENE_CAROLINA_CAROLINA_COMPARISONS_OF_SURVIVAL_AND_OVERWINTERING_ECOLOGY.

Farnsworth SD, Seigel RA (2013). Responses, movements, and survival of relocated box turtles during construction of the intercounty connector highway in Maryland. Transp Res Rec 2362: 1-8. doi: 10.3141/2362-01.

Fijan N, Matasin Z, Petrinec Z et al. (1991). Isolation of an iridovirus-like agent from the green frog (*Rana esculenta* L.). Vet Arch Zagreb 61: 151-158.

Forzán M, Wood J (2013). Low detection of ranavirus DNA in wild postmetamorphic green frogs, *Rana* (*Lithobates*) *clamitans*, despite previous or concurrent tadpole mortality. J Wildl Dis 49: 879-886.

Fox SF, Greer AL, Tores-Cervantes R et al. (2006). First case of ranavirus-associated morbidity and mortality in natural populations of the South American frog, *Atelognathus patagonicus*. Dis Aquat Organ 72: 87-92.

Francis-Floyd R (1992). Comparative hematology for largemouth bass (*Micropterus salmoides*) and black crappie (*Pomoxis nigromaculatus*) from Lake Weir, Lake Holy, and Newman's Lake. Final report. Florida Freshwater Game and Fish Commission, Tallahassee.

Gahl MK, Calhoun AJK (2010). The role of multiple stressors in ranavirus-caused amphibian mortalities in Acadia National Park wetlands. Can J Zool 88: 108-118.

Galli L, Pereira A, Márquez A et al. (2006). Ranavirus detection by PCR in cultured tadpoles (*Rana catesbeiana* Shaw, 1802) from South America. Aquaculture 257: 78-82.

Geng Y, Wang KY, Zhou ZY et al. (2011). First report of a ranavirus associated with morbidity and mortality in farmed Chinese giant salamanders. J Comp Pathol 145: 95-112.

George, MR, John KR, Mansoor MM, Saravanakumar R, Sundar P, Pradeep V (2014). Isolation and characterization of a ranavirus from koi, *Cyprinus carpio* L., experiencing mass mortalities in India. J Fish Dis. doi: 10. 1111/jfd. 12246.

Glenney JW, Julian JT, Quartz WM (2010). Preliminary amphibian health survey in the Delaware Water Gap National Recreation Area. J Aquat Anim Health 22: 102-114.

Gobbo F, Cappellozza E, Pastore MR et al. (2010). Susceptibility of black bullhead *Ameiurus melas* to a panel of ranavirus isolates. Dis Aquat Organ 90: 167-174.

Goldberg TL (2002). Largemouth bass virus: an emerging problem for warmwater fisheries? In: Philipp DP, Ridgway MS (eds) Black bass: ecology, conservation and management. American Fisheries Society Symposium, Bethesda.

Goodman RM, Miller DL, Ararso YT (2013). Prevalence of ranavirus in Virginia turtles as detected by tail-clip sampling versus oral-cloacal swabbing. Northeast Nat 20: 325-332.

Granoff A, Came PE, Rafferty KA (1965). The isolation and properties of viruses from *Rana pipiens*: their possible relationship to the renal adenocarcinoma of the leopard frog. Ann N Y Acad Sci 126: 237-255.

Grant EC, Inendino KR, Love WJ et al. (2005). Effects of practices related to catch-and-release angling on mortality and viral transmission in juvenile largemouth bass infected with largemouth bass virus. J Aquat Anim Health 17: 315-322.

Gray MJ, Miller DL, Schmuster AC et al. (2007). Frog virus 3 prevalence in tadpole populations at cattle-access and non-access wetlands in Tennessee, USA. Dis Aquat Organ 77: 97-103.

Gray MJ, Miller DL, Hoverman JT (2009a). Ecology and pathology of amphibian ranaviruses. Dis Aquat Organ 87: 243-266.

Gray MJ, Miller DL, Hoverman JT (2009b). First report of ranavirus infecting lungless salamanders. Herpetol Rev 40: 316-319.

Gray MJ, Brunner JL, Earl JE, Ariel E (2015). Design and analysis of ranavirus studies: surveillance and assessing risk. In: Gray MJ, Chinchar VG (eds) Ranaviruses: lethal pathogens of ectothermic vertebrates. Springer, New York.

Green DE, Converse KA (2005). Diseases of frogs and toads. In: Majumadar SK,

Huffman JE, Brenner FJ et al. (eds) Wildlife diseases: landscape epidemiology, spatial distribution and utilization of remote sensing techniques. The Pennsylvania Academy of Science, Easton, pp 89–117.

Green DE, Converse KA, Schrader AK (2002). Epizootiology of sixty-four amphibian morbidity and mortality events in the USA, 1996-2001. Ann N Y Acad Sci 969: 323–339.

Greer AL, Berrill M, Wilson PJ (2005). Five amphibian mortality events associated with ranavirus infection in south central Ontario, Canada. Dis Aquat Organ 67: 9–14.

Greer AL, Brunner JL, Collins JP (2009). Spatial and temporal patterns of *Ambystoma tigrinum virus* (ATV) prevalence in tiger salamanders *Ambystoma tigrinum nebulosum*. Dis Aquat Organ 85: 1–6.

Grizzle JM, Brunner CJ (2003). Review of largemouth bass virus. Fisheries 28: 10–14.

Grizzle JM, Altinok I, Fraser WA et al. (2002). First isolation of largemouth bass virus. Dis Aquat Organ 50: 233–235.

Groocock GH, Grimmett SG, Getchell RG et al. (2008). A survey to determine the presence and distribution of largemouth bass virus in wild freshwater bass in New York State. J Aquat Anim Health 20: 158–164.

Haislip NA, Gray MJ, Hoverman JT et al. (2011). Development and disease: how susceptibility to an emerging pathogen changes through anuran development. PLoS One 6: e22307.

Hamed MK, Gray MJ, Miller DL (2013). First report of ranavirus in plethodontidae salamanders from the Mount Rogers National Recreation Area, Virginia, USA. Herpetol Rev 44: 455–457.

Hanson LA, Petrie-Hansons L, Meals KO et al. (2001). Persistence of largemouth bass virus infection in a northern Mississippi reservoir after a die-off. J Aquat Anim Health 13: 27–34.

Harp EM, Petranka JW (2006). Ranavirus in wood frogs (*Rana sylvatica*): potential sources of transmission within and between ponds. J Wildl Dis 42: 307–318.

Heard MJ, Smith KF, Ripp KJ et al. (2013). The threat of disease increases as species move towards extinction. Conserv Biol 24: 1378–1388.

Hedrick RP, McDowell TS (1995). Properties of iridoviruses from ornamental fish.

Vet Res 26: 423-427.

Heldstab A, Bestetti G (1982). Spontaneous viral hepatitis in a spur-tailed Mediterranean land tortoise. J Zoo Anim Med 13: 113-120.

Holopainen R, Ohlemeyer S, Schuetze H et al. (2009). Ranavirus phylogeny and differentiation based on major capsid protein, DNA polymerase and neurofilament triplet H1-like protein genes. Dis Aquat Organ 85: 81-91.

Homan RN, Bartling JR, Stenger RJ et al. (2013). Detection of ranavirus in Ohio, USA. Herpetol Rev 44: 615-618.

Hoverman JT, Gray MJ, Miller DL (2010). Anuran susceptibility to ranaviruses: role of species identity, exposure route and a novel virus isolate. Dis Aquat Organ 98: 97-107.

Hoverman JT, Gray MJ, Haislip NA et al. (2011). Phylogeny, life history, and ecology contribute to differences in amphibian susceptibility to ranaviruses. Ecohealth 8: 301-319.

Hoverman JT, Gray MJ, Haislip NA et al. (2012a). Widespread occurrence of ranavirus in pond breeding amphibian populations. Ecohealth 9: 36-48.

Hoverman JT, Mihaljevic JR, Richgels LD et al. (2012b). Widespread co-occurrence of virulent pathogens within California amphibian communities. Ecohealth 9: 288-292.

Huang Y, Huang X, Liu H et al. (2009). Complete sequence determination of a novel reptile iridovirus isolated from soft-shelled turtle and evolutionary analysis of *Iridoviridae*. BMC Genomics 10: 224.

Huang SM, Tu C, Tseng CH et al. (2011). Genetic analysis of fish iridoviruses isolated in Taiwan during 2001-2009. Arch Virol 156: 1505-1515.

Hyatt AD, Gould AR, Zupanovic Z et al. (2000). Comparative studies of piscine and amphibian iridoviruses. Arch Virol 145: 301-331.

Hyatt AD, Williamson M, Coupar BEH et al. (2002). First identification of a ranavirus from green pythons (*Chondropython viridis*). J Wildl Dis 38: 239-252.

Iwanowicz L, Densmore C, Hahn C et al. (2013). Identification of largemouth bass virus in the introduced Northern snakehead inhabiting the Chesapeake Bay Watershed. J Aquat Anim Health 25: 191-196.

Jancovich JK, Davidson EW, Morado JF et al. (1997). Isolation of a lethal virus from the endangered tiger salamander *Ambystoma tigrinum stebbinsi*. Dis Aquat

Organ 31: 161–167.

Jancovich JK, Davidson EW, Seiler A et al. (2001). Transmission of the *Ambystoma tigrinum virus* to alternative hosts. Dis Aquat Organ 46: 159–163.

Jancovich JK, Davidson EW, Parameswaran N et al. (2005). Evidence for emergence of an amphibian iridoviral disease because of human-enhanced spread. Mol Ecol 14: 213–224.

Jancovich JK, Bremont M, Touchman JW et al. (2010). Evidence for multiple recent host species shifts among the ranaviruses (family Iridoviridae). J Virol 84: 2636–2647.

Jancovich JK, Steckler N, Waltzek TB (2015). Ranavirus taxonomy and phylogeny. In: Gray MJ, Chinchar VG (eds) Ranaviruses: lethal pathogens of ectothermic vertebrates. Springer, New York.

Johnson AJ (2006). Iridovirus infections of captive and free-ranging chelonians in the United States. Ph. D. Thesis, Veterinary Medicine. University of Florida, Gainesville, p 149.

Johnson AJ, Pessier AP, Jacobson ER (2007). Experimental transmission and induction of ranaviral disease in western ornate box turtles (*Terrapene ornata ornata*) and red-eared sliders (*Trachemys scripta elegans*). Vet Pathol 44: 285–297.

Johnson AJ, Pessier AP, Wellehan JFX et al. (2008). Ranavirus infection of free-ranging and captive box turtles and tortoises in the United States. J Wildl Dis 44: 851–863.

Johnson AJ, Wendland L, Norton TM et al. (2010). Development and use of an indirect enzyme linked immunosorbent assay for detection of iridovirus exposure in gopher tortoises (*Gopherus polyphemus*) and eastern box turtles (*Terrapene carolina carolina*). Vet Microbiol 142: 160–167.

Juhúsz T, Woynarovichne LM, Csaba G et al. (2013). Isolation of ranavirus causing mass mortality in brown bullheads (*Ameiurus nebulosus*) in Hungary. Magyar Allatorvosok Lapja 135: 763–768.

Kanchanakhan S (1998). An ulcerative disease of the cultured tiger frog, *Rana tigrina*, in Thailand: virological examination. AAHRI Newsl 7: 1–2.

Kik M, Martel A, Spitzen-van der Sluijs A et al. (2011). Ranavirus associated mass mortality in wild amphibians in the Netherlands, 2010: a first report. Vet J 190: 284–286.

Kik M, Stege M, Boonyarittichaikij R et al. (2012). Concurrent ranavirus and *Batrachochytrium dendrobatidis* infection in captive frogs (*Phyllobates* and *Dendrobates* species), the Netherlands, 2012: a first report. Vet J 194: 247–249.

Kimble SJ, Karna AK, Johnson AJ et al. (2014). Mosquitoes as a potential vector of ranavirus transmission in terrestrial turtles. EcoHealth. doi: 10. 1007/s10393-014-0974-3.

Kolby JE, Smith KM, Berger L et al. (2014). First evidence of amphibian chytrid fungus (*Batrachochytrium dendrobatidis*) and ranavirus in Hong Kong amphibian trade. PLoS One 9: e90750.

Landsberg JH, Kiryu Y, Tabuchi M et al. (2013). Co-infection by alveolate parasites and frog virus 3-like ranavirus during and amphibian larval mortality event in Florida, USA. Dis Aquat Organ 105: 89–99.

Langdon, JS (1989). Experimental transmission and pathogenicity of epizootic haematopoietic necrosis virus (ehnv) in redfin perch *Perca fluviatilis* l and 11 other teleosts. *Journal of Fish Diseases*, 12, 295–310.

Langdon JS, Humphrey JD (1987). Epizootic haematopoietic necrosis a new viral disease in redfin perch *Perca fluviatilis* L. in Australia. J Fish Dis 10: 289–298.

Langdon JS, Humphrey JD, Copland J et al. (1986a). The disease status of Australian salmonids: viruses and viral diseases. J Fish Dis 9: 129–135.

Langdon JS, Humphrey JD, Williams LM et al. (1986b). First virus isolation from Australian fish: an iridovirus-like pathogen from redfin perch *Perca fluviatilis* L. J Fish Dis 9: 263–268.

Langdon JS, Humphrey JD, Williams LM (1988). Outbreaks of an EHNV-like iridovirus in cultured rainbow trout, *Salmo gairdneri* Richardson, in Australia. J Fish Dis 11: 93–96.

Lei X-Y, Ou T, Zhu R-L et al. (2012). Sequencing and analysis of the complete genome of *Rana grylio virus* (RGV). Arch Virol 157: 1559–1564.

Leimbach S, Schütze H, Bergmann SM (2014). Susceptibility of European sheatfish *Silurus glanis* to a panel of ranaviruses. J Appl Ichthyol 30: 93–101.

Ma J, Zeng L, Zhou Y et al. (2014). Ultrastructural morphogenesis of an amphibian iridovirus isolated from Chinese giant salamander (*Andrias davidianus*). J Comp Pathol 150: 325–331.

Majji S, LaPatra S, Long SM et al. (2006). *Rana catesbeiana* virus Z (RCV-Z): a novel pathogenic ranavirus. Dis Aquat Organ 73：1-11.

Mao J, Hedrick RP, Chinchar VG (1997). Molecular characterization, sequence analysis, and taxonomic position of newly isolated fish iridoviruses. Virology 229：212-220.

Mao JD, Green E, Fellers G et al. (1999a). Molecular characterization of iridoviruses isolated from sympatric amphibians and fish. Virus Res 63：45-52.

Mao JH, Wang J, Chinchar GD et al. (1999b). Molecular characterization of a ranavirus isolated from largemouth bass *Micropterus salmoides*. Dis Aquat Organ 37：107-114.

Marschang RE, Becher P, Posthaus H et al. (1999). Isolation and characterization of an iridovirus from Hermann's tortoises (*Testudo hermanni*). Arch Virol 144：1909-1922.

Marschang RE, Braun S, Becher P (2005). Isolation of a ranavirus from a gecko (*Uroplatus fimbriatus*). J Zoo Wildl Med 36：295-300.

Marschang RE, Stöhr AC, Blahak S et al. (2013). Ranaviruses in snakes, lizards, and chelonians. In：Proceedings of second international symposium on ranaviruses, Knoxville, 27-29 July.

Marsh IB, Whittington RJ, O'Rourke B et al. (2002). Rapid differentiation of Australian, European and American ranaviruses based on variation in major capsid protein gene sequence. Mol Cell Probes 16：137-151.

Martel A, Fard MS, van Rooij P et al. (2012). Road-killed common toads (*Bufo bufo*) in Flanders (Belgium) reveal low prevalence of ranavirus and *Batrachochytrium dendrobatidis*. J Wildl Dis 48：835-839.

Mavian C, Lopez-Bueno A, Balseiro A et al. (2012). The genome sequence of the emerging common midwife toad virus identifies an evolutionary intermediate within ranaviruses. J Virol 86：3617-3625.

Mazzoni R, José de Mesquita A, Fleury LFF et al. (2009). Mass mortality associated with frog virus 3-like ranavirus infection in farmed tadpoles, *Rana catesbeiana*, from Brazil. Dis Aquat Organ 86：181-191.

Miller DL, Rajeev S, Gray MJ et al. (2007). Frog virus 3 infection, cultured American bullfrogs. Emerg Infect Dis 13：342-343.

Miller DL, Rajeev S, Brookins M et al. (2008). Concurrent infection with *Ranavirus*, *Batrachochytrium dendrobatidis*, and *Aeromonas* in a captive amphibian colony. J Zoo Wildl Med 39：445-449.

Miller DL, Gray MJ, Rajeev S et al. (2009). Pathological findings in larval and juvenile anurans inhabiting farm ponds in Tennessee, USA. J Wildl Dis 45: 314-324.

Miller D, Gray M, Storfer A (2011). Ecopathology of ranaviruses infecting amphibians. Viruses 3: 2351-2373.

Miller DL, Pessier AP, Hick P, Whittington RJ (2015). Comparative pathology of ranaviruses and diagnostic techniques. In: Gray MJ, Chinchar VG (eds) Ranaviruses: lethal pathogens of ectothermic vertebrates. Springer, New York.

Monson-Collar K, Hazard L, Dolcemascolo P (2013). A Ranavirus-related mortality recent and the first report of Ranavirus in New Jersey. Herpetol Rev 44: 263-265.

Moody NJG, Owens L (1994). Experimental demonstration of pathogenicity of a frog virus, Bohle iridovirus, for fish species, barramundi *Lates calcifer*. Dis Aquat Organ 18: 95-102.

Müller M, Zangger N, Denzler T (1988). Iridovirus-epidemie bei der griechischen landschildkröte (*Testudo hermanni hermanni*). Verhandl Ber 30. Int Symp Erkr Zoo-und Wildtiere, Sofia, pp 271-274.

Murali S, Wu MF, Guo IC et al. (2002). Molecular characterization and pathogenicity of a grouper iridovirus (GIV) isolated from yellow grouper, *Epinephelus awoara* (Temminck & Schlegel). J Fish Dis 25: 91-100.

Neal JW, Eggleton MA, Goodwin AE (2009). The effects of largemouth bass virus on a quality largemouth bass population in Arkansas. J Wildl Dis 45: 766-771.

O'Bryan CJ, Gray MJ, Brooks CS (2012). Further presence of ranavirus infection in amphibian populations of Tennessee, USA. Herpetol Rev 43: 293-295.

OIE (2008). Aquatic animal health code. Section 2. 4 diseases of amphibians. http://www. oie. int/ international-standard-setting/aquatic-code/. Accessed 20 March 2014.

Paetow LJ, Pauli BD, McLaughlin JD et al. (2011). First detection of ranavirus in *Lithobates pipiens* in Quebec. Herpetol Rev 42: 211-214.

Pasmans F, Blahak S, Martel A et al. (2008). Ranavirus-associated mass mortality in imported red tailed knobby newts (*Tylototriton kweichowensis*): a case report. Vet J 175: 257-259.

Pearman PB, Garner TWJ (2005). Susceptibility of Italian agile frog populations to an emerging strain of Ranavirus parallels population genetic diversity. Ecol Lett

8: 401-408.

Pearman PB, Garner TWJ, Straub M et al. (2004). Response of the Italian agile frog (*Rana latastei*) to a *Ranavirus*, frog virus 3: a model for viral emergence in naïve populations. J Wildl Dis 40: 660-669.

Petranka JW, Murray SM, Kennedy CA (2003). Response of amphibians to restoration of a southern Appalachian wetland: perturbations confounded post-restoration assessment. Wetlands 23: 278-290.

Picco AM, Collins JP (2008). Amphibian commerce as a likely source of pathogen pollution. Conserv Biol 22: 1582-1589.

Picco AM, Brunner JL, Collins JP (2007). Susceptibility of the endangered California tiger salamander, *Ambystoma californiense*, to Ranavirus infection. J Wildl Dis 43: 286-290.

Picco AM, Karam AP, Collins JP (2010). Pathogen host switching in commercial trade with management recommendations. Ecohealth 7: 252-256.

Plumb JA, Hanson LA (2011). Health maintenance and principal microbial diseases of cultured fishes. Wiley, Hoboken.

Plumb JA, Zilberg D (1999a). Survival of largemouth bass iridovirus in frozen fish. J Aquat Anim Health 11: 94-96.

Plumb JA, Zilberg D (1999b). The lethal dose of largemouth bass virus in juvenile largemouth bass and the comparative susceptibility of striped bass. J Aquat Anim Health 11: 246-252.

Plumb JA, Grizzle JM, Young HE et al. (1996). An iridovirus isolated from wild largemouth bass. J Aquat Anim Health 8: 265-270.

Plumb JA, Noyes AD, Graziono S et al. (1999). Isolation and identification of viruses from adult largemouth bass during a 1997 - 1998 survey in the southeastern United States. J Aquat Anim Health 11: 391-399.

Pozet F, Morand M, Moussa A et al. (1992). Isolation and preliminary characterization of a pathogenic icosahedral deoxyribovirus from the catfish *Ictalurus melas*. Dis Aquat Organ 14: 35-42.

Prasankok P, Chutmongkonkul M, Kanchankhan S (2005). Characterization of iridovirus isolated from diseased marbled sleepy goby, *Oxyeleotris marmoratus*. In: Walker PJ, Lester RG, Bondad-Reantaso M (eds) Diseases in Asian Aquaculture V. Asian Fisheries Society, Manila.

Price, SJ, Garner TWJ, Nichols RA, et al. (2014). Collapse of amphibian communities due to an introduced *Ranavirus*. Curr Biol 24: 2586–2591. http://www.cell.com/current-biology/pdfExtended/S0960-9822(14).01149-X.

Qin QW, Chang SF, Ngoh-lim GH et al. (2003). Characterization of a novel ranavirus isolated from grouper *Epinephelus tauvina*. Dis Aquat Organ 53: 1–9.

Richter SC, Drayer AN, Strong JR et al. (2013). High prevalence of ranavirus infection in permanent constructed wetlands in eastern Kentucky, USA. Herpetol Rev 44: 464–466.

Ridenhour BJ, Storfer AT (2008). Geographically variable selection in *Ambystoma tigrinum* virus (Iridoviridae) throughout the western USA. J Evol Biol 21: 1151–1159.

Rothermel BR, Travis ER, Miller DL et al. (2013). High occupancy of stream salamanders despite high *Ranavirus* prevalence in a southern Appalachians watershed. Ecoheath 10: 184–189.

Ruder MG, Allison AB, Miller DL et al. (2010). Pathology in practice. J Am Vet Med Assoc 237: 783–785.

Russell DM, Goldberg CS, Sprague L et al. (2011). Ranavirus outbreaks in amphibian populations of northern Idaho. Herpetol Rev 42: 223–225.

Schloegel LM, Picco AM, Kilpatrick AM et al. (2009). Magnitude of the US trade in amphibians and presence of *Batrachochytrium dendrobatidis* and ranavirus infection in imported North American bullfrogs (*Rana catesbeiana*). Biol Conserv 142: 1420–1426.

Schloegel LM, Daszak P, Cunningham AA et al. (2010). Two amphibian diseases, chytridiomycosis and ranaviral disease, are now globally notifiable to the World Health Organization for Animal Health (OIE): an assessment. Dis Aquat Organ 92: 101–108.

Schock DM, Bollinger TK, Chinchar VG et al. (2008). Experimental evidence that amphibian ranaviruses are multi-host pathogens. Copeia 1: 133–143.

Schock DM, Ruthig GR, Collins JP et al. (2010). Amphibian chytrid fungus and ranaviruses in the Northwest Territories, Canada. Dis Aquat Organ 92: 231–240.

Schramm HL Jr, Davis JG (2006). Survival of largemouth bass from populations infected with largemouth bass virus and subjected to simulated tournament conditions. N Am J Fish Manag 26: 826–832.

Sharifian-Fard M, Pasmans F, Adriaensen C et al. (2011). Ranaviruses in invasive bullfrogs, Belgium. Emerg Infect Dis 17: 2371–2372.

Shaver K (2012). In Md., fear for the turtles. http://www.washingtonpost.com/local/commuting/2012/02/01/gIQA5OOZ9Q_story.html. Accessed 31 July 2014.

Sim RR, Wack AN, Allender AC, Murphy KJ, Bronson E (2012). Management of a concurrent ranavirus and herpesvirus epizootic event incaptive eastern box turtles (*Terrapene carolina carolina*). Proceedings Association of Reptilian and Amphibian Veterinarians.

Southard GM, Fries LT, Terre DR (2009). Largemouth bass virus in Texas: distribution and management issues. J Aquat Anim Health 21: 36–42.

Souza MJ, Gray MJ, Colclough P et al. (2012). Prevalence of infection by *Batrachochytrium dendrobatidis* and *Ranavirus* in eastern Hellbenders (*Cryptobranchus alleganeiensis alleganeiensis*). J Wildl Dis 48: 560–566.

Speare R, Smith JR (1992). An iridovirus-like agent isolated from the ornate burrowing frog, *Limnodynastes* in northern Australia. Dis Aquat Organ 14: 51–57.

Speare R, Freeland WJ, Bolton SJ (1991). A possible iridovirus in erythrocytes of *Bufo marinus* in Costa Rica. J Wildl Dis 27: 457–462.

St Amour V, Wong WM, Garner TWJ et al. (2008). Anthropogenic influence on prevalence of 2 amphibian pathogens. Emerg Infect Dis 12: 1175–1176.

Stark T, Laurijssens C, Weterings M et al. (2014). Death in the clouds: Ranavirus associated mortality in an assemblage of cloud forest amphibians in Nicaragua. Acta Herpetol 9: 125–127.

Stöhr AC, Hoffmann A, Papp T et al. (2013a). Long-term study of an infection with ranaviruses in a group of edible frogs (*Pelophylax* kl. *esculentus*) and partial characterization of two viruses based on four genomic regions. Vet J 197: 238–244.

Stöhr AC, Blahak S, Heckers KO et al. (2013b). Ranavirus infections associated with skin lesions in lizards. Vet Res 44: 84.

Stöhr AC, Fleck J, Mutchmann F et al. (2013c). Ranavirus infection in a group of wild caught Lake Urmia newts *Neurergus crocatus* imported from Iraq to Germany. Dis Aquat Organ 103: 185–189.

Stöhr AC, Heckers KO, Ball L et al. (2013d). Coinfection with multiple viruses in European pond turtles (*Emys orbicularis*). In: Proceedings of Association of

Reptilian and Amphibian Veterinarians, Indianapolis.

Stöhr AC, López-Bueno A, Blahak S, Caeiro MF, Rosa GM, et al. (2015). Phylogeny and Differentiation of Reptilian and Amphibian Ranaviruses Detected in Europe. PLoS ONE 10 (2).: e0118633. doi: 10. 1371/journalpone. 0118633.

Storfer A, Alfaro ME, Ridenhour BJ et al. (2007). Phylogenetic concordance analysis shows an emerging pathogen is novel and endemic. Ecol Lett 10: 1075-1083.

Tapiovaara H, Olesen NJ, Linden J et al (1998). Isolation of an iridovirus from pikeperch *Stizostedion lucioperca*. Dis Aquat Organ 32: 185-193.

Teacher AGF, Cunningham AA, Garner TWJ (2010). Assessing the long-term impacts of *Ranavirus* infection on wild common frog populations. Anim Conserv 13: 514-522.

Todd-Thompson M (2010). Seasonality, variation in species prevalence and localized disease for *Ranavirus* in Cades Cove (Great Smoky Mountains National Park) amphibians. M. Sc. Thesis, University of Tennessee, Knoxville.

Torrence SM, Green DE, Benson CJ et al. (2010). A new ranavirus isolated from *Pseudacris clarkii* tadpoles in Playa wetlands in the southern High Plains, Texas. J Aqaut Anim Health 22: 65-72.

Une Y, Nakajinma K, Taharaguchi S et al. (2009a). Ranavirus infection outbreak in the salamander (*Hynobius nebulosus*) in Japan. J Comp Pathol 141: 310.

Une Y, Sakuma A, Matsueda H et al. (2009b). Ranavirus outbreak in North American bullfrogs (*Rana catesbeiana*), Japan. Emerg Infect Dis 15: 1146-1147.

USFWS (2011). National wild fish health survey database. USFWS, Washington, DC. www. fws. gov/wildfishsurvey/database/nwfhs. Accessed 20 May 2014.

USGS (2005). USGS National Wildlife Health Center Quarterly Wildlife Mortality Report April 2005 to June 2005. http: //www. nwhc. usgs. gov/publications/quarterlyreports/2005qtr2.F jsp.

USGS (2008). USGS National Wildlife Health Center Quarterly Wildlife Mortality Report October 2008 to December 2008 http: //www. nwhc. usgs. gov/publications/quarterlyreports/2008qtr4. jsp.

Uyehara IK, Gamble T, Cotner S (2010). The presence of *Ranavirus* in anuran populations at Itasca State Park, Minnesota, USA. Herpetol Rev 41: 177-179.

Waltzek TB, Miller DL, Gray MJ et al. (2014). New disease records for hatchery-reared sturgeon: expansion of host range of frog virus 3 into pallid sturgeon, *Scaphirhynchus albus*. Dis Aquat Organ 111(3).: 219–227.

Weir RP, Moody NJG, Hyatt AD (2012). Isolation and characterization of a novel Bohle-like virus from two frog species in the Darwin rural area, Australia. Dis Aquat Organ 99: 169–177.

Weng SP, He JG, Wang XH et al. (2002). Outbreaks of an iridovirus disease in cultured tiger frog, *Rana tigrina rugulosa*, in southern China. J Fish Dis 25: 423–427.

Westhouse RA, Jacobson ER, Harris RK et al. (1996). Respiratory and pharyngo-esophageal iridovirus infection in a gopher tortoise (*Gopherus polyphemus*). J Wildl Dis 32: 682–686.

Whitfield SM, Geerdes E, Chacon I et al. (2013). Infection and co-infection by the amphibian chytrid fungus and ranavirus in wild Costa Rican frogs. Dis Aquat Organ 104: 173–178.

Whittington RJ, Reddacliff GL (1995). Influence of environmental temperature on experimental infection of redfin perch (*Perca fluviatilis*) and rainbow trout (*Oncorhynchus mykiss*) with epizootic haematopoietic necrosis virus, an Australian iridovirus. Aust Vet J 72: 421–424.

Whittington RJ, Philbey A, Reddacliff GL et al. (1994). Epidemiology of epizootic haematopoietic necrosis virus (EHNV) infection in farmed rainbow trout, *Oncorhynchus mykiss* (Walbaum): findings based on virus isolation, antigen capture ELISA and serology. J Fish Dis 17: 205–218.

Whittington RJ, Kearns C, Hyatt AD et al. (1996). Spread of epizootic haematopoietic necrosis virus (EHNV) in redfin perch (*Perca fluviatilis*) in southern Australia. Aust Vet J 73: 112–114.

Whittington RJ, Reddacliff LA, Marsh I et al. (1999). Further observations on the epidemiology and spread of epizootic haematopoietic necrosis virus (EHNV) in farmed rainbow trout *Oncorhynchus mykiss* in southeastern Australia and a recommended sampling strategy for surveillance. Dis Aquat Organ 35: 125–130.

Whittington RJ, Becker JA, Dennis MM (2010). Iridovirus infections in finfish—critical review with emphasis on ranaviruses. J Fish Dis 33: 95–122.

Wolf K, Bullock GL, Dunbar CE et al. (1969). Tadpole edema virus: a viscerotropic pathogen for anuran amphibians. J Infect Dis 118: 253–262.

Woodland JE, Brunner CJ, Noyes AD et al. (2002a). Experimental oral transmission of largemouth bass virus. J Fish Dis 25: 669–672.

Woodland JE, Noyes AD, Grizzle JM (2002b). A survey to detect largemouth bass virus among fish from hatcheries in the southeastern USA. Trans Am Fish Soc 131: 308–311.

Xu K, Zhu D, Wei Y et al. (2010). Broad distribution of ranavirus in free-ranging *Rana dybowskii*, in Heilongjiang, China. Ecohealth 7: 18–23.

Zhang Q, Li Z, Jiang Y et al. (1996). Preliminary studies on virus isolation and cell infection from diseased frog *Rana grylio*. Acta Hydrobiol Sin 4: 390–392.

Zhang QY, Xiao F, Li ZQ et al. (2001). Characterization of an iridovirus from the cultured pig frog, *Rana grylio*, with lethal syndrome. Dis Aquat Organ 48: 27–36.

Zhou ZY, Geng Y, Liu XX et al. (2013). Characterization of a ranavirus isolated from the Chinese giant salamander (*Andrias davidianus*, Blanchard, 1871) in China. Aquaculture 384- 387: 66–73.

Zilberg D, Grizzle JM, Plumb JA (2000). Preliminary description of lesions in juvenile largemouth bass injected with largemouth bass virus. Dis Aquat Organ 39: 143–146.

Zupanovic Z, Lopez G, Hyatt AD et al. (1998a). Giant toads, *Bufo marinus*, in Australia and Venezuela have antibodies against ranaviruses. Dis Aquat Organ 32: 1–8.

Zupanovic Z, Musso C, Lopez G et al. (1998b). Isolation and characterization of iridoviruses from the giant toad *Bufo marinus* in Venezuela. Dis Aquat Organ 33: 1–9.

第二章 蛙病毒的分类学及种系发生

詹姆斯·K. 贾柯维奇[1]，纳塔利·K. 斯特克勒和托马斯·B. 华尔兹克[2]

1 引言

59 核质巨 DNA 病毒（nucleocytoplasmic large DNA viruses，NCLDVs）是一类单源集群的病毒（monophyletic cluster of viruses），该类病毒在世界范围内，感染从单细胞生物到人类之间的真核生物。核质巨 DNA 病毒群涵盖 6 个病毒科，即痘病毒科（*Poxviridae*）、非洲猪瘟病毒科（*Asfarviridae*）、虹彩病毒科（*Iridoviridae*）、囊泡病毒科（*Ascoviridae*）、拟菌病毒科（*Mimiviridae*）和藻类 DNA 病毒科（*Phycodnaviridae*）（Yutin 和 Koonin，2012；Yutin 等人，2009；图 1）。此外，马赛病毒分离株（Marseillevirus isolates）可归类为核质巨 DNA 病毒的成员，当我们对这一重要而复杂的双链 DNA 病毒类群的认识进一步加深时，有可能会有更多的病毒分离株和病毒科加入 NCLDV 集群。最近，负责监管病毒分类的国际病毒分类委员会（ICTV）收到一个建议（Colson 等人，2012，2013），这个建议就是将核质巨 DNA 病毒归类为一个新目，命名为原病毒目（Megavivales），将 NCLDV 归类为一个明确的分类等级会为大双链 DNA 病毒提供一个必要的分类结构。因此，这一改变分类的建议，在不久的将来，极有可能被接受，到那时，我们将把这一病毒群称为 NCLDV。NCLDV 的一些成员具有已知最大的病毒基因组。例如，拟菌病毒科的成员，其基因组大小有
60 1 200 kbp，编码 1 000 个以上的病毒基因（Raoult 等人，2004）。这些病毒在感染细胞的细胞质里进行复制，然而 NCLDV 有一些成员，如虹彩病毒科在其复

---

[1] J. K. Jancovich /Department of Biological Sciences, California State University, 333 S. Twin Oaks Valley Rd, San Marcos, CA 92096, USA/e-mail: jjancovich@ csusm. edu.

[2] N. K. Steckler, T. B. Waltzek/Department of Infectious Diseases and Pathology, College of Veterinary Medicine, University of Florida, Gainesville, FL 32611, USA.

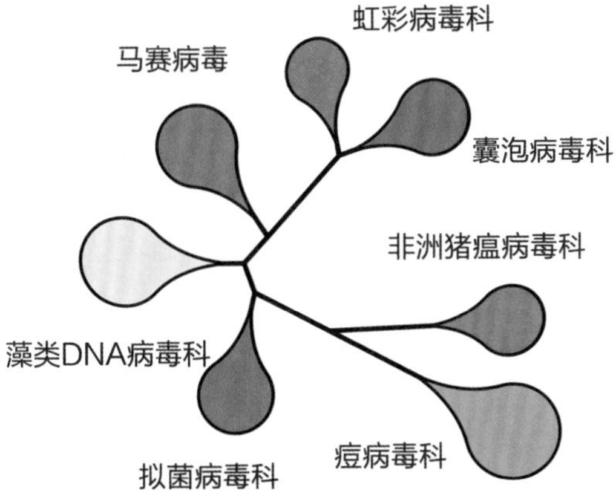

图 1 NCLDV 成员的种系发生示意图。图形所表示的进化树由 Yutin 等人（2009）最先发表的、依据 DNA 多聚酶基因保守区 263 氨基酸的系统发生绘制的。进化树没有根据比例进行绘制。

制周期中还包括了一个核阶段。结果，NCLDV 的成员包含许多在细胞质中复制需要的编码基因，但是这些病毒成员仍然要完全依靠宿主的翻译机制。NCLDV 基因组的比较分析显示，一套 50 个病毒核心基因，在 NCLDV 之间是保守的（Yutin 和 Koonin，2012）。这一发现，有力地支持了这个病毒群来源于同一祖先的假说。在 NCLDV 中最好鉴定的科是痘病毒科，该科包括一种人类主要致病源，即天花病毒（smallpox virus）。在最近几十年中，我们对 NCLDV 的其他科，特别是虹彩病毒科的成员，在分子生物学、生态学以及感染动态学领域的认识不断加深。

虹彩病毒科由 5 个属组成，其中虹彩病毒属（*Iridovirus*）和绿虹彩病毒属（*Chloriridovirus*）这两个属的病毒成员感染无脊椎动物宿主。除这两个属以外，虹彩病毒科还包括肿大细胞病毒属（*Megalocytivirus*）、淋巴囊肿病毒属（*Lymphocystivirus*）和蛙病毒属（*Ranavirus*），这 3 个属的病毒成员主要感染冷血脊椎动物（Jancovich 等人，2012）。虹彩病毒具有线性双链 DNA 基因组，该基因组会周期性地突变并使末端变得冗长（Goorha 和 Murti，1982），虹彩病毒基因组的大小在虹彩病毒科内具有很大的差异，变化范围为 140 kbp ~303 kbp。然而，由于基因组末端冗长，单位长度基因组大小即每一个独特基因大小之和则较小，其范围为 105 kbp~212 kbp（Jancovich 等人，2012）。虹彩病毒科的病毒成员共享 26 个核心基因（Eaton 等人，2007），这一核心基因簇包括病毒结构蛋白以及涉及基因表达调控、病毒复制和毒力有关的蛋白（Jancovich 等人，2015；Grayfer 等人，2015）。26 个核心基因的序列分析已被用来产生高分辨率的进化树（见图 2），进而用于分析虹彩病毒科以及蛙病毒属的病毒成员（Jancovich 等人，2012）。

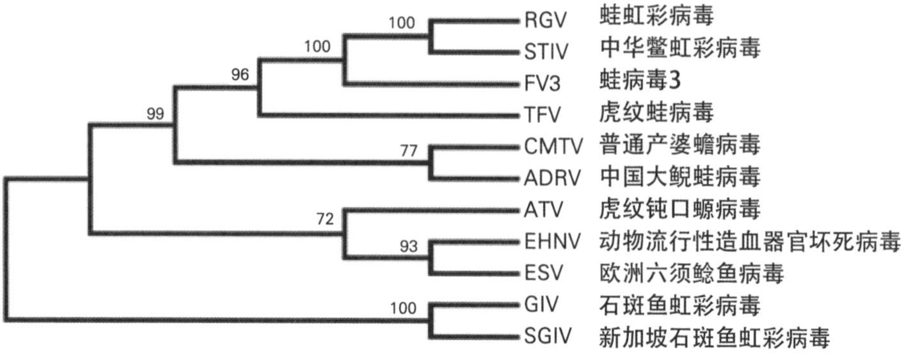

图 2 描述 11 个全序列蛙病毒之间进化关系的进化分支图。其是建立在对串联在一起的 26 个虹彩病毒保守基因推导氨基酸序列进行序列比对基础之上的，这些基因是由 Eaton 等人（2007）所界定的。数据集包括 13 287 个比对氨基酸位点，运用 MEGA6 进行了最大似然分析（Tamura 等人，2013）。各分支上的数字为自展值（1 000 次重复抽样检验）。见表 1 分类单元的缩写。

2 蛙病毒的分类学

61 蛙病毒属病毒能够感染一系列广泛不同的冷血脊椎动物宿主，包括鱼类、两栖类和爬行类（Marschang，2011；Miller 等人，2011；Whittington 等人，2010）。另外，有学者提出假设，在进化史上，蛙病毒的宿主已从鱼类跳跃到了两栖类和爬行类（Jancovich 等人，2010；Mavian 等人，2012a）。这一广泛的宿主范围已成为很多蛙病毒研究的焦点，从而使得许多研究者试图弄清楚蛙病毒是如何感染多种多样宿主的（Brenes 等人，2014）。同时，在进化史上，研究者还试图弄清楚，蛙病毒的宿主究竟是什么时候从鱼类跳跃到其他冷血脊椎动物的（Chen 等人，2013；Jancovich 等人，2010；Mavian 等人，2012a），以及什么样的遗传元件导致蛙病毒宿主范围如此之大及其致病机理（Jancovich 等人，2015）。

目前，蛙病毒属有 6 个病毒种得到了国际病毒分类委员会认可（ICTV recognization）（Jancovich 等人，2012）。这些病毒种包括蛙病毒 3（FV3）和其他 5 个病毒种。FV3 是蛙病毒属的典型病毒种，并且也是虹彩病毒科最好鉴定的病毒成员。其他 5 个病毒种分别为虎纹钝口螈病毒（ATV）、博乐虹彩病毒（BIV）、动物流行性造血器官坏死病毒（EHNV）、欧洲鲶鱼病毒（ECV）和桑堤-库珀蛙病毒（SCRV）（Jancovich 等人，2012）。此外，还有其他几种遗传距离比较远的蛙病毒还没有得到国际病毒分类委员会虹彩病毒研究组的认可，这些病毒包括新加坡石斑鱼虹彩病毒（Singapore grouper iridovirus，SGIV）（Song 等人，2004）、石斑鱼虹彩病毒（Grouper iridovirus，GIV）、食用蛙病毒（Rana esculenta virus，REV）（Holopainen 等人，2009）、普通产婆蟾病毒（CMTV）（Mavian 等人，2012a）、中国大鲵蛙病毒（Andrias davidianus ranavirus，ADRV，也称中国大鲵虹彩病毒）（Chen 等人，2013）、鳕鱼虹彩病 62
毒（Cod iridovirus，CoIV）（Ariel 等人，2010）、短鳍鳗蛙病毒（Short-finned eel ranavirus，SERV）（Holopainen 等人，2009）、梭鲈虹彩病毒（Pike-perch iridovirus，PPIV）（Holopainen 等人，2009）以及大菱鲆蛙病毒（Ranavirus maxima，Rmax）（Ariel 等人，2010）。有多个标准来描述蛙病毒属的成员，其中包括基因组 DNA 限制性内切酶片段长度多态性（restriction endonuclease fragment length polymorphism，RFLP）图谱、病毒蛋白图谱、DNA 序列分析以及宿主的特异性（Jancovich 等人，2012）。除这些标准以外，利用全基因组序列信息进行 dot plot 分析以及利用单个和串联基因序列进行种系发生分析，这些依据都为蛙病毒的分类提供了全新的认识（Eaton 等人，2007；Jancovich 等人，2010；Mavian 等人，2012a；Tan 等人，2004；Wang 等人，2014）。dot plot 分析对蛙病毒基因组的组织结构进行了总体描述，同时为鉴定病毒基因组内的插入（insertions）、缺失（deletions）以及倒位（inversions）提供了一个可视途径。dot plot 研究清楚地表明，尽管蛙病毒共享着它们大多数基因，但是基因的顺序不是保守的，而且可以作为一种方法来区分进化亲缘关系远近的病毒株或病毒种。例如，基因顺序在 FV3、虎纹蛙病毒以及中华鳖病毒（Soft-shelled turtle virus）之间是保守的，从而可以区别于 ATV 与 EHNV 中所见到的基因顺序（Jancovich 等人，2015）。

运用已经全部测序的蛙病毒 26 个核心基因进行系统发生分析，已经确立 4 个不同的病毒系（lineages）（见图 2 和表 1）：（1）TFV/FV3/BIV 样病毒；（2）CMTV/ADRV 样病毒；（3）ATV/EHNV 样病毒；（4）SGIV/GIV 样病毒。如果按照 MCP 分析所提出的，SCRV 有可能组成第 5 个病毒系（见图 3 和表 2）。此外，还有其他一些蛙病毒基因组全序列已被测序，特别是那些存在于鱼类中的蛙病毒，未来很有可能还有其他病毒系加入进来。蛙病毒系没有明确界定的宿主范围，它包括那些只将鱼类作为靶宿主的病毒（例如，GIV 样病毒和 SCRV 样病毒），只感染两栖类的病毒（例如，CMTV/ADRV 样病毒），既感染两栖类也感染鱼类的病毒（例如，ATV/EHNV 样病毒）以及使两栖类、鱼类和爬行类三类动物都感染的病毒（TFV/FV3/BIV 样病毒，见图 3）。因此，系统发生分析将会被研究者用来对新发现的蛙病毒进行鉴定和归类。

研究者通过对一个或多个病毒基因测序，将新分离到的蛙病毒归类到相应的病毒系中。例如，新分离到蛙病毒株的系统发生分析以及分类学上归类已集中在一个独特的且高度保守的基因上（例如 MCP 基因）（Allender 等人，2013；Duffu 和 Andrews，2013；Geng 等人，2011；George 等人，2014；Kolby 等人，2014；Marsh 等人，2002；Waltzek 等人，2014）或集中在一组串联组成的多个病毒基因上（Holopainen 等人，2009；Iwanowicz 等人，2013）。虽然 MCP 基因的分析非常方便，但是这一蛋白高度保守的性质会掩饰不同病毒分离株之间的差异。总的来讲，无论哪一种方法都提供了一个有用的起点，从而有利于蛙病毒分离株的鉴定和分类。此外，现有的一系列不同蛙病毒株全基因组序列信息有助于开发更加快速、敏感和独特的方法对新发现的蛙病毒进行检测和分类。例如，在基因组内设计鉴定高变区（hypervariable regions）两侧的引物，会使得区分病毒分离株更加容易。

目前，已有 12 个完整测序的蛙病毒（见表 1）。除此之外，同一病毒多个
64 病毒株如 FV3（Morrison 等人，2014）以及亲缘关系相近的病毒（He 等人，2002；Huang 等人，2009；Lei 等人，2012）全基因组序列信息也是现成的。经过完整测序的蛙病毒基因组 dot plot 比较分析将在本书的另外章节作详细的讨论（Jancovich 等人，2015）。在蛙病毒基因组中，目前有 4 个独特基因组的组织结构已被鉴定（Chen 等人，2013；Eaton 等人，2007；Jancovich 等人，2003，2010；Mavian 等人，2012a，b；Song 等人，2004；Tan 等人，2004；Tsai 等人，2005）。SCRV 可能代表着第 5 个类群（JK Jancovich 和 TB Waltzek，未发表资
66 料），其他蛙病毒基因组的结构还有待发现。有趣的是，整个基因组的 dot plot 分析显示，用 26 个核心基因进行系统发生分析，具有相似基因组结构的蛙病毒会聚类。因此，蛙病毒基因结构与 26 个核心基因的系统发生之间似乎存在着直接关联。

至今还不清楚，为什么在蛙病毒之间的整个基因组结构如此不同。说到底，在 NCLDV 的其他病毒成员中，也没有如此不同的基因组结构。例如，所有痘病毒的基因组都展示出保守的中央核心区、可变区以及末端反向重复区（inverted terminal repeat regions）。核心区包括所有痘病毒共同的复制基因（replicative genes），而末端反向重复区的编码基因则影响宿主的特异性和致病性（Gubser 等人，2004；Upton 等人，2003）。对比之下，在上述 4 个蛙病毒系

表 1　基因组已测序的蛙病毒属成员

蛙病毒	缩写	宿主	基因组大小（kb）	GC 百分比（%）	预测的 ORFs	登录号	参考文献
中国大鲵蛙病毒	ADRV	蝾螈	106 719	55	101	KF033124	Wang et al.（2014）
			106 734			KC865735	Chen et al.（2013）
虎纹钝口螈病毒	ATV	蝾螈	106 332	54	96	AY150217	Jancovich et al.（2003）
普通产婆蟾病毒	CMTV	蛙	106 878	55	104	JQ231222	Mavian et al.（2012a）
动物流行性造血器官坏死病毒	EHNV	鱼	127 011	54	100	FJ433873	Jancovich et al.（2010）
欧洲六须鲶病毒	ESV	鱼	127 732	54	136	JQ724856	Mavian et al.（2012b）
蛙病毒 3	FV3	蛙	105 903	55	98	AY548484	Tan et al.（2004）
石斑鱼虹彩病毒	GIV	鱼	139 793	49	139	AY666015	Tsai et al.（2005）
蛙虹彩病毒	RGV	蛙	105 791	55	106	JQ654586	Lei et al.（2012）
新加城石斑鱼虹彩病毒	SGIV	鱼	140 131	48	162	AY521625	Song et al.（2004）
中华鳖虹彩病毒	STIV	龟	105 890	55	105	EU627010	Huang et al.（2009）
缅因州斑点钝口螈病毒	SSME	蝾螈	105 070	55	95	KJ1751441	Morrison et al.（2014）
虎纹蛙病毒	TFV	蛙	105 057	55	105	AF389451	He et al.（2002）

斜体蛙病毒名表示蛙病毒属中公认的病毒种。

中，虽然基因保守，但是基因顺序（gene order）存在差别。也许基因组结构的不同反映了它们内在的高频率重组（Chinchar 和 Granoff，1986），高频率重组可导致病毒基因组明显的重排。因此，假如随着时间的推移，病毒基因组重组增加，那么序列趋异的蛙病毒就会同样表现出较低的序列共线性（sequence collinearity）。鉴于此，未来的研究工作应侧重于弄清楚蛙病毒之间基因组的可变性和多样性，以及病毒生态学、宿主范围和致病性之间的关系。

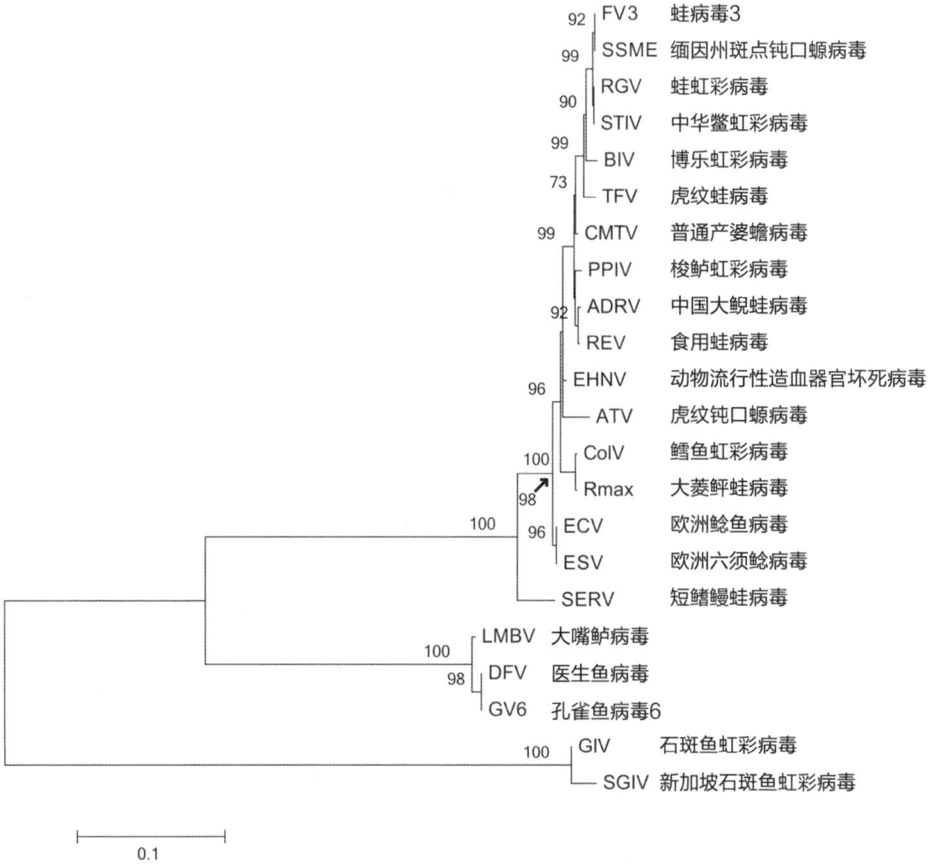

图 3 描述虹彩病毒科（*Iridoviridae*）22 种病毒之间进化关系的系统发生树。建树的依据是病毒主衣壳蛋白基因（major capsid gene）核苷酸序列全长的序列比对。数据集包括 1 392 个核苷酸位点的比对，运用 MEGA6 进行了最大似然分析（Tamura 等人，2013）。各分支上的数字为自展值（1 000 次重复抽样检验）。见表 1 和表 2 的分类缩写。分支长度基于推断替换的数值，用尺度条表示。

3　蛙病毒分类的未来：我们从这里出发将走向何方？

展望蛙病毒分类的未来，我们必须弄清楚国际病毒分类委员会是如何界定不同层次的病毒。国际病毒分类委员会将一个病毒种定义为“按多个标准来区分，在性质上有别于其他种类的一个单系发育的病毒类群”（Adams 等人，2013）。一个病毒种的标准是由国际病毒分类委员会单独研究小组界定的，其中包括自然和实验的宿主范围、细胞及组织嗜性（tropism）、致病性、抗原性及其基因组或基因的相关度（Adams 等人，2013）。但是，关键的部分就在于，一个病毒种的界定不是按单一标准，而是按多个标准来区分的。除此之外，国际病毒分类委员会还将一个属定义为“具有特定共同标准的一个物种群”（Adams 等人，2013）。

正如上面讨论过的，蛙病毒分类依据基因组 DNA 限制性内切酶片段长度
多态性图谱、病毒蛋白图谱（virus protein profiles）、DNA 序列分析以及宿主的
特异性（Jancovich 等人，2012）。不幸的是，为了描述一个病毒种有别于其他
病毒种，这些标准不允许我们定量区分种内和种间的不同。然而，最近几年，
我们通过单个病毒基因的测序和病毒全基因组的测序，对蛙病毒多样性的认识
进一步加深。因此，虹彩病毒科研究小组（Iridoviridae Study Group）必须对该 67
病毒科内属及种的命名标准作出必要的重新评估。例如，石斑鱼虹彩病毒
（GIV）和新加坡石斑鱼虹彩病毒（SGIV），在目前蛙病毒属的病毒分离株之
间，似乎在亲缘关系上是最远的病毒（见图 2 和图 3）。整个基因组的 dot plot
分析显示，GIV 和 SGIV 存在共线性关系（collinearity），但是石斑鱼虹彩病毒
与其他蛙病毒只有少数几个共线性区域（Jancovich 等人，2015）。此外，GIV/
SGIV 缺乏在其他蛙病毒中所见到的 DNA 甲基转移酶基因，因而 GIV/SGIV 基
因组属于不具备甲基化的基因组（Song 等人，2004；Tsai 等人，2005）。因此，
GIV/SGIV 必须考虑归为一个新属，最低限度要划为蛙病毒属中的一个不同
的种。

类似的有 SCRV 样病毒，它是一个包括医生鱼病毒（DFV）、大嘴鲈病毒（LMBV）和孔雀鱼病毒 6（GV6）的类群，也是另外一个亲缘关系较近的病毒类群，该病毒类群也必须考虑归为虹彩病毒科中的一个新属（见图 3）。现有的 LMBV 基因组序列允许我们作出更全面的比较，同时还有助于对蛙病毒的 SCRV 类群分类地位作出描述。目前，我们即将完成 LMBV、DFV 和 GV6 基因组的测序。一旦完成测序，我们就能够对这一病毒类群进行更全面的分析，

表 2　蛙病毒主衣壳蛋白序列的比较

蛙病毒	缩写	宿主	登录号	同源性(%)	参考文献
蛙病毒 3	FV3	蛙	FJ459783	100	Holopainen et al. (2009)
缅因州斑点钝口螈病毒	SSME	蝾螈	KJ175144	100	Morrison et al. (2014)
蛙虹彩病毒	RGV	蛙	JQ654586	99.8	Lei et al. (2012)
中华鳖虹彩病毒	STIV	龟	EU627010	99.8	Huang et al. (2009)
博乐虹彩病毒	BIV	蛙	AY187046	98.7	Marsh et al. (2002)
普通产婆蟾病毒	CMTV	蛙	JQ231222	98.6	Mavian et al. (2012a)
虎纹蛙病毒	TFV	蛙	AF389451	98.6	He et al. (2002)
食用蛙病毒	REV	蛙	FJ358611	98.3	Holopainen et al. (2009)
梭鲈虹彩病毒	PPIV	鱼	FJ358610	98.3	Holopainen et al. (2009)
中国大鲵蛙病毒	ADRV	蝾螈	KC865735	98.3	Chen et al. (2013)
动物流行性造血器官坏死病毒	EHNV	鱼	FJ433873	97.8	Jancovich et al. (2010)
欧洲鲶鱼病毒	ECV	鱼	FJ358608	97.3	Holopainen et al. (2009)
欧洲六须鲶病毒	ESV	鱼	JQ724856	97.3	Mavian et al. (2012b)
鳕鱼虹彩病毒	CoIV	鱼	GU391284	97.2	Ariel et al. (2010)
大菱鲆蛙病毒	Rmax	蛙	GU391285	97.1	Ariel et al. (2010)
虎纹钝口螈病毒	ATV	蝾螈	AY150217	96.3	Jancovich et al. (2003)
短鳍鳗蛙病毒	SERV	鱼	FJ358612	94.1	Holopainen et al. (2009)
孔雀鱼病毒 6	GV6	鱼	FR677325	78.1	Unpublished
医生鱼病毒	DFV	鱼	FR677324	78.1	Unpublished
大嘴鲈病毒	LMBV	鱼	FR682503	78.0	Unpublished
石斑鱼虹彩病毒	GIV	鱼	AY666015	70.0	Tsai et al. (2005)
新加坡石斑鱼虹彩病毒	SGIV	鱼	AY521625	69.4	Song et al. (2004)

蛙病毒 3 主衣壳蛋白全长序列(1 392 个字母)与其他蛙病毒核苷酸的比较。斜体蛙病毒名表示蛙病毒属中公认的病毒种。

注:大嘴鲈病毒、孔雀鱼病毒 6 和医生鱼病毒全部被认为是桑堤-库珀蛙病毒的同义名。本表对 Waltzek 等人(2014)的表进行了改进。

并确定它们是否应该归类为虹彩病毒科的一个单一属。

其他被鉴定的鱼类蛙病毒包括鳕鱼虹彩病毒和大菱鲆蛙病毒（Ariel 等人，2010）、短鳍鳗蛙病毒（Holopainen 等人，2009）和梭鲈虹彩病毒（Tapiovaara 等人，1998）。虽然对上述病毒已进行初步测序（Ariel 等人，2010；Holopainen 等人，2009），但是我们还需要对这些病毒的全基因组进行测序，以便确定它们是属于蛙病毒属，还是需要建立一个新属。因此，蛙病毒分类的未来也会反映出需要把目前认可的病毒种（例如，ATV/EHNV，TFV/FV3/BIV 以及 CMTV/ADRV）划分为一个单一复合病毒种，同时补充一些新种（例如，SGIV/GIV 和 LMBV/DFV/GV6），或者将这些种劈开，分成不同的属。为达到这一目的，在最后作出分类改变之前，虹彩病毒科研究小组还需要对这些变化的可能结果作出重新评估。

4　最后的一些想法

蛙病毒的分类正持续地在世界范围内展开，特别是当一个新的病毒分离株被发现时。新发现病毒分离株的分类学归类，主要是依据单个和多个病毒基因以及宿主、蛋白、血清学和形态学特征。但是，单基因的分类分析不可能像全基因组分析或使用蛙病毒 26 个核心基因进行系统发生比较分析所得出的结果一样可靠。随着越来越多的病毒全基因组被测序，我们对蛙病毒分类学的多样性和复杂性的认识将描绘出新的篇章。

致谢

我们要感谢 V. 格雷戈里·钦察尔（V. Gregory Chinchar）和特雷弗·威廉 68
姆斯（Trevor Williams）对本稿件作出的评审。本著作的发行，部分资金得到了美国国家卫生研究院（National Institutes of Health）项目的赞助（项目编号：1R15AI101889-01）以及加州州立大学圣马科斯分校（J. K. J.）和佛罗里达大学（N. S. 和 T. B. W.）的赞助。

开放存取出版物的发行得到了田纳西大学（林业、野生动物与渔业系，研究与交流办公室，农业研究所）、华盛顿州立大学图书馆、戈登州立大学（学术事务办公室）、两栖爬行动物兽医协会以及两栖爬行动物保护协会的资金赞助。

开放存取

参考文献

Adams MJ, Lefkowitz EJ, King AMQ, Carstens EB (2013). Recently agreed changes to the International Code of Virus Classification and Nomenclature. Arch Virol 158: 2633–2639.

Allender MC, Bunick D, Mitchell MA (2013). Development and validation of TaqMan quantitative PCR for detection of frog virus 3-like virus in eastern box turtles (*Terrapene carolina carolina*). J Virol Methods 188: 121–125.

Ariel E, Holopainen R, Olesen NJ, Tapiovaara H (2010). Comparative study of ranavirus isolates from cod (*Gadus morhua*) and turbot (*Psetta maxima*) with reference to other ranaviruses. Arch Virol 155: 1261–1271.

Brenes R, Gray MJ, Waltzek TB, Wilkes RP, Miller DL (2014). Transmission of ranavirus between ectothermic vertebrate hosts. PLoS One 9: e92476.

Chen ZY, Gui JF, Gao XC, Pei C, Hong YJ, Zhang QY (2013). Genome architecture changes and major gene variations of *Andrias davidianus ranavirus* (ADRV). Vet Res 44: 101–114.

Chinchar VG, Granoff A (1986). Temperature-sensitive mutants of frog virus 3: biochemical and genetic characterization. J Virol 58: 192–202.

Colson P, de Lamballerie X, Fournous G, Raoult D (2012). Reclassification of giant viruses composing a fourth domain of life in the new order Megavirales. Inter-virology 55: 321–332.

Colson P, De Lamballerie X, Yutin N, Asgari S, Bigot Y, Bideshi DK, Cheng XW, Federici BA, Van Etten JL, Koonin EV, La Scola B, Raoult D (2013). "Megavirales", a proposed new order for eukaryotic nucleocytoplasmic large DNA viruses. Arch Virol 158: 2517–2521.

Duffus AL, Andrews AM (2013). Phylogenetic analysis of a frog virus 3-like ranavirus found at a site with recurrent mortality and morbidity events in southeastern Ontario, Canada: partial major capsid protein sequence alone is not sufficient for fine-scale differentiation. J Wildl Dis 49: 464–467.

Eaton HE, Metcalf J, Penny E, Tcherepanov V, Upton C, Brunetti CR (2007). Comparative genomic analysis of the family *Iridoviridae*: re-annotating and defining the core set of iridovirus genes. Virol J 4: 11–28.

Geng Y, Wang KY, Zhou ZY, Li CW, Wang J, He M, Yin ZQ, Lai WM (2011). First report of a ranavirus associated with morbidity and mortality in farmed Chinese giant salamanders (*Andrias davidianus*). J Comp Pathol 145: 95–102.

George MR, John KR, Mansoor MM, Saravanakumar R, Sundar P, Pradeep V (2014). Isolation and characterization of a ranavirus from koi, *Cyprinus carpio* L., experiencing mass mortalities in India. J Fish Dis. doi: 10. 1111/jfd. 12246.

Goorha R, Murti KG (1982). The genome of frog virus-3, an animal DNA virus, is circularly permuted and terminally redundant. Proc Natl Acad Sci USA 79: 248–252.

Grayfer L, Edholm E-S, De Jesús Andino F, Chinchar VG, Robert J (2015). Ranavirus host immunity and immune evasion. In: Gray MJ, Chinchar VG (eds) Ranaviruses: lethal pathogens of ectothermic vertebrates. Springer, New York.

Gubser C, Hue S, Kellam P, Smith GL (2004). Poxvirus genomes: a phylogenetic analysis. J Gen Virol 85: 105–117.

He JG, Lu L, Deng M, He HH, Weng SP, Wang XH, Zhou SY, Long QX, Wang XZ, Chan SM (2002). Sequence analysis of the complete genome of an iridovirus isolated from the tiger frog. Virology 292: 185–197.

Holopainen R, Ohlemeyer S, Schutze H, Bergmann SM, Tapiovaara H (2009). Ranavirus phylogeny and differentiation based on major capsid protein, DNA polymerase and neurofilament triplet H1-like protein genes. Dis Aquat Organ 85: 81–91.

Huang YH, Huang XH, Liu H, Gong J, Ouyang ZL, Cui HC, Cao JH, Zhao YT, Wang XJ, Jiang YL, Qin QW (2009). Complete sequence determination of a novel reptile iridovirus isolated from soft-shelled turtle and evolutionary analysis of Iridoviridae. BMC Genomics 10: 224–238.

Iwanowicz L, Densmore C, Hahn C, McAllister P, Odenkirk J (2013). Identifi cation of largemouth bass virus in the introduced Northern Snakehead inhabiting the Chesapeake Bay watershed. J Aquat Anim Health 25: 191–196.

Jancovich JK, Mao J, Chinchar VG, Wyatt C, Case ST, Kumar S, Valente G, Subramanian S, Davidson EW, Collins JP, Jacobs BL (2003). Genomic sequence of a ranavirus (family *Iridoviridae*) associated with salamander mortalities in North America. Virology 316: 90–103.

Jancovich JK, Bremont M, Touchman JW, Jacobs BL (2010). Evidence for multiple recent host species shifts among the ranaviruses (family *Iridoviridae*). J Virol 84: 2636–2647.

Jancovich JK, Chinchar VG, Hyatt A, Myazaki T, Williams T, Zhnag QY (2012). Family *Iridoviridae*. In: King AMQ (ed) Ninth report of the International Committee on Taxonomy of Viruses. Elsevier, San Diego.

Jancovich JK, Qin Q, Zhang Q-Y, Chinchar VG (2015). Ranavirus replication: molecular, cellular, and immunological events. In: Gray MJ, Chinchar VG (eds) Ranaviruses: lethal pathogens of ectothermic vertebrates. Springer, New York.

Kolby JE, Smith KM, Berger L, Karesh WB, Preston A, Pessier AP, Skerratt LF (2014). First evidence of amphibian chytrid fungus (*Batrachochytrium dendrobatidis*) and ranavirus in Hong Kong amphibian trade. PLoS One 9: e90750.

Lei XY, Ou T, Zhu RL, Zhang QY (2012). Sequencing and analysis of the complete genome of Rana grylio virus (RGV). Arch Virol 157: 1559–1564.

Marschang RE (2011). Viruses infecting reptiles. Viruses 3: 2087–2126.

Marsh IB, Whittington RJ, O'Rourke B, Hyatt AD, Chisholm O (2002). Rapid differentiation of Australian, European and American ranaviruses based on variation in major capsid protein gene sequence. Mol Cell Probes 16: 137–151.

Mavian C, Lopez-Bueno A, Balseiro A, Casais R, Alcami A, Alejo A (2012a). The genome sequence of the emerging common midwife toad virus identifies an evolutionary intermediate within ranaviruses. J Virol 86: 3617–3625.

Mavian C, Lopez-Bueno A, Fernandez Somalo MP, Alcami A, Alejo A (2012b). Complete genome sequence of the European sheatfish virus. J Virol 86: 6365–6366.

Miller D, Gray M, Storfer A (2011). Ecopathology of ranaviruses infecting amphibians. Viruses 3: 2351–2373.

Morrison EA, Garner S, Echaubard P, Lesbarreres D, Kyle CJ, Brunetti CR (2014). Complete genome analysis of a frog virus 3 (FV3) isolate and sequence comparison with isolates of differing levels of virulence. Virol J 11: 46–59.

Raoult D, Audic S, Robert C, Abergel C, Renesto P, Ogata H, La Scola B, Suzan M, Claverie JM (2004). The 1. 2-megabase genome sequence of Mimivirus.

Science 306：1344–1350.

Song WJ，Qin QW，Qiu J，Huang CH，Wang F，Hew CL（2004）. Functional genomics analysis of Singapore grouper iridovirus：complete sequence determination and proteomic analysis. J Virol 78：12576–12590.

Tamura K，Stecher G，Peterson D，Filipski A，Kumar S（2013）. MEGA6：molecular evolutionary genetics analysis version 6. 0. Mol Biol Evol 30：2725–2729.

Tan WGH，Barkman TJ，Chinchar VG，Essani K（2004）. Comparative genomic analyses of frog virus 3，type species of the genus *Ranavirus*（family *Iridoviridae*）. Virology 323：70–84.

Tapiovaara H，Olesen NJ，Linden J，Rimaila-Parnanen E，von Bonsdorff CH（1998）. Isolation of an iridovirus from pike-perch *Stizostedion lucioperca*. Dis Aquat Organ 32：185–193.

Tsai CT，Ting JW，Wu MH，Wu MF，Guo IC，Chang CY（2005）. Complete genome sequence of the grouper iridovirus and comparison of genomic organization with those of other iridoviruses. J Virol 79：2010–2023.

Upton C，Slack S，Hunter AL，Ehlers A，Roper RL（2003）. Poxvirus orthologous clusters：toward defining the minimum essential poxvirus genome. J Virol 77：7590–7600.

Waltzek TB，Miller DL，Gray MJ，Drecktrah B，Briggler JT，MacConnell B，Hudson K，Hopper L，Friary J，Yun SC，Malm KV，Weber ES，Hedrick RP（2014）. New disease records for hatchery-reared sturgeon. I. Expansion of the host range of frog virus 3 into hatchery-reared pallid sturgeon *Scaphirhynchus albus*. Dis Aquat Organ 111：219–227.

Wang N，Zhang M，Zhang L，Jing H，Jiang Y，Wu S，Lin X（2014）. Complete genome sequence of a ranavirus isolated from Chinese giant salamander（*Andrias davidianus*）. Genome Announc 2：e01032–e01013.

Whittington RJ，Becker JA，Dennis MM（2010）. Iridovirus infections in finfish—critical review with emphasis on ranaviruses. J Fish Dis 33：95–122.

Yutin N，Koonin EV（2012）. Hidden evolutionary complexity of nucleo-cytoplasmic large DNA viruses of eukaryotes. Virol J 9：161–179.

Yutin N，Wolf YI，Raoult D，Koonin EV（2009）. Eukaryotic large nucleo-cytoplasmic DNA viruses：clusters of orthologous genes and reconstruction of viral genome evolution. Virol J 6：223–236.

第三章 蛙病毒生态学和进化：从流行病学到灭绝

杰西·L. 布伦纳，安德鲁·斯托福①，马修·J. 格雷②和杰森·T. 霍韦尔曼③

1 引言

71 蛙病毒在北方豹蛙（*Lithobates pipiens*）原代肾细胞中被偶然发现后的数十年间，一直被认为对鱼类和两栖类影响很小（Granoff 等人，1966；Chinchar 等人，2009；Williams 等人，2005）。越来越多的证据显示，蛙病毒是好几种鱼类和两栖类出现广泛传播的流行病和死亡的原因，从而使得上述观点发生了彻底的改变（Ahne 等人，1997；Chinchar 2002，Williams 等人，2005）。随着全世界范围内蛙病毒地理分布范围的明显扩大以及不同变温脊椎动物种群数量不断下降的证据增多，全世界研究者对这一病毒属产生了浓厚兴趣（Duffus 等人，2015）。蛙病毒在野外和人工饲养环境引起的大规模死亡常常以快速发病和高死亡率为标志，但奇怪的是，在自然种群中发生的蛙病毒感染症状不太明显。总的说来，蛙病毒的流行可以影响种群动态变化，这种影响从明显感染开始一直延续到局部种群的灭绝。在这一章，我们将对蛙病毒的流行病学作一综述，其侧重点是蛙病毒感染的各种影响因素。然后，我们讨论蛙病毒在种内和种间的传播以及蛙病毒病流行的结果。我们同样也考察蛙病毒的进化，并重点讨论

---

① J. L. Brunner · A. Storfer/School of Biological Sciences, Washington State University, Pullman, WA 99164, USA/e-mail: jesse. brunner@ wsu. edu; astorfer@ wsu. edu.

② M. J. Gray/Center for Wildlife Health, Department of Forestry, Wildlife and Fisheries, University of Tennessee, 274, Ellington Plant Sciences Building, Knoxville, TN 37996, USA/e-mail: mgray11@ utk. edu.

③ J. T. Hoverman/Department of Forestry and Natural Resources, Purdue University, West Lafayette, IN 47907, USA/e-mail: jhoverm@ purdue. edu/M. J. Gray, V. G. Chinchar (eds.), Ranaviruses, DOI 10. 1007/978-3-319-13755-1_4.

蛙病毒的局部适应和毒力。越来越多的证据显示，蛙病毒能通过人类活动传播 72
到世界各地，这一点对于认识蛙病毒的流行病学很重要。本章的最后，我们将讨论蛙病毒对其宿主的影响，并考察蛙病毒是否会造成宿主的灭绝。

2　蛙病毒的流行病学

我们所知道的有关蛙病毒流行病学、地理学和宿主范围的大多数知识来自对大规模死亡事件的调查、小种群数量和时间点的零星监测研究以及重点对少数具有经济重要性或保护价值的物种进行的较大规模的监测研究工作（Grizzle 和 Brunner，2003；Gray 等人，2009a；Whittington 等人，2010；Miller 等人，2011；Duffus 等人，2015）。达弗斯等人（2015）对全世界蛙病毒的分布和宿主范围进行了综述、总计，在 6 大洲的 32 个国家，公认的有 6 个蛙病毒种，能造成至少 175 种变温脊椎动物感染或明显发病（Duffus 等人，2015）。

2.1　两栖类动物蛙病毒的流行病学

蛙病毒感染两栖类动物和引起的相关死亡事件，在北美洲、南美洲、欧洲、非洲和亚洲都已报道（Duffus 等人，2015）。在北美洲，两栖类动物 43%~57% 的死亡归因于假定性蛙病毒感染（Green 等人，2002；Muths 等人，2006）。这些流行病通常发生在中夏和晚夏，并涉及发育晚期的蝌蚪和刚变态的蝌蚪（见图 1；Green 等人，2002）。死亡来得比较突然，常常一天之内有成千上万的正常幼体死亡，在几天之内，死亡率高达 90% 以上（见图 2；Green 等人，2002）。最近有报道称，至少有 200 000 个幼体在短短 24 小时之内全部死亡（Wheelwright 等人，2014）。在欧洲（Ariel 等人，2009a；Kik 等人，2011）、南美洲（Fox 等人，2006；Stark 等人，2014）以及亚洲（Une 等人，2009）的野生两栖类种群中，也有类似快速发病和季节性暴发的报道。与此同时，还有很多报道称，人工饲养的种群呈现类似的感染方式（Duffus 等人，2015）。最近，普赖斯（Price）等人（2014）报道，西班牙在引入新型蛙病毒以后，多个地点出现了两栖类种群数量的下降。然而，最近有几篇报道虽然没有关于纵向资料收集，并且错过了一些死亡事件（Gray 等人，2015），但是报道称蛙病毒仍然存在于两栖类动物幼体和变态后的两栖类动物中，这些两栖动物没有明显发病或发生死亡（Duffus 等人，2015）。值得指出的是，蛙病毒引起的亚致死性感染可以影响与健康相关的性状，如生长与发育（Echaubard 等人，2010）。

与两栖类动物幼体蛙病毒暴发常见模式对比而言，在英国，林蛙（*Rana* 73

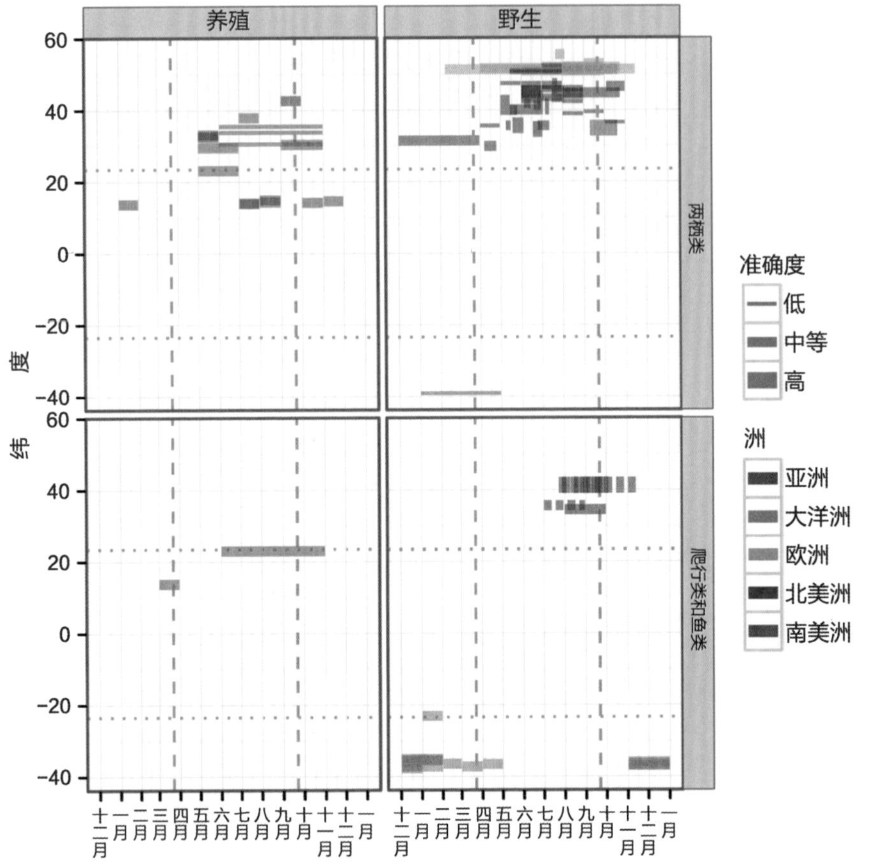

图 1 两栖类、爬行类和鱼类（虚线）养殖种群和野生种群中由蛙病毒引发的大规模死亡的季节性定时与所对应的纬度图。大多数大规模死亡事件常常始于并结束于夏季月份。纵向虚线表示春夏与夏秋临界点，横向虚线表示南北回归线。数据涵盖了 40 篇文献中报道的 109 个大规模死亡事件，其中包括定时发生和局部发生的大规模死亡。报道中只提供月份名的，我们做了包括整个月份的处理。提供了期限的，我们在时间上也对结束日期做了调整，从而与时间期限相配。月中指每月的第 15 天。对日期精确度的描述，其范围是从低（当只提供部分信息的时候）到高（例如，大规模死亡事件的发生而不是结束）（当提供准确日期的时候）。当有的报道只提供国家、州或省份的数据时，许多大规模死亡事件发生的纬度有时会不准确，在这些情况或例子中，我们就使用这个地区的中点。

temporaria）蛙病毒的暴发，大多数只限于成蛙（Cunningham 等人，1993；Teacher 等人，2010；Duffus 等人，2013）。达弗斯（Duffus）等人（2013）在一年之内所采集的 288 条蝌蚪中，只检测出 1 条蝌蚪感染了蛙病毒，3 年之内，

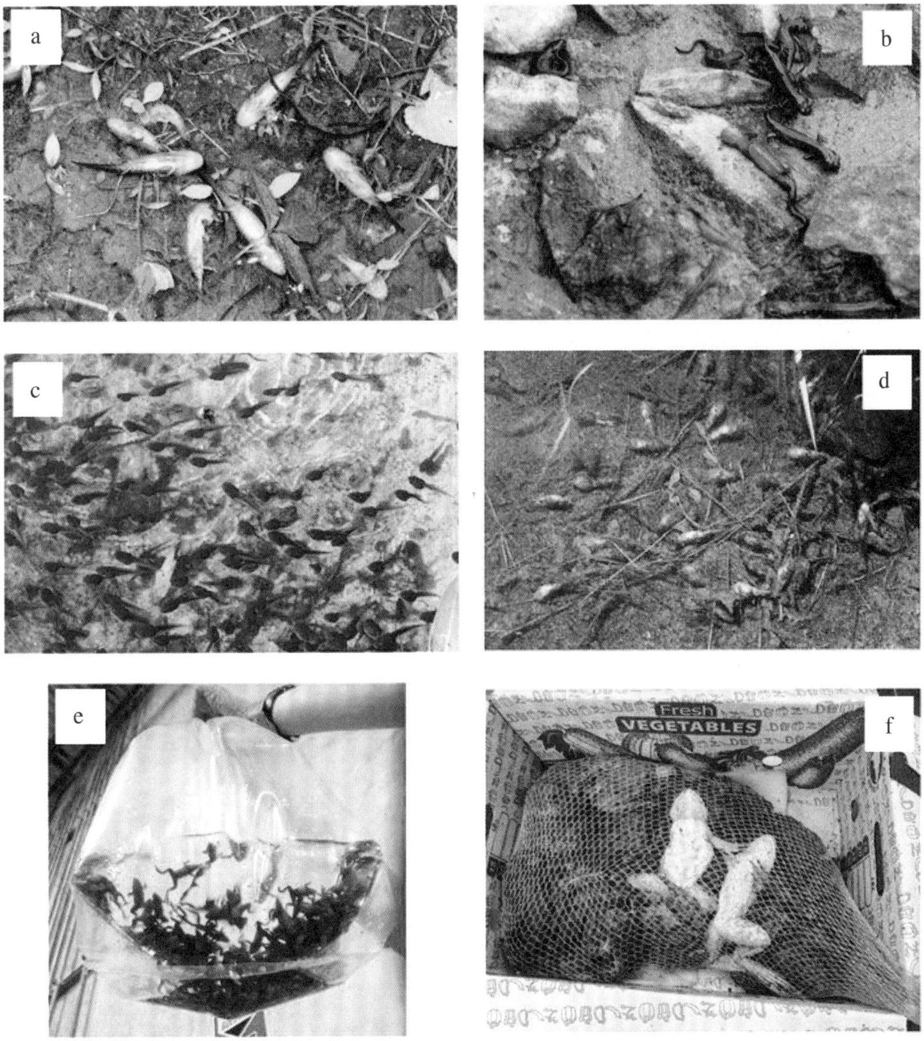

图 2　两栖类动物蛙病毒大规模死亡，包括幼龄（a. 照片由 Matthew Niemiller 提供）和成年（b. 照片由 Ana Balseiro 提供）两个年龄类别。蛙病毒暴发后，在数日之内，从没有明显死亡（c. 照片由 Nathaniel Wheelwright 提供）到完全死亡（d. 照片由 Nathaniel Wheelwright 提供）的进程很快。全球转运亚临床症状感染的动物个体可以导致蛙病毒感染的发生（e. 照片由 onathan Kolby 提供）。在人工饲养条件下，大量宿主的高接触率以及运输可以导致蛙病毒毒力的进化（f. 照片由 Jonathan Kolby 提供）。

在 120 条成年林蛙中检测出 32 只蛙感染了蛙病毒。在英国，蛙病毒虽然使成年林蛙种群数量下降缓慢，但是传播很广泛（Teacher 等人，2010）。

2.2 鱼类蛙病毒的流行病学

74 许多蛙病毒与养殖鱼类的死亡事件有关联（Ahne 等人，1997；Chua 等人，1994；Deng 等人，2011；Langdon 等人，1988；Prasankok 等人，2005；Qin 等人，2003）。但是对于这些水产养殖系统的流行病学和生态学极少有文献作详细的
75 记载。大多数关于鱼类蛙病毒的流行病学和生态学的资料都来自澳大利亚动物流行性造血器官坏死病毒的研究以及美国东南部（Whittington 等人，2010）对大嘴鲈病毒（简称 LMBV，该病毒是桑堤-库珀蛙病毒的一个病毒株）的研究。在早夏，EHNV 的暴发造成了成千上万条感染蛙病毒的红鲈鲈（*Perca fluviatilis*）稚鱼快速（2～3 周）大规模死亡，并伴随着少量成鱼的感染（Langdon 等人，1986；Langdon 和 Humphrey，1987）。在原先没有 EHNV 感染病史的隔离池中，成鱼（1 龄以上）发生了大规模死亡，提示先前曾接触过病毒或是种系发生的作用。此外，剧烈的大规模死亡以及大规模死亡持续超过 30 年的，如果在深水中和边缘地区进行不确定的检测，那么，EHNV 流行病发生的频度就很难评估（Whittington 等人，2010）。在澳大利亚，EHNV 也同样造成养殖的虹鳟鱼（*Oncorhynchus mykiss*）死亡事件，在该死亡事件中，病毒感染不是很严重，但是病毒的毒力很强（Langdon 等人，1988；Whittington 等人，1994，1999）。

大多数大嘴鲈的大规模死亡发生在夏天，并涉及大规格（>30 cm）的大嘴鲈（Grizzle 和 Brunner，2003）。此外，LMBV 有时会与数千条大鱼死亡有关联（Plumb 等人，1996；Hanson 等人，2001）。同时，有文献记录，在 20 世纪 90 年代，有些湖中个体较大、年龄较老的大嘴鲈种群数量下降也与之有关（Maceina 和 Grizzle，2006），但大嘴鲈发病率或死亡率很低，不容易被注意到（Grizzle 和 Brunner，2003）。而且，LMBV 在没有发生大规模死亡病史的地点，还常常见于在临床上表现正常的动物中（Hanson 等人，2001；Grizzle 等人，2002；Grizzle 和 Brunner，2003）。如果要找出这两种鱼的蛙病毒的共同点的话，那就是都发生在应激条件下（例如，暖期）。其流行病学主要涉及特定敏感的生活史阶段，EHNV 主要发生于稚鱼阶段，LMBV 则主要发生在成鱼阶段。

在鱼类蛙病毒中，代表蛙病毒属不同系统发生的病毒系（Qin 等人，2003；Huang 等人，2011），值得一提的就是石斑鱼虹彩病毒（Grouper iridovirus，GIV）和新加坡石斑鱼虹彩病毒（Singapore grouper iridovirus，SGIV）。自 20 世纪 90 年代以来，这两种病毒在东南亚的石斑鱼（*Epinephelus spp.*）和其他海洋养殖的有鳍鱼类中造成了大量死亡和损失（Chua 等人，1994；Qin 等人，2003；Harikrishnan 等人，2010）。有人对这一病毒流行病学作了详情报道，而

且这一流行病会持续数周（Nagasawa 和 Cruz-Lacierda，2004），造成高达 90%的死亡率，应激因素（例如操作应激、水质应激）在加重临床症状和死亡方面起着很重要的作用（Chua 等人，1994）。

2.3　爬行类动物蛙病毒的流行病学

在爬行类动物当中，已零星地检测出蛙病毒感染，主要存在于康复设施和其他人工饲养环境的动物当中（Ariel，2011；Allender 等人，2013a；Chinchar 和 Waltzek，2014）。同时，还在自由放养的龟鳖类动物中观察到了好几次蛙病毒的大暴发（Allender 等人，2006；Johnson 等人，2008；Belzer 和 Seibert，2011；Farnsworth 和 Seigel，2013）。在自由生活的其他爬行类动物种群当中，还没有 76
发现蛙病毒的报道。至少在北美洲，不像鱼类和两栖类，龟鳖类动物蛙病毒的暴发更为分散，常常连续几年，在整个活动季节都可观察到龟鳖类动物死亡的发生，例如乌龟在冬天没有进行冬眠的时候就是如此（Belzer 和 Seibert，2011；Farnsworth 和 Seigel，2013），但是大多数记录的病例都发生在夏季（见图 1）。现有的证据提示，蛙病毒在龟鳖类动物中能造成急性而快速的致死性感染（Johnson 等人，2007）。大多数被感染的个体会在几周之内死亡，然而，在龟鳖类动物发生死亡的任何一个特定的时间点，蛙病毒感染的流行率都比较低。比如，在美国宾夕法尼亚州的一个私人自然野生动物救助所，在蛙病毒引起的大规模死亡期间，71 只东方箱龟（*Terrapene carolina carolina*）中有 15 只死亡，死亡率为 21%（Johnson 等人，2008；Belzer 和 Seibert，2011），第 2 年，在 6 只存活的东方箱龟当中，只有 1 只是血清学阳性，这表明，绝大多数被感染的个体已经死亡（Johnson 等人，2010）。类似地，阿蓝德等人（2013a）发现，在美国田纳西州自由放养的东方箱龟，连续 3 年蛙病毒感染的流行率都相当低，309 只东方箱龟当中仅有 1 只感染，感染率为 0.3%。在美国东南部，来自 5 个州自由放养的穴居沙龟（*Gopherus polyphemus*）的血清学阳性率也非常低，只有 1.5%（Johnson 等人，2010），这一结果与高毒力、急性感染相一致。有证据显示，更多水生龟类感染的蛙病毒可能没有致病性（请参阅第 7 节）。与这一假设相一致的是，在美国弗吉尼亚州高度水生的东部锦龟（*Chrysemys picta picta*）蛙病毒的流行率比较高，63 只中有 11 只感染，流行率为 17%（Goodman 等人，2013）。

依据现有的资料，在龟鳖类动物中，似乎不太可能出现由自身维持的蛙病毒流行造成零星的死亡事件。龟鳖类动物蛙病毒的感染呈急性，且常常是致死性的，这样就没有时间让感染龟与未感染龟接触。此外，现存的大多数龟鳖类种群密度都比较低，这就进一步限制了病毒传播的机会。因此，龟鳖类动物的大规模死亡很可能是由倾倒感染蛙病毒的其他动物引发的。

2.4 蛙病毒流行病学小结

虽然在自然种群中围绕着蛙病毒的流行病学，仍然存在着大量的不确定性，但是存在着两种明显的模式。第一种模式，蛙病毒的流行或死亡事件，至少在鱼类和两栖类动物中，存在着共同的时间模式（见图 1），表现出宿主动物大规模死亡，一般发生在夏季月份，常常进展迅速，但这一模式也存在着一些例外，例如，牛蛙（*Lithobates catesbeianus*，以前的学名为 *Rana catesbeiana*）在日本和美国东南部大规模死亡，是发生在十月（Hoverman 等人，2012；Une 等人，2009）。第二种模式，在不同的种群和地理位置之间，蛙病毒的流行结果存在着很大的变异性，从没有或者没有明显死亡到几乎没有存活的大规模死亡。下一节，我们将从蛙病毒流行病学的发生时间开始，探讨针对这两种模式提出的几种假设。

3 对蛙病毒流行具有明显季节性的解释

77 蛙病毒的流行常常发生在晚春或夏季月份，从发病到结束只有几周的时间（Langdon 和 Humphrey，1987；Green 等人，2002；Grizzle 和 Brunner，2003）。目前，存在着四种不互相排斥的假设，来解释蛙病毒导致的大规模死亡具有明显的季节性并且发生迅猛的原因。第一种假设，这种流行模式可能是假的，其原因可能是检测偏差；第二种假设，蛙病毒引起的大规模死亡，只简单地反映出蛙病毒在年初引入种群以后基本的流行病动态学；第三种假设，宿主在某些发育阶段，对蛙病毒更加易感，这些发育阶段与夏季月份相一致；最后一种假设，这些事件都发生在夏季温度升高的时候。接下来的内容，我们将评估这些假设及其基本机制。

3.1 检测偏差

通常对蛙病毒感染变态两栖类和发育晚期幼体（Green 等人，2002）引起的大规模死亡的观察，常常限于进入浅水区完成变态的群体，因为只有在浅水区，才容易观察到它们。类似地，大量的稚鱼倾向于在人们常去的近岸聚群，这可能是在红鳍鲈中首先检测到 EHNV 的原因（Whittington 等人，2010）。大规模死亡事件，常常是因为发生在没有人活动的地方以及种类处于隐性感染而没有引起注意。例如，由蛙病毒引起的乌龟发病和死亡，会因这些动物神秘的自然属性，而在观察时常常被错过（Farnsworth 和 Seigel，2013）。然而，夏季所暴发的许多蛙病毒病，在研究比较深入且观察频度比较高的动物种群当中，常常呈现快速死亡模式（Brunner 等人，2011；Langdon 和 Humphrey，1987；Petranka 等人，2007；Wheelwright 等人，2014），从而提示，检测偏差对观察到的死亡事件发生时间不是一个普遍接受的解释。

3.2　蛙病毒感染的季节性引入与发病率

蛙病毒引发大规模死亡的快速发生和季节性只简单地反映出病毒在年初被引入以后，感染的发生率迅速增加，一个假设就是，两栖类动物蛙病毒流行病发生于亚致死性感染的成年动物返回到繁殖场以后（Brunner 等人，2004）。成年动物可以直接将疾病传染给同种幼体，假如它们和其他物种在时间和空间上重叠在一起，还可传染给其他物种，或者是当它们再度感染死亡以及采食感染 78
病毒的幼体而受到间接传染。布伦纳（Brunner）等人（2004）发现，在美国的西南部，返回到繁殖场的成年横带虎斑钝口螈（*Ambystoma mavortium nebulosum*）感染了虎纹钝口螈病毒。类似地，最近的一个调查发现，在美国东部，返回到繁殖场的雄性林蛙，有 39% 发生了亚临床感染（JLB，EJ Crespi 和 S. Hall，华盛顿州立大学；S. Duncan，NM Mattheus 和 L. Rissler，阿拉巴马大学，未发表资料）。倾倒成体或尸体会引发随后的幼体感染。因此，蛙病毒感染的动态学类似于许多其他传染病（Keeling 和 Rohani，2008），蛙病毒在引入以后传播非常缓慢，这是因为感染比较稀少，而且只有在流行建立以后感染才会加速。由于蛙病毒感染在两栖类动物中常常是致死性的，动物通常在实验室感染接种后的几天到几周就会发生死亡（Gray 等人，2009a；Hoverman 等人，2011）。学者们都认为，用死亡来追踪感染，时间会稍稍延后。在野外，实际死亡的过程似乎更快速、简单。因为在流行早期，死亡的少数几个个体很难被检测出来，这是由于它们死亡个体很少、分解快速或者被吞食。只有存在许多死亡动物的时候（也许食腐动物可暴食），感染事件才变得明显。此外，传染性尸体的增多也会促进传播（Pearman 等人，2004；Harp 和 Petranka，2006；Brunner 等人，2007），增加流行的速度，在两栖类动物中，存在着支持这一假设的一些证据。托德-汤普森（Todd-Thompson，2010）对栖息在美国阿巴拉契亚山脉碟形湿地的两栖类种群进行蛙病毒的纵向调查研究时，观察到了一个典型的流行病学曲线。蛙病毒直到 4 月末才能被检测出来，此时，感染的幼体达 20%，20 天后，感染率增加到 80%，这一现象与钝口螈幼体的大规模死亡过程相一致。在托伦斯（Torrence 等人，2010）所采集的斑点合唱蛙（*Pseudacris clarkii*）蝌蚪中，流行病感染症状的快速增加，同样也符合这一模式

(Torrence 等人，2010)。的确如此，死亡事件通常伴随着高感染的流行(Bollinger 等人，1999；Greer 等人，2005；Fox 等人，2006；Kik 等人，2011；Hoverman 等人，2012；Homan 等人，2013；Titus 和 Green，2013)。

在某些蛙病毒-宿主系统中，感染的流行与疾病的发生没有关联。例如，格里尔（Greer 等人，2009）发现，在亚利桑那州北部的三个池塘中，虎纹钝口螈 ATV 的感染率，升高到接近 50% 的一个峰值，但没有观察到虎纹钝口螈发病和死亡。类似地，达弗斯（Duffus）等人（2008）发现，在早夏，有 20%～32% 的林蛙蝌蚪感染，但没有表现出疾病的临床症状。同样也常常发现，在不发病和死亡的情况下，LMBV 也很常见（Hanson 等人，2001；Grizzle 和 Brunner，2003；Groocock 等人，2008；Southard 等人，2009）。因此，在多个时间点采集宿主种群的流行病学和地理统计学数据，对于弄清楚蛙病毒对其宿主种群所造成的影响是至关重要的（Gray 等人，2015）。

3.3 整个发育期间的易感性

79 除频繁观察到稚体比成体对蛙病毒更加易感以外（Cullen 等人，1995；Ariel，1997；Cullen 和 Owens，2002；Bang Jensen 等人，2011b），宿主还会在某些特定的发育阶段，变得容易发生蛙病毒感染，这些发育阶段与夏季月份相对应。在两栖类动物中，所报道的由蛙病毒引起的大规模死亡事件，常常发生在接近变态或正在变态的个体中（Speare 和 Smith，1992；Green 和 Converse，2005；Greer 等人，2005）。两栖类动物的变态需要消耗能量，而且会造成一段天然免疫抑制期（Rollins-Smith，1998；Carey 等人，1999）。有人推测，正在变态的无尾两栖类动物特别容易发生蛙病毒感染。这种现象可以解释蛙病毒引起两栖类动物大规模死亡及其感染加速的原因（Gahl 和 Calhoun，2010）。沃恩（Warne）等人（2011）研究发现，接触过蛙病毒的林蛙蝌蚪，会使林蛙每个发育阶段的死亡概率增加 1.7 倍（Gosner，1960）。在发育晚期，林蛙幼体对蛙病毒易感性更高，这一结果得到了一个流行病学模型的支持。该模型运用发育阶段特异的易感性，准确地预测了死亡事件发生的时间（见图 3；Warne 等人，2011）。假如所有的发育阶段具有相同的易感性，那么林蛙种群的死亡也会发生在春季，而发生在春季的死亡是不会被观察到的。海斯利普（Haislip）等人（2011）同样发现，两栖类动物的不同发育阶段，对蛙病毒的易感性存在着差别，但是变态期并不经常是最易感的发育阶段。因此，变态的发生，不能单独用来解释两栖类种群中蛙病毒暴发的时间原因。

3.4　温度与易感性

夏天较高的温度可能是导致定时暴发蛙病毒的另外一个因素。格里兹尔和布伦纳（2003）推测，在夏季，LMBV 引发大嘴鲈大规模死亡，是由于大嘴鲈在较高的温度条件下对病毒的易感性增高，或者是由其他与温度相关的应激因子引起的，例如水体的低氧浓度（Goldberg，2002），这一观点得到了实验接种感染结果的支持（Grant 等人，2003）。在早夏，红鳍鲈 EHNV 流行病的发生，主要是在稚鱼中（Langdon，1989），这也有可能与温度对易感性的影响有关。稚鲈在早夏喜欢在浅而较温暖的水域觅食，从而促进了 EHNV 的感染和疾病的发生（Whittington 和 Reddacliff，1995；Ariel 等人，2009b）。而成年鲈鱼喜欢在较深并且较冷的水域觅食，可能因此避免或清除了感染（Whittington 和 Reddacliff，1995；Whittington 等人，2010）。在澳大利亚，虹鳟 ENHV 的暴发似乎与温度无关，但与不良的饲养和高鱼群密度有关（Whittington 等人，1999，2010）。

温度既可影响宿主-寄生生物（host-parasite）互相关系的动态学，也可作为宿主的一个应激因子。首先，病原的复制速率与宿主免疫反应的动态学对温 80
度是敏感的（Altizer 等人，2013）。阿里尔（Ariel）等人（2009b）报道，随着温度上升到某个最适温度值，蛙病毒在细胞培养中的复制速率会随之增加，病毒复制的最适温度会随着病毒分离株和细胞系的不同而有所不同，一般来说，这个最适温度为 24 ℃～28 ℃。但是，从生存在冷水中的鳗鱼中分离到的短鳍鳗蛙病毒（Short-finned eel ranavirus，SERV）复制的最适温度为 20 ℃或更低，从而提示，病毒对宿主环境存在着适应（Ariel 等人，2009b）。格兰特（Grant）等人（2003）同样也找到了 LMBV 的复制存在着宿主特异温度模式的一些证据，LMBV 在 30 ℃生长要比在 25 ℃ 生长快得多，但是当温度升高到 81
35 ℃时，细胞培养 1 天以后，病毒不再复制，可能就是大嘴鲈所栖息的典型水生系统，其温度没有超过 30 ℃的缘故（Eaton 和 Scheller，1996；Grant 等人，2003）。因此，强有力的证据显示，蛙病毒的复制喜高温，并且具有宿主依赖性（Speare 和 Smith，1992；Grant 等人，2003；Rojas 等人，2005；Ariel 等人，2009b）。从离体的复制速率来表示在体的复制速率这个意义上说，病毒的复制速率和宿主的死亡率一般来说会随着温度的升高而增加。来自好几个研究的结果与这一观点相吻合。在实验水浴接种成年红鳍鲈时，水温保持在 12 ℃～21 ℃时所有鱼全部死亡，而水温保持在 6 ℃～10 ℃时那些鱼既没有感染，也没有快速清除感染（Whittington 和 Reddacliff，1995）。用 EHNV 实验接种欧洲红鳍鲈和虹鳟鱼群，同样也发现死亡率会随着温度（15 ℃～20 ℃）的升高而

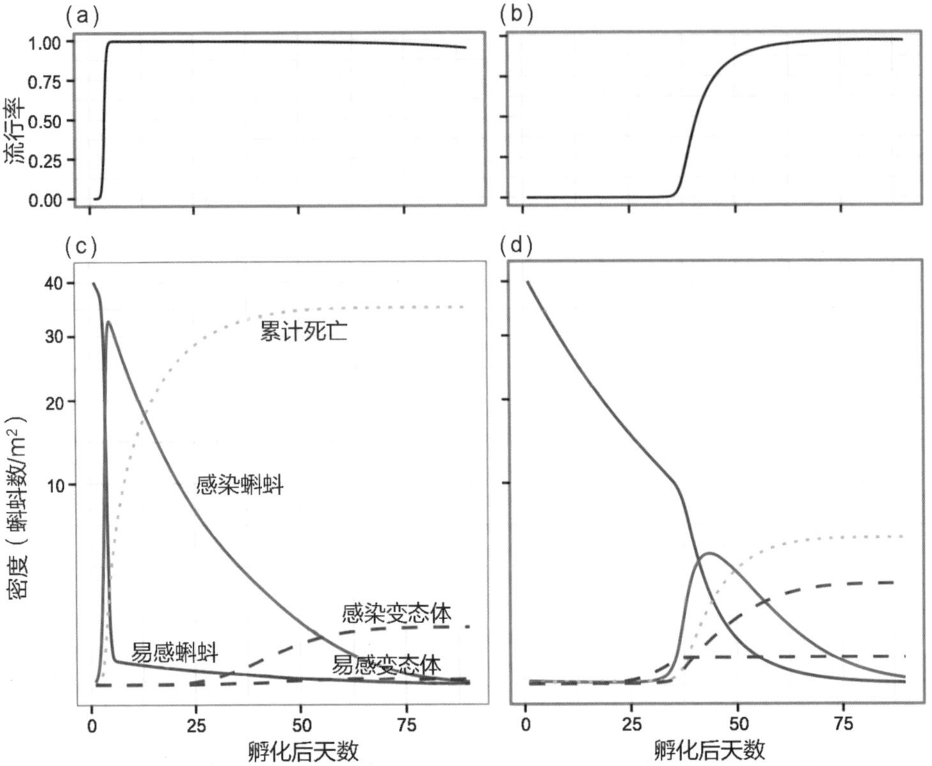

图 3 一个正在发育的林蛙蝌蚪种群蛙病毒流行病学的“易染-感染-易染”模型的流行（a，b）和动态学（c，d）。其前提条件是所有的蝌蚪具有同等的易染性（a，c）或者是蝌蚪接近变态时变得更加易感（b，d）。该模型包括一个每天 0.04 条的恒定背景死亡率（该估计值来自 DeBenedictis，1974）以及一个每天 0.25 阶段的发育率，模型中的发育阶段起于 Gosner（1960）阶段 20（孵化），止于阶段 41（变态），等同于平均 60 天的幼体发育期。传播率的估计值来自林蛙蝌蚪流行病学的中试（JLB，华盛顿州立大学，未发表资料）。刚孵化蝌蚪的起始密度为 40 条/m^2，该值是自然密度范围（例如自然密度为 26 条/m^2～790 条/m^2；Petranka 等人，2003）的最低值，但是，当密度为 400 条/m^2 时，也没有发生定性上的变化。对感染动物死亡率（0.0331 条/天）或恢复率（0.0169 条/天）的估计均来自 Reeve 等人（2013）。值得注意的是，在这个模型中，恢复后的蝌蚪会再次变得易感，因为我们缺乏这些蝌蚪有免疫记忆的证据。通过 Gosner 发育阶段特异性的感染率计算，发育阶段特异性的易感性包括了多重传播时期，感染率的估计可见 Warne 等人（2011）的一个 LD50 研究。

增加，而在 10 ℃（Ariel 和 Jensen，2009）则没有见到明显的死亡。

类似的情况还有贝利（Bayley）等人（2013）报道，林蛙蝌蚪在 20 ℃时，接种感染 FV3（Frog virus 3）或 REV（Rana esculenta virus）以后，死亡率在96%以上。但在 15 ℃接种感染蛙病毒时，其死亡率则小于 32%。然而，在用 ATV 感染虎纹钝口螈时，则观察到了相反的死亡模式（Rojas 等人，2005），当虎纹钝口螈幼体接种感染 ATV 以后，饲养在 10 ℃或 18 ℃温度条件下的幼体，死亡率在 80%以上。而饲养在 26 ℃温度条件下的幼体，死亡率则小于 38%。在 10 ℃时的病毒滴度高于在 18 ℃时的病毒滴度，提示较低的温度可以抑制宿主对 ATV 的免疫反应。在龟类中，同样也有证据显示，温度影响着感染结果，在 22 ℃感染 FV3 的红耳龟（*Trachemys scripta elegans*），死亡率为 100%，而在 28 ℃感染 FV3 的红耳龟，死亡率则为 50%（Allender 等人，2013b）。此外，22 ℃ 感染的龟与 28 ℃感染的龟相比，其死亡时间要短得多，而且病毒载量也大得多。因而阿蓝德（Allender）等人（2013b）提出一个假设，在较高的温度条件下，细胞免疫或体液免疫能使龟清除感染。

除把重点放在病毒复制的动态学和宿主的免疫反应以外，温度就其本身而言就是一个应激因子。贝利等人（2013）发现，林蛙蝌蚪感染 FV3、PPIV 和 REV 以后，饲养在 20 ℃要比饲养在 15 ℃的死亡率高得多，但在 20 ℃时，没有感染病毒的对照组动物死亡率同样也比较高，这就说明，较高的温度一般来说对动物造成了应激。接种感染好几种鱼类的结果显示，在鱼类的温度极限附近，死亡率会增加（Whittington 和 Reddacliff，1995；Grant 等人，2003）。阿里尔和詹森（Jensen，2009）注意到，虹鳟感染EHNV以后，在 20 ℃时，死亡率最高，这一温度已经超出了该种类的自然温度范围，且很有可能产生应激或免疫抑制。然而，本研究中的红鳍鲈，在中间温度时却产生了更高的死亡率，在 15 ℃时的死亡率是在 20 ℃时的 2 倍，而在 10 ℃时死亡率极低，因此，在红鳍鲈中，温度诱发的应激似乎很重要。也许，病毒复制的适宜温度是 15 ℃，而不是宿主免疫系统的适宜温度，宿主免疫系统的适宜温度在 20 ℃时占优势（Ariel 和 Jensen，2009）。类似地，伊查巴特（Echaubard）等人（2014）发现，用北方豹纹蛙和林蛙蝌蚪进行实验流行病学研究时，在 14 ℃时，对照组和那 82
些感染两种 FV3 样病毒的死亡率，要比在 22 ℃时高得多，从而提供了另外的证据，即温度可能影响蛙病毒病的流行，还可能依赖于宿主和病毒类型（Echaubard 等人，2014）。综上所述，温度有可能通过多个机制影响着蛙病毒的感染。考虑到气候变化导致全球气温上升，所以有必要把更多的研究重点放在温度及其变化（Raffel 等人，2006；还可参阅 Terrell 等人，2013 的文献）对宿主与病原生理学所产生的影响上（Altizer 等人，2013）。

4 面对其他自然应激因子和人为应激因子的易感性

蛙病毒复杂的流行病学特征之一就是流行结果的差异，可以从没有（或没有表现出）死亡到大规模死亡。这一明显的随机性导致学者们作出这样的假设，环境因子存在着较大的时间和空间上的变化，因而，在疾病的暴发中起着很重要的作用（Gray 等人，2009a）。人们普遍认为，自然的和人为的应激因子可以抑制宿主的免疫功能，从而使宿主个体在应激环境中变得更加易感和易发病（见 Martin，2009 综述；Blaustein 等人，2012）。虽然，许多研究者都用应激这一术语来表示任何不良的、一般来说不可预测的条件，这些条件能使生物感染，但是对于“应激-诱导的易感性”假设来说，实质上有着其生理学的基础。这种假设认为，糖皮质应激激素的缓慢升高，对宿主的免疫系统有着不良的影响，比如，糖皮质激素能降低循环系统中淋巴细胞群体数量、减少细胞因子的产生或者抑制细胞介导的免疫应答（Sheridan 等人，1994；Haddad 等人，2002；Dhabhar，2009）。

然而，界定应激和确定应激因子以及推导它们所引起的反应，常常是比较困难的，在缺乏生态环境、发育和进化背景的情况下尤其如此（Martin，2009）。例如，沃恩等人（2011）发现，变态前的林蛙蝌蚪接种感染蛙病毒以后，糖皮质激素的浓度要高于对照组，并且还经历了更快速的发育和更迅速的重量减轻。由于糖皮质激素负责动用资源（例如，对感染产生应答反应）以及促使变态前的蝌蚪加速变态，作者提出这样的假设，糖皮质激素这一峰值的作用主要是维持能量的均衡。只有那些具有足够能量储备的个体，才能同时支持快速的发育和体内强健的免疫应答（Warne 等人，2011）。很显然，即使有相当大的背景，要搞清楚与糖皮质激素浓度升高有关联的免疫力，还存在着很大的难度。因此，毫不奇怪，所报道的自然和人为应激因子对宿主病毒易感性的影响是高度可变的。

4.1 捕食者和其他自然应激因子

83 数十年的研究，特别是对两栖类动物的研究，已显示出捕食者能够改变个体和种群的生理学、行为学以及形态学（Tollrian 和 Harvell，1999）。由于捕食的威胁，两栖类动物的蝌蚪可以改变体内应激激素皮质酮的生成（Fraker 等人，2009），进而改变它们的免疫力。最近，好几个研究用笼养的捕食者来考查捕食风险对疾病结果的影响，这些捕食者能够散发化学信息素（例如：利它信息素，kairomones），可以防止捕食者接触并杀死猎物。克尔比（Kerby）等人（2011）研究发现，与对照组相比，当感染 ATV 的虎纹钝口螈与蜻蜓（*Anax junius*）幼体捕食者所释放的化学信息素接触时，虎纹钝口螈传染病的流行率和死亡率都增加。然而，海斯利普等人（2012）研究发现，在用绿蛙（*L. clamitans*）、林蛙（*L. sylvaticus*）、高地合唱团青蛙（*P. feriarum*）和灰树蛙（*Hyla chrysoscelis*）4 种无尾两栖动物的幼体与伟蜓属未定种（*Anax* sp.）和北美巨型水蝽（*Belostoma flumineum*）2 种捕食者进行的一组类似实验中，捕食者的信息素对发病和死亡并没有影响。类似地，里夫（Reeve）等人（2013）发现，在实验室和中试条件下，笼养的捕食者对林蛙蝌蚪的死亡率根本没有影响，在对照组与捕食者信息素处理组之间，糖皮质激素的浓度并没有差别。这就说明，捕食者应激既不能普遍促进皮质酮的生成，也不能使这种作用随着时间而消散。因此，结论是，虽然还有对其他物种的研究提供有帮助的证据，但是捕食者不能使无尾两栖类动物对蛙病毒更易感。此外，也可能存在其他推断的应激因子，如饵料水平的下降显著地增加糖皮质激素的浓度，因而使得蝌蚪不可能变得具有传染性或经历蛙病毒诱导的死亡（Reeve 等人，2013）。因此，即使是面临能量挑战和生理应激，两栖类也有可能对蛙病毒不易感。

4.2　人为应激因子

有人提出，人为应激因子相对于自然的应激因子来说，对蛙病毒疾病的影响更大（Reeve 等人，2013）。事实也确实如此，许多新出现的传染性疾病都与人类活动密切相关，包括土地利用的改变以及污染（Daszak 等人，2001）。斯特-奥马尔（St-Amour）等人（2008）发现，邻近工业和人类居住区，绿蛙种群蛙病毒感染的流行会增加，但其机制目前还不清楚。好几个研究发现，用于放牧牛的湿地，蛙病毒的流行会增加，其中的原因可能是显露的植被减少，导致两栖类幼体集聚，水质下降（Gray 等人，2007；Greer 和 Collins，2008；Hoverman 等人，2012）。研究发现（Gahl 和 Calhoun，2008，2010），蛙病毒的暴发虽然很弱，但如果在具有较高浓度的铝离子、含量较低的钙、较高的温度的池塘中以及那些高海拔的地方，蛙病毒暴发的概率会增加。在美国大烟山国 84
家公园进行的一项长期研究发现，在较高的海拔高度，红背蝾螈（plethodontid salamanders）蛙病毒的患病率更大，作者认为这一现象是由较高的水温、更多的人类进入以及蛙病毒粒子（virions）流向下游（Gray 等人，2009b；Sutton 等人，2014）引起的。虽然相关研究取得了这些吸引人的成果，但是也有必要进行进一步的实验研究，探讨人为应激因子的相对重要性，并确定其基本机制。

杀虫剂是另外一个影响并产生蛙病毒病的人为因素。水体可以从杀虫剂的直接应用、陆地径流或随风漂流而接受杀虫剂（Davidson 等人，2002）。此外，许多杀虫剂对野生动物具有免疫抑制作用（Marcogliese 和 Pietrock，2011）。虎纹钝口螈幼体接触除草剂莠去津（Atrazine）以后，外周血液中的白细胞计数会显著降低，并且对 ATV 的易感性增加（Forson 和 Storfer，2006b）。杀虫药毒死蜱（Chlorpyrifos）和西维因（Carbaryl）同样也增加虎纹钝口螈 ATV 感染的死亡率（Kerby 和 Storfer，2009；Kerby 等人，2011）。然而，弗尔森（Forson）和斯托福（Storfer，2006a）的研究发现，水中的莠去津可以降低长趾钝口螈（*Ambystoma macrodactylum*）幼体的 ATV 感染。因此，作者提出杀虫药可以使病毒失活，也可以刺激宿主的免疫系统。在 4 个研究中，接触杀虫药和蛙病毒都是同时进行的。由于杀虫药的免疫抑制要花好几天才能表现出来，因此，在实验设计中，添加病毒以前要先接触杀虫药，这样才会显示出更为一致的结果。

5 蛙病毒的种内传播

蛙病毒可以通过接触感染的个体、水体或污染物，如塘底基质，以及吞食感染动物的一部分或全部而进行传播（Langdon 等人，1988；Reddacliff 和 Whittington，1996；Jancovich 等人，1997；Plumb 和 Zilberg，1999b；Woodland 等人，2002b；Pearman 等人，2004；Harp 和 Petranka，2006；Brunner 等人，2007；Cunningham 等人，2007a；Robert 等人，2011；Brenes 等人，2014a）。非洲爪蟾（*Xenopus laevis*）幼体和成体在接触病毒水体后 3 小时就可在肠道中检出病毒 mRNA，然后病毒会扩散到其他组织。这一结果显示，两栖类动物的肠道是水源性病毒进入体内的第一场所（Robert 等人，2011），而皮肤有可能不是变态两栖类病毒感染的常见场所，其原因主要是两栖类动物能分泌抗菌肽布满皮肤之上，这些抗菌肽能使数种病原失活，其中包括 FV3 和 ATV（Chinchar 等人，2001，2004；Sheafor 等人，2008；Rollins-Smith，2009）。然而，布伦纳等人（2007）的研究显示，虎纹钝口螈幼体离开水体 1 秒钟进行皮肤与皮肤的接触，足以感染 ATV，有临床症状的虎纹钝口螈的幼体可以将病毒传播给未感染的幼体。因此，病毒至少在幼体期间，有可能通过未感染幼体的皮肤上皮而进入体内。

85 根据我们的研究推测，两栖类动物在幼体期间吞食被病毒感染的组织是常见和重要的传播途径。观察发现，两栖类动物和爬行类动物吞食同类是一种常见现象（Crump，1983；Polis 和 Myers，1985），即使无尾目动物的蝌蚪也会同类相残和吞食同类（Altig 等人，2007）。有两个研究发现，林蛙和拉塔斯斯蛙（*Rana latastei*）蝌蚪，当允许它们吞食 FV3 感染死亡的同类时，死亡率会大大增加（Pearman 等人，2004；Harp 和 Petranka，2006）。类似地，布伦纳等人（2007）研究发现，ATV 感染的虎纹钝口螈幼体，当它们临近死亡很容易被吞食时，极具传染性。

腹腔注射 EHNV 的鱼类，其胃肠道缺乏病灶，但自然感染的鱼类胃肠道会发生病灶，从而提示，在自然状态下，存在着口服感染的传播途径（Langdon 等人，1988；Reddacliff 和 Whittington，1996）。LMBV 同样也可以通过水体（Plumb 和 Zilberg，1999b）和吞食感染的猎物而进行传播（Woodland 等人，2002a）。由于 LMBV 已从皮肤黏液中检测出来，因而也存在直接接触而进行传播的可能。伍德兰（Woodland）等人（2005）所做的实验中，LMBV 在小水族箱中，从感染的鱼传染给稚鱼时，无论是否阻隔直接接触，感染都非常有效，因而提示，通过水体传播是主要的传播途径。值得注意的一点就是，LMBV 病毒病的暴发主要涉及成年动物，而大多数研究使用的是稚鱼。

龟类动物的传播途径，目前还不清楚。约翰逊（Johnson）等人（2007）采用肌肉注射缅甸星龟蛙病毒（Burmese star tortoise ranavirus）的方法，仅仅在一只箱龟和几只红耳龟中诱发了感染，但是那些口服同样剂量病毒的龟却保持未感染状态。然而，最近，布雷内斯（Brenes）等人（2014a）的研究显示，从白鲟（*Scaphirhynchus albus*）（Waltzek 等人，2014）得到的 FV3 样病毒分离株，经水体传播给了红耳龟，其中用水浴接种病毒的龟，有 20% 发生了感染。与 FV3 样病毒感染的可普灰树蛙一起共养的红耳龟，有 30% 的龟发生了感染。还应值得注意的是，通过节肢动物媒介进行的传播并没有排除在外。蛙红细胞病毒（Frog erythrocytic virus）好像是一种虹彩病毒（Gruia-Gray 等人，1989），该病毒通过惊骇库蚊（*Culex territans*）和蠓（*Forcipomyia fairfaxensis*）在变态后的牛蛙之间进行机械性的传播（Gruia-Gray 和 Desser，1992）。阿蓝德等人（2006）推测，蛙病毒也有可能通过媒介在龟类之间进行传播，这是因为研究者发现病毒存在于龟类血液循环的血细胞之中。金布尔（Kimble）等人（2014）最近在东方箱龟感染蛙病毒的地点，从蚊中检测到了蛙病毒。蚊虫的传播有助于解释蛙病毒是如何持续在很少相遇的龟之间传播以及感染后很快死亡的原因。

不同的传播途径有可能导致个体感染不同量的病毒，从而对感染的概率和结果产生巨大的影响。鱼类和两栖类动物剂量反应实验显示，接种更大剂量的蛙病毒导致感染概率的增加，从感染到死亡的生存时间缩短（Plumb 和

Zilberg，1999b；Pearman 等人，2004；Brunner 等人，2005；Deng 等人，2011；Warne 等人，2011）。吞食被感染的组织，很可能使宿主获得更大剂量的病毒，86 与动物之间单个接触或在污染的水中游动相比，会大大增加感染的机会。与这一观点相吻合的是霍韦尔曼（Hoverman）等人（2010）的发现，口服 FV3 样病毒的蝌蚪与那些通过水浴接种而感染的蝌蚪相比，其死亡的发生要快得多。虽然极少评估与生态相关的背景，但是对于造成感染和发病，一定存在着一个最小剂量和一个临界值。例如虎纹钝口螈幼体的最小剂量为 10^2PFU mL^{-1}。超过临界值，剂量就变得不重要了，例如病毒感染量 >10^4 PFU mL^{-1}（Brunner 等人，2005）。值得引起格外关注的是，即使是高标准的接种感染，其感染的过程和结果变化也非常大。

最后需要指出的是，我们还没有见到关于野生种群蛙病毒的传播率或动态学的公开发表资料。实质上，针对传播的每一个研究都集中在蛙病毒的传播途径上，忽略了宿主行为、密度和接触率的关键作用，而这些因素在构成传播动态学方面起着重要作用。目前，还存在的一个问题就是缺乏纵向研究数据，也就是说，追踪蛙病毒感染的发病率和死亡率随着时间的变化几乎是不可能的（Gray 等人，2015）。未来的研究应侧重于参数化流行病学模型的数据收集，从而显著地加快我们对蛙病毒生态学的认识（Gray 等人，2015）。

6 蛙病毒的种间传播

目前已很清楚，现在蛙病毒能感染一系列广泛不同的宿主（Duffus 等人，2015），但是，直到现在，还不清楚这些病毒是否严格限于亲缘关系相近的分类群中，或者是否能在变温脊椎动物不同类别之间进行传播。来自野外的轶事证据（anecdotal evidence）表明，蛙病毒存在种间传播的可能性。例如，已有报道，病鱼和病龟与蛙病毒感染引起的大规模死亡存在着关联（Mao 等人，1999；Farnsworth 和 Seigel，2013）。此外，好几个实验室的研究显示，从同一脊椎动物类群分离到的 BIV 和 FV3 样病毒，能够实验感染另一类群的动物（Moody 和 Owens，1994；Ariel 和 Owens，1997；Bang Jensen 等人，2009，2011b；Gobbo 等人，2010；Bayley 等人，2013；Brenes 等人，2014b）。最近，布雷内斯等人（2014a）进行了成对比较实验，将接种过蛙病毒的和未接种蛙病毒的宿主分别置于一个细目筛网的两侧，防止它们直接接触，但可允许水体和病毒粒子通过。他们的实验结果显示，刚出壳的红耳龟和西部食蚊鱼（*Gambusia affinis*）都能将 FV3 样病毒传播给可普灰树蛙蝌蚪，分别导致了 50% 和 10% 的死亡率。反过来，可普灰树蛙蝌蚪引起了30%的红耳龟感染，但红耳龟在实验的28天中没有发生死亡，同时，没有引起西部食蚊鱼感染。这一研究显示，存在着不同动物类别间的传播，这是很重要的。同时，更为重要的问题就是搞清楚不同动物类别的个体是如何以不同的方式进行相互接触（如吞食），促进直接传播的，或者是如何在时间和空间发生重叠，从而产生间接传播的（Gray等人，2009a）。

7　不同物种对蛙病毒的易感性

研究两栖类物种对蛙病毒的相对易感性和发病的实验在不断增加（Cullen等人，1995；Cullen和Owen，2002；Schock等人，2008；Hoverman等人，2010，2011；Haislip等人，2011）。但是，涉及爬行类动物的比较研究一直很缺乏，直到最近，布雷内斯等人（2014a，b）才开展了一些研究。而涉及鱼类蛙病毒的易感性研究，在很大程度上则局限于EHNV（Becker等人，2013；Langdon，1989；Brenes等人，2014b）。然而，值得一提的是，最近好几个研究，将不同种的鱼接种从鱼类和两栖类动物中分离到的多个蛙病毒株。因此，从总体上来讲，有关鱼类对蛙病毒相对易感性的研究资料正在积累（Bang Jensen等人，2009，2011a；Gobbo等人，2010；Brenes等人，2014b）。 87

从这些研究中，可以归纳出两种模式。第一种模式，物种对任何特定蛙病毒的易感性变化非常大。例如，在澳大利亚，贝克尔（Becker）等人（2013）用EHNV接种感染了12种具有经济和生态重要性的淡水鱼类，其中，有4种鱼对蛙病毒易感，有1种鱼即东方食蚊鱼（*G. holbrooki*）是隐性带毒者，有7种鱼既没有被感染，也没有再次获得感染（Becker等人，2013）。类似地，对来自北美洲的14种无尾两栖类动物和5种有尾两栖类动物进行蛙病毒接种感染实验，实验结束时，其结果变化非常大，有的种类全部死亡，有的种类则没有检测出感染（Hoverman等人，2010，2011；Haislip等人，2011）。用龟类、鱼类、无尾两栖类动物分离到的3种FV3样病毒进行接种感染实验，结果导致5种鱼中只有2种鱼感染，其中包括西部食蚊鱼；3种高度水生的龟类中，有2种动物即佛罗里达鳖（*Apalone ferox*）和密西西比地图龟（*Graptemys seudogeographica kohni*）发生了感染（Brenes等人，2014b）。除宿主之间的差别以外，在不同的发育阶段和环境条件下，很显然，易感性也有所变化（见3.3节）。因此，当我们经常谈及鱼类和蛙类对蛙病毒的易感性时，以相同的方式来对待所有的蛙病毒，因而越来越清楚，如此宽泛的陈述掩盖了大量重要的变化。对这些变化的解释仍面临着挑战。

随着研究不断向前迈进，霍韦尔曼等人（2011）运用比较系统发生的研究方法对 7 科 19 种两栖类动物进行了研究，评估其对蛙病毒 3（FV3）易感性的可能相关关系，例如系统发生的关联性、生活史和生态学等。某些科如蛙科（Ranidae）的动物，相对于其他科如雨蛙科（Hylidae）和钝口螈科（Ambystomatidae）的动物来说更加易感。但是，同样也存在着与生态学和生活史相关的易感模式。在半永久湿地繁殖的物种，例如一些稀有物种，在分布上受到限制，对蛙病毒更加易感。此外，还有证据显示，具有快速发育幼体阶段的物种也比较易感。这一结果可能是受到生活史平衡的驱动，如此一来，用于生长和发育上的投入，会以防御病原为代价，这种平衡见于感染吸虫的两栖类动物之中（Johnson 等人，2012）。很清楚，蛙病毒感染，就生长和发育而言，能转嫁成本（Echaubard 等人，2010）。

88 鉴于易感性的种间差异，如果存在排毒率和行为上的差异，那么群落的组成很有可能影响蛙病毒病暴发的概率、动态学以及结果。能增大病原传播的物种会增加蛙病毒病暴发的概率（Paull 等人，2012）。此外，宿主物种接种感染的次序也会改变蛙病毒病暴发的结果。布雷内斯（2013）的研究结果显示，在水生环境中进行中试试验，结果显示，如果林蛙蝌蚪首先感染接种蛙病毒，则比首先感染接种蛙病毒的三锯拟蝗蛙（*P. feriarum*）或者首先感染接种蛙病毒的斑点钝口螈（*A. maculatum*）幼体，在群落水平的死亡率要高得多。另外，假如群落由 3 个高度易感的物种组成，那么其死亡率要比只有一个高度易感物种组成的群落死亡率高得多（Brenes，2013）。

第二种易感性的常见模式就是，尽管它们一般来说有着广泛的宿主范围，但蛙病毒对分离病毒的同类别动物有更好的感染效果。特别值得一提的是鱼类，甚至还有一些爬行类动物，相对于两栖类动物来说，对 ATV 和 FV3 样病毒的易感性要差（Jancovich 等人，2001；Picco 等人，2010；Allender 等人，2013b；Brenes 等人，2014a）。有好几个研究显示，ATV（Jancovich 等人，2001；Picco 等人，2010）和鱼类 FV3 对鱼类几乎没有或完全没有传播性（Ariel 等人，2010；Gobbo 等人，2010；Bang Jensen 等人，2011a；Bang Jensen 等人，2011b）。与之类似的是，两栖类动物可能会对鱼类蛙病毒没有易感性。贝利等人（2013）研究发现，欧洲林蛙（European common frog）在蝌蚪和成体期都能感染 FV3 和 REV，而且蝌蚪还可感染 PPIV，但不能感染其他几种鱼类病毒，如医生鱼病毒、欧洲六须鲶病毒、孔雀鱼病毒 6、EHNV 和 SERV。即使是在蛙病毒种内也存在着宿主范围的差别。例如，布雷内斯等人（2014b）用从鱼和龟中分离到的 FV3 样病毒使龟产生了亚临床感染，但是从蛙中分离到的病毒株没有使龟发生感染。然而，我们提请注意，不要对这些模式进行太广泛的推断。研究者对蛙病毒的宿主范围还只是勾画出一个大体轮廓，因而，仍保留了一个公开课题——上述模式是否普遍存在？假如是，其理由是什么？

8 蛙病毒在环境和携毒者中的存活时间

蛙病毒的存活时间存在两种潜在的机制，即在环境中耐受，或者在亚致死性感染的宿主（贮藏宿主，reservoirs）中耐受。从历史的观点来说，蛙病毒能够抵抗环境的降解。例如，EHNV 能在-20 ℃和-70 ℃保存的鱼组织中持续存活 2 年以上，在 4 ℃时，至少能存活 7 天（Langdon，1989）。EHNV 在蒸馏水中，15 ℃条件下，97 天后，病毒滴度没有下降，同时还观察到，EHNV 在无菌塑料培养皿中的干组织培养基中，15 ℃避光保存 113~200 天（Langdon，1989），能长期存活。类似地，LMBV 能在冷冻组织中存活达 155 天（Plumb 和 Zilberg，1999a）。然而，蛙病毒在比较接近真实生态环境条件下，降解更快。有研究报道，LMBV 在其感染的水体中，24 小时降解了 90%，但是，该病毒在水体中保留可检测水平至少有 7 天时间（Grizzle 和 Brunner，2003）。纳 89
齐尔（Nazir）等人（2012）的研究发现，从蛙、龟和壁虎中分离到的 4 种 FV3 样病毒，20 ℃条件下，在未消毒池塘中存活时间的 T-90 值为 22~34 天，而在 4 ℃时，存活时间可以增加到 72 天，假如不跨年度的话，这一时间足以允许病毒在一个流行区的环境中持续传播。然而，来自细菌直接作用而进行实验分离到的病毒粒子，可能存在一定问题，这是因为细菌和其他微生物不能吞食和使病原失活。为解决这一问题，约翰逊和布伦纳（2014）从 5 个池塘采集了水样，并对这些水样进行了过滤消毒、紫外消毒，对剩下的水样未作处理，然后将 FV3 样病毒直接加入到水中。他们发现，在过滤消毒的水样中，T-90 值为 8 天，而在未作处理的池塘水样中，T-90 值仅为 1 天，并且池塘水样中还存在水生微生物。类似地，在春季，将普通的浮游动物蚤状溞（*Daphnia pulex*）加入已接种病毒的水体中，结果观察到蛙病毒发生了快速降解。此外，约翰逊和布伦纳（2014）得出结论，蛙病毒可能对不良环境条件如干燥、冷冻有抵抗力，同时，它们有可能在水体通过自然产生的微生物和浮游动物进行快速的降解，从而提示，直接传播途径，如接触、吞食，可能比水源性传播更为重要。一个很重要的警示就是，所有这些研究所使用的病毒都是在细胞培养中生长的蛙病毒，而在自然条件下，蛙病毒常通过分泌黏液排毒或通过皮肤脱落排毒等，有可能使蛙病毒避免在微生物和环境中降解。

纳齐尔等人（2012）同样也做了蛙病毒在土壤中存活时间的测试，结果T-90值为30~48天，从而引起了人们对蛙病毒通过污染的土壤而进行传播的关注（Harp和Petranka，2006）。然而，布伦纳等人（2007）的研究发现，当对池塘底泥进行干燥处理时，ATV会变得没有传染性。因此，水体或它们的基质含水，是传染的关键所在。

蛙病毒能在死的或者活的感染宿主体内持续存活。现在已很清楚，蛙病毒能在冷冻的尸体中存活很长时间（Langdon，1989）。在环境中，如果在发生大规模死亡以后尸体被很快冷冻，那么冷冻的尸体可能会是来年蛙病毒重要的传染源（Bollinger等人，1999）。此外，某些物种的个体在生活史的某些阶段是病毒的携带者，即仍保留感染状态，并在没有清除或死于感染的情况下长期传染其他个体，从而作为更为易感物种或发育阶段的病毒储藏宿主（Haydon等人，2002）。总的来讲，在没有明显感染的状况下，只有一小部分个体能够存活几周或几个月（Langdon，1989；Cullen和Owens，2002；Brunner等人，2004；Robert等人，2007；Haislip等人，2011；Hoverman等人，2011；Brenes，2013；Brenes等人，2014b）。例如，霍韦尔曼等人（2011）、海斯利普等人（2011）、布雷内斯（2013）与其他学者（2014b）所做实验当中，43个变温脊椎动物感染接种FV3样病毒28天以后，在感染和死亡之间只有大约85%的相关性。此外，在所做的实验当中，只是少数几个亚临床感染的个体将蛙病毒传播给了更为易感的个体或物种，从而启动蛙病毒的暴发（Brunner等人，2004）。另外，
90 罗伯特（Robert）等人（2007）的研究结果显示，通常能抵抗FV3感染的非洲爪蟾（*X. laevis*）可能是无症状的病毒携带者。在后来的一个研究中，莫拉莱斯（Morales）等人（2010）发现，腹腔巨噬细胞能使静止的FV3感染潜伏至少达3周。无症状的蛙病毒感染会在γ射线造成免疫功能不全的动物中重新激活（Robert等人，2007）。这些感染了病毒却无明显症状的非洲爪蟾或者感染了病毒的其他物种个体能否存活，并保留更长时间的感染，以及它们如何将感染传播给其他个体，仍然是值得公开研究的课题（请参阅3.2节）。

9 蛙病毒及其宿主的选择与共进化

蛙病毒对它们的宿主种群具有很强的选择力，常常造成疾病的流行，从而造成种群的剧烈波动，甚至造成局部种群的灭绝。欧洲林蛙种群就是一个很好的例子，蛙病毒在过去的选择中，似乎喜好MHC（major histocompatibility complex，主要组织相容性复合体）类型I等位基因，该等位基因与病毒识别和

提供抗原性有关。研究发现，特定的 MHC 类型 I 等位基因，在有蛙病毒感染病史的种群出现的频率要比没有蛙病毒感染病史的种群高得多（Teacher 等人，2009）。此外，还观察到杂合度和亲缘关系（relatedness）的下降，从而提示，蛙病毒病的流行会导致交配模式（mating patterns）的改变（Teacher 等人，2009）。

有充分的证据显示，两栖类种群对蛙病毒的易感性存在差异性，推测其原因，有可能是基础遗传差异。实验室的试验普遍显示，不同种群之间的死亡率和感染率存在着巨大差别（Pearman 等人，2004；Brunner 等人，2005；Pearman 和 Garner，2005；Schock 等人，2008；Brunner 和 Collins，2009；Echaubard 等人，2014）。有一个很好的例子，皮尔曼（Pearman）等人（2004）的研究发现，拉塔斯蛙具有低的遗传多样性的 2 个种群，在感染接种 FV3 后，仅 5 天就全部死亡。而具有较高遗传多样性的 4 个其他种群，死亡率只有 40%～70%。类似地，非洲爪蟾近交系（inbred lines）的蝌蚪相对于远交系（outbred lines）的蝌蚪来说，其存活时间更短（Gantress 等人，2003）。

鉴于比较大的可能是遗传的易感性变异以及由蛙病毒给宿主群带来的强烈选择压力，从总体上，我们认为，蛙病毒及其宿主存在着共进化。这种共进化潜力，可以以虎纹钝口螈-ATV 系统作为例证，因为在发现虎纹钝口螈时，常常没有其他两栖类动物，而蛙病毒的流行却很常见（Brunner 等人，2004）。在这一系统中，有三点证据提示，宿主-病毒存在着共进化。

第一，在美国亚利桑那州虎纹钝口螈种群中，发病的频度和吞食同类的频度之间存在负相关关系（Pfennig 等人，1991）。虽然同类残食的动物可通过捕食同类带来性能上的优势（Reilly 等人，1992），但是同类残食的动物捕食，其代价主要是会增加从同类中获得 ATV 的风险，因而减少捕食同类有可能阻止疾病的传播（Pfennig 等人，1991；Bolker 等人，2008）。普通花园实验（Common Garden Experiments）提示，这些模式在遗传上是基本的，因而有可能是进化选择的结果（Parris 等人，2005）。在 ATV 处理组与未处理组之间，动物同类相食表型的产生是不可塑的，并且在野外所观察到的同类相食频度的差别，可以在实验室复制（Parris 等人，2005）。

第二，病毒的存活时间与宿主的生活史紧密耦合，从而提示两者存在共进 91
化史。幼体、有鳃成体以及变态后的成体都对 ATV 易感，但幼体相对变态体来说更易复发（Brunner 等人，2004）。在池塘中，蛙病毒流行病主要发生在幼体期，但是许多变态体离开池塘后，处于亚临床感染状态，冬天一过，来年便返回池塘进行繁殖（Brunner 等人，2004）。因此，全年作为种群间或种群内病毒的重要传染源，虎纹钝口螈本身就是 ATV 的重要储藏宿主。

第三，系统发生一致性的分子遗传分析表明，在蝾螈种群与地方 ATV 病毒株之间存在着共进化（Storfer 等人，2007）。排除三宿主转换（three host switches）所引起的感染蝾螈作为垂钓鱼饵的搬运（Jancovich 等人，2005）以外，蝾螈与病毒两者之间的进化树（phylogenetic trees）具有完全的一致性（Storfer 等人，2007）。此外，进化分枝的长度（nodal depths），或者假定物种形成的时间点，或者分枝事件的时间点，与 ATV 病毒株及其关联的虎纹钝口螈种群紧密相关（Storfer 等人，2007）。不同虎纹钝口螈种群分离到的不同 ATV 病毒株，在分子进化上，似乎存在着地方性选择。空间分布不同的虎纹钝口螈种群之间以及与宿主免疫逃避（Jancovich 和 Jacobs，2011）相关联的 ATV 基因已进化出独特的氨基酸差异，而不依赖于它们系统发生的进化关系（Ridenhour 和 Storfer，2008）。总之，这些发现为 ATV 及其宿主虎纹钝口螈共进化史的研究提供了强有力的支持。

10　蛙病毒毒力的进化

虽然蛙病毒常常被认为具有很强的毒力，但是在病毒株与宿主物种之间，其毒力变化很大。例如，布伦纳和柯林斯（Brunner 和 Collins，2009）用 ATV 的 9 个病毒株感染接种虎纹钝口螈幼体，结果发现，在不同的病毒分离株之间，其毒力（以接种到死亡的时间来评估）变化非常大，而且这种毒力明显是可遗传的。这一结果以及许多其他研究毒力的类似结果提出了一个问题：为什么一些病毒或病毒株的毒力要比另外一些的毒力强呢？

病毒毒力变异的一个公认而重要的假设被认为是传播与毒力之间的一种平衡，而毒力一般被认为是病毒在宿主体内复制以及来自宿主传播的一种不可避免的附带结果（Lenski 和 May，1994；Alizon 等人，2009）。强毒力的病原相对于弱毒力的病原来说，具有更少的传播机会，这是因为它们能快速地杀死宿主。然而，毒力的最佳水平，甚至是平衡的存在，在很大程度上取决于宿主的生态（如密度、死亡背景）、病原（如宿主体内病原的竞争和宿主的免疫反
92 应）以及它们之间的相互作用（如传播方式和发病原因）（Day，2001，2002；Day 和 Proulx，2004；Ebert，1999）。适宜快速传播的条件（如宿主种群的密度过大、未成年宿主的大举涌入）会缩短感染期（如高的背景死亡率、通过免疫系统快速的清除），或者导致宿主体内不同病毒株之间的竞争（如多重感染）。

这些条件一般来说，适宜毒力更强的病原（May 和 Nowak，1994；Ebert 和

Mangin，1997；Williams 和 Day，2001；Cooper 等人，2002；Gandon 等人，2002；Day，2003；Restif 和 Koella，2003；Alizon 和 van Baalen，2008）。耐受感染的宿主如使适应度最小化，而不是抵抗或清除感染的宿主，被认为是规避了对宿主毒力增强的进化（Roy 和 Kirchner，2000；Restif 和 Koella，2003）。然而，对宿主-病原系统的详细了解，常常驱动着对预期的进化动态学进行研究（Day，2001，2002；Sabelis 和 Metz，2002；Day 和 Proulx，2004）。因此，我们必须小心翼翼地将毒力学理论应用到蛙病毒-宿主系统中。

然而，一个越来越清楚的模式就是，从人工饲养环境如水产养殖和蛙养殖场、鱼饵商店中分离到的蛙病毒，相对于野生病毒株来说，毒力更强。例如，从鱼饵商店分离到的一株 ATV 对虎纹钝口螈的毒力显著高于本地病毒株，能够造成更大的死亡（Storfer 等人，2007）。类似地，在美国爱达荷州，从牛蛙养殖场分离到的一株 FV3 即 RCV-Z（Majji 等人，2006），在体生长显著加快，并且在牛蛙和斑点蛙（*Lithobates luteiventris*）蝌蚪中造成的死亡率显著高于从华盛顿野生牛蛙种群中分离到的一株病毒（AS，KC，华盛顿州立大学，未发表资料）。霍韦尔曼等人（2011）的研究同样也发现，从牛蛙人工养殖场分离到的一株 FV3 样病毒，与 FV3 典型病毒株相比，毒力更强，在实验室试验中造成 8 个无尾两栖类物种的幼体平均死亡率在 51% 以上。从白鲟孵化场分离到的一株 FV3 样病毒，被证实是目前为止所确定的毒力最强的病毒株（Waltzek 等人，2014）。

在人工饲养环境中，对蛙病毒毒力增强的进化，存在着好几种假设：第一，毒力成本（在传播之前，宿主的死亡就已经发生）有可能在人工饲养环境中降低，这是因为高的背景死亡率或者环境条件促进了病毒的快速传播。高的种群密度可以增加接触率，并造成排出病毒粒子的积聚，这两者都会增加传播率。另外，宿主的死亡并不是蛙病毒传染期的结束。在养殖场中，易感动物由于吞食没有被及时清除的感染的尸体，因而发生感染。而在自然界中，其他吞食者和促进分解的生物有可能丰度比较高，这些感染的尸体很快就会消失。第二，在流行的最初阶段，当易感宿主不受限制时，快速传播和强毒力应是最佳选择（Day 和 Proulx，2004；Bolker 等人，2010）。如果新的易感动物不断被引入一个正在经历流行病的种群，就像在一些出售虎纹钝口螈鱼饵商店里发生的一样（Picco 和 Collins，2008）。这样就有利于增殖更快、毒力更强的病毒株的形成。当宿主受到限制时，更有利于这些病毒的形成。最后，在一个宿主内部病原之间的竞争也有利于病毒毒力的增强，甚至是以减少传播为代价（例如，
de Roode 等人，2005）。人工饲养环境，常常会从许多不同的来源接受动物， 93
从而隐藏着好几个共循环的蛙病毒株，进而对选择毒力最强的病毒株产生影响

(Antia 等人，1994；Bull，1994)。然而，所有这些假设仍有待验证。从总体上讲，这些学者从逻辑上建议，需要对促进蛙病毒的传播、混合以及存活的人工饲养设施条件和措施作出改变。

这些结果同样也提出了值得关注的问题，感染蛙病毒的鱼饵虎纹钝口螈或美国牛蛙的引进，有可能把毒力强的病毒株带到从未感染的宿主（Naïve hosts）栖息区，或带到原先感染过，但已经适应其他蛙病毒的宿主栖息区。特别值得一提的是，牛蛙构成了国际贸易两栖类的主体。施勒格尔（Schloegel）等人（2009）报道，在2000—2005年，超过2 800 000只两栖类动物进口到了美国，其中蛙病毒的感染率达8.5%。养殖场养殖的和野生的牛蛙以及其他两栖类动物在全球的贸易量巨大（Altherr 等人，2011），从而使得国际贸易成为蛙病毒跨地区传播和引入的重要途径（图2，Schloegel 等人，2010）。因此，一个关键问题就是，是所引入的强毒力蛙病毒株的持续存在，还是竞争，形成了毒力超过本地野生型的蛙病毒株。

11 蛙病毒带来的灭绝风险

卡斯罗特和博尔科路（Castrot 和 Bolker，2005）对病原造成宿主灭绝的三个理论机制进行了综述。第一，小的而被隔离的种群可能存在因疾病带来随机灭绝的风险。高度传染性、强毒力的病原如蛙病毒，同样能使最初的高密度种群数量下降到因地理统计学波动或相关遗传影响所导致的随机灭绝种群数量水平。第二，以非密度依赖方式传播的病原能够持续传播，从而引起宿主密度的下降，甚至能下降到宿主灭绝点。一个常见的非密度依赖传播例子就是通过性传播的传染病。但是不管宿主密度如何，任何形式的聚集，包括局部采食地点或繁殖地点的聚集，都能维持高的传播率。即使是短暂的非密度依赖传播，例如繁殖，也能引起病原驱动的灭绝（Ryder 等人，2007）。第三，具有生物（其他物种）或非生物（环境）储藏宿主的病原，无论宿主的密度如何，都会持续感染宿主，因而具有造成宿主灭绝的潜在风险。

蛙病毒似乎满足宿主灭绝所需要的条件，通过这三个机制的任何一个或全部机制造成宿主灭绝（Miller 等人，2011）。广泛的宿主范围以及区域与国际贸易造成蛙病毒频繁跨地区传播。不难想象一株强毒力和高传染性的蛙病毒是如何被引入到小的或从未感染的种群（naïve populations）中的。另外在实验室，ATV 感染虎纹钝口螈幼体的试验发现，传播速率会随着被感染幼体的密度增加而增快，很快达到饱和，而且即使虎纹钝口螈幼体在密度最低时，也会有相

当高的感染率（Greer 等人，2008）。此外，有许多变温脊椎动物在采食、栖息
以及交配时会集聚在一起，这样就会导致至少短期的非密度依赖性传播。最 94
后，正如上面所指出的，蛙病毒可以在环境中存活足够长的时间，以传染从未感染的宿主（naïve hosts）。更重要的是鱼类、两栖类动物以及爬行类动物常常与其他不易感或丰度更高的物种在分布上互相重叠，而这些重叠物种是所研究物种的病毒贮藏宿主。总体而言，有足够的理由认为，蛙病毒可以导致宿主的灭绝（Miller 等人，2011）。

尽管存在着蛙病毒造成宿主灭绝的这一潜在风险，但是要找到蛙病毒导致区域或全球物种灭绝或者造成种群数量下降的具体例子，还是比较困难。存在着几个长期跟踪野生变温脊椎动物种群的数据集，这些种群反复发生蛙病毒感染，并引起大规模死亡。至少对两栖类动物来说，需要长期的数据集来确定种群数量是否下降（Alford 和 Richards，1999）。从某种程度上说，这可以通过这样一个事实得到解释，即蛙病毒常常感染处于发育阶段的幼体或稚体（见 2.1 和 3.3 节），对于许多分类群来说，宿主的这些发育阶段有可能在地理统计学上没有成体重要（Biek 等人，2002）。由此我们推测，感染蛙病毒成体最多的种群或物种可能会显示出种群数量下降的风险最大（见 Earl 和 Gray，2014）。的确如此，蒂彻（Teacher）等人（2010）报道，在英国池塘中的林蛙成体种群丰度的中值下降了 81%，这是由于已知的蛙病毒病反复发作引起了大规模死亡。再次重申，在英国，蛙病毒似乎主要感染成体（Cunningham 等人，1993；Teacher 等人，2010；Duffus 等人，2013）。类似地，在成龟中观察到的明显由蛙病毒感染造成的死亡，最终也将这些种群置于灭绝风险的境地（Belzer 和 Seibert，2011；Farnsworth 和 Seigel，2013）。

在两栖类幼体中，蛙病毒反复流行同样会导致种群数量的下降以及区域性的种群灭绝。例如，彼得兰卡（Petranka）等人（2003，2007）报道，在几个新构建的湿地当中，由于每年蛙病毒病引起大规模死亡，最小种群恢复期在 8 年以上。最近，厄尔（Earl）和格雷（Gray）（2014）运用矩阵种群预测模型（matrix population-projection model）展示，假如一个封闭林蛙种群幼体或变态体每隔 5 年感染一次蛙病毒，那么这个种群就会存在局部灭绝的可能性。对于每年都感染的种群来说，灭绝的时间快达 5 年（Earl 和 Gray，2014）。此外，敏感性分析显示，对于林蛙来说，变态前期的存活要比变态后期的存活更加重要。所提供的初步证据显示，蛙病毒可引起幼体的显著死亡，并导致种群的灭绝（Earl 和 Gray，2014）。在限制扩散的破碎化景观中，易感性物种相对于前面所公认的灭绝风险来说，可能会存在更大的灭绝风险（Collins 和 Crump，2009）。

蛙病毒反复流行导致两栖类种群数量下降最有说服力的例子，来自西班牙欧罗巴山皮科斯国家公园（Picos de Europa）（Price 等人，2014）。在该研究中，对两栖类种群监测了 6 年，并且在此期间引进了蛙病毒，在接下来的几年里，种群数量显著下降。什么原因致使这些病毒-宿主种群的相互作用有别于其他种群数量没有下降的种群，目前仍不清楚。很明显，有必要对蛙病毒病暴发在种群水平上的影响作更多的研究，特别是要对反复发生蛙病毒感染并引起
95 大规模死亡的地点做长期跟踪研究（Gray 等人，2015）。此外，将宿主和病毒传播（如复合种群动态学）纳入数学模型和研究之中，以搞清楚蛙病毒施加给变温脊椎动物宿主的风险显得尤为必要。

致谢

我们要感谢布伦纳（Brunner）实验室成员对早期稿件的评审意见，特别要感谢杰克·克尔比（Jake Kerby）和戴维·勒萨勒雷（David Lesbarrères）仔细认真的评审以及提出的有帮助性建议，还要感谢编辑人员，没有他们巨大的努力，这一章和这本书就不可能存在。同时，还要感谢国家科学基金会资金项目 DEB 1316549 提供的支持。

开放存取出版物的发行得到了田纳西大学（林业、野生动物与渔业系，研究与交流办公室，农业研究所）、华盛顿州立大学图书馆、戈登州立大学（学术事务办公室）、两栖爬行动物兽医协会以及两栖爬行动物保护协会的资金赞助。

开放存取

参考文献

Ahne W, Bremont M, Hedrick RP, Hyatt AD, Whittington RJ (1997). Special topic review: iridoviruses associated with epizootic haematopoietic necrosis (EHN) in aquaculture. World J Microbiol Biotechnol 13: 367-373.

Alford RA, Richards SJ (1999). Global amphibian declines: a problem in applied ecology. Annu Rev Ecol Syst 30: 133-165.

Alizon S, van Baalen M (2008). Multiple infections, immune dynamics, and the evolution of virulence. Am Nat 172: E150-E168.

Alizon S, Hurford A, Mideo N, Van Baalen M (2009). Virulence evolution and the trade-off hypothesis: history, current state of affairs and the future. J Evol Biol 22: 245-259.

Allender MC, Fry MM, Irizarry AR, Craig L, Johnson AJ, Jones M (2006). Intracytoplasmic inclusions in circulating leukocytes from an eastern box turtle (*Terrapene carolina carolina*) with iridoviral infection. J Wildl Dis 42: 677-684.

Allender MC, Mitchell MA, McRuer D, Christian S, Byrd J (2013a). Prevalence, clinical signs, and natural history characteristics of frog virus 3-like infections in eastern box turtles (*Terrapene carolina carolina*). Herpetol Conserv Biol 8: 308-320.

Allender MC, Mitchell MA, Torres T, Sekowska J, Driskell EA (2013b). Pathogenicity of frog virus 3-like virus in red-eared slider turtles (*Trachemys scripta elegans*) at two environmental temperatures. J Comp Pathol 149: 356-367.

Altherr S, Goyenechea A, Schubert DJ (2011). Canapés to extinction: the international trade in frogs' legs and its ecological impact. A report by Pro Wildlife, Defenders of Wildlife and Animal Welfare Institute, Washington, DC.

Altig R, Whiles MR, Taylor CL (2007). What do tadpoles really eat? Assessing the trophic status of an understudied and imperiled group of consumers in freshwater habitats. Freshw Biol 52: 386-395.

Altizer S, Ostfeld RS, Johnson PT, Kutz S, Harvell CD (2013). Climate change and infectious diseases: from evidence to a predictive framework. Science 341: 514-519.

Antia R, Levin BR, May RM (1994). Within-host population dynamics and the evolution and maintenance of microparasite virulence. Am Nat 144: 457-472.

Ariel E (1997). Pathology and serological aspects of Bohle iridovirus infections in six selected water-associated reptiles in North Queensland. Dissertation, James Cook University, North Queensland.

Ariel E (2011). Viruses in reptiles. Vet Res 42: 12.

Ariel E, Jensen BB (2009). Challenge studies of European stocks of redfin perch, *Perca fluviatilis* L., and rainbow trout, *Oncorhynchus mykiss* (Walbaum), with epizootic haematopoietic necrosis virus. J Fish Dis 32: 1017-1025.

Ariel E, Owens L (1997). Epizootic mortalities in tilapia *Oreochromis mossambicus*.

Dis Aquat Organ 29: 1-6.

Ariel E, Kielgast J, Svart HE, Larsen K, Tapiovaara H, Bang JB, Holopainen R (2009a). Ranavirus in wild edible frogs *Pelophylax kl. esculentus* in Denmark. Dis Aquat Organ 85: 7-14.

Ariel E, Nicolajsen N, Christophersen MB, Holopainen R, Tapiovaara H, Jensen BB (2009b). Propagation and isolation of ranaviruses in cell culture. Aquaculture 294: 159-164.

Ariel E, Holopainen R, Olesen NJ, Tapiovaara H (2010). Comparative study of ranavirus isolates from cod (*Gadus morhua*) and turbot (*Psetta maxima*) with reference to other ranaviruses. Arch Virol 155: 1261-1271.

Bang Jensen B, Ersboll AK, Ariel E (2009). Susceptibility of pike *Esox lucius* to a panel of *Ranavirus* isolates. Dis Aquat Organ 83: 169-179.

Bang Jensen B, Reschova S, Cinkova K, Ariel E, Vesely T (2011a). Common carp (*Cyprinus carpio*) and goldfish (*Carassius auratus*) were not susceptible to challenge with ranavirus under certain challenge conditions. Bull Eur Assoc Fish Pathol 31: 112-118.

Bang Jensen B, Holopainen R, Tapiovaara H, Ariel E (2011b). Susceptibility of pike-perch *Sander lucioperca* to a panel of ranavirus isolates. Aquaculture 313: 24-30.

Bayley AE, Hill BJ, Feist SW (2013). Susceptibility of the European common frog *Rana temporaria* to a panel of ranavirus isolates from fish and amphibian hosts. Dis Aquat Organ 103: 171-183.

Beck BH, Bakal RS, Brunner CJ, Grizzle JM (2006). Virus distribution and signs of disease after immersion exposure to largemouth bass virus. J Aquat Anim Health 18: 176-183.

Becker JA, Tweedie A, Gilligan D, Asmus M, Whittington RJ (2013). Experimental infection of Australian freshwater fish with epizootic haematopoietic necrosis virus (EHNV). J Aquat Anim Health 25: 66-76.

Belzer W, Seibert S (2011). A natural history of *Ranavirus* in an eastern box turtle population.Turtle Tortoise Newsl 15: 18-25.

Biek R, Funk WC, Maxell BA, Mills LS (2002). What is missing in amphibian decline research: insights from ecological sensitivity analysis. Conserv Biol 16: 728-734.

Blaustein AR, Gervasi SS, Johnson PT, Hoverman JT, Belden LK, Bradley PW,

Xie GY (2012). Ecophysiology meets conservation: understanding the role of disease in amphibian population declines. Philos Trans R Soc Lond B Biol Sci 367: 1688–1707.

Bolker BM, de Castro F, Storfer A, Mech S, Harvey E, Collins JP (2008). Disease as a selective force precluding widespread cannibalism: a case study of an iridovirus of tiger salamanders, *Ambystoma tigrinum*. Evol Ecol Res 10: 105–128.

Bolker BM, Nanda A, Shah D (2010). Transient virulence of emerging pathogens. J R Soc Interface 7: 811–822.

Bollinger TK, Mao J, Schock D, Brigham RM, Chinchar VG (1999). Pathology, isolation, and preliminary molecular characterization of a novel iridovirus from tiger salamanders in Saskatchewan. J Wildl Dis 35: 413–429.

Brenes R (2013). Mechanisms contributing to the emergence of ranavirus in ectothermic vertebrate communities. Dissertation, University of Tennessee, Knoxville.

Brenes R, Gray MJ, Waltzek TB, Wilkes RP, Miller DL (2014a). Transmission of ranavirus between ectothermic vertebrate hosts. PLoS One 9: e92476.

Brenes R, Miller DL, Waltzek TB, Wilkes RP, Tucker JL, Chaney JC, Hardman RH, Brand MD, Huether RR, Gray MJ (2014b). Susceptibility of fish and turtles to three ranaviruses isolated from different ectothermic vertebrate classes. J Aquat Anim Health 26: 118–126.

Brunner JL, Collins JP (2009). Testing assumptions of the trade-off theory of the evolution of parasite virulence. Evol Ecol Res 11: 1169–1188.

Brunner JL, Schock DM, Collins JP, Davidson EW (2004). The role of an intraspecific reservoir in the persistence of a lethal ranavirus. Ecology 85: 560–566.

Brunner JL, Richards K, Collins JP (2005). Dose and host characteristics influence virulence of ranavirus infections. Oecologia 144: 399–406.

Brunner JL, Schock DM, Collins JP (2007). Transmission dynamics of the amphibian ranavirus *Ambystoma tigrinum* virus. Dis Aquat Organ 77: 87–95.

Brunner JL, Barnett KE, Gosier CJ, McNulty SA, Rubbo MJ, Kolozsvary MB (2011). Ranavirus infection in die-offs of vernal pool amphibians in New York, USA. Herpetol Rev 42: 76–79.

Bull JJ (1994). Perspective: virulence. Evolution 48: 1423–1437.

Carey C, Cohen N, Rollins-Smith L (1999). Amphibian declines: an immunological perspective. Dev Comp Immunol 23: 459–472.

Chinchar VG (2002). Ranaviruses (family *Iridoviridae*): emerging cold-blooded killers. Arch Virol 147: 447–470.

Chinchar VG, Waltzek TB (2014). Ranaviruses: not just for frogs. PLoS Pathog 10: e1003850.

Chinchar VG, Wang J, Murti G, Carey C, Rollins-Smith L (2001). Inactivation of frog virus 3 and channel catfish virus by esculentin-2P and ranatuerin-2P, two antimicrobial peptides isolated from frog skin. Virology 288: 351–357.

Chinchar VG, Bryan L, Silphadaung U, Noga E, Wade D, Rollins-Smith L (2004). Inactivation of viruses infecting ectothermic animals by amphibian and piscine antimicrobial peptides. Virology 323: 268–275.

Chinchar VG, Hyatt AD, Miyazaki T, Williams T (2009). Family *Iridoviridae*: poor viral relations no longer. In: Van Etten JL (ed) Current topics in microbiology and immunology, vol 328, Lesser known large dsDNA viruses. Springer, Berlin.

Chua FHC, Ng ML, Ng KL, Loo JJ, Wee JY (1994). Investigation of outbreaks of a novel disease, 'Sleepy Grouper Disease', affecting the brown-spotted grouper, *Epinephelus tauvina* Forska. J Fish Dis 17: 417–427.

Collins JP, Crump ML (2009). Extinction in our times: global amphibian decline. Oxford University Press, Oxford.

Cooper VS, Reiskind MH, Miller JA, Shelton KA, Walther BA, Elkinton JS, Ewald PW (2002). Timing of transmission and the evolution of virulence of an insect virus. Proc R Soc Lond B Biol Sci 269: 1161–1165.

Crump ML (1983). Opportunistic cannibalism by amphibian larvae in temporary aquatic environments. Am Nat 121: 281–289.

Cullen BR, Owens L (2002). Experimental challenge and clinical cases of Bohle iridovirus (BIV) in native Australian anurans. Dis Aquat Organ 49: 83–92.

Cullen CR, Owens L, Whittington RJ (1995). Experimental infection of Australian anurans (*Limnodynastes terraereginae* and *Litoria latopalmata*) with Bohle iridovirus. Dis Aquat Organ 23: 83–92.

Cunningham AA, Langton TES, Bennet PM, Drury SES, Gough RE, Kirkwood JK (1993). Unusual mortality associated with poxvirus-like particles in frogs (*Rana temporaria*). Vet Rec 133: 141–142.

Cunningham AA, Hyatt AD, Russell P, Bennett PM (2007a). Emerging epidemic diseases of frogs in Britain are dependent on the source of ranavirus agent and the route of exposure. Epidemiol Infect 135：1200–1212.

Cunningham AA, Hyatt AD, Russell P, Bennett PM (2007b). Experimental transmission of a ranavirus disease of common toads (*Bufo bufo*) to common frogs (*Rana temporaria*). Epidemiol Infect 135：1213–1216.

Daszak P, Cunningham AA, Hyatt AD (2001). Anthropogenic environmental change and the emergence of infectious diseases in wildlife. Acta Trop 78：103–116.

Davidson C, Shaffer HB, Jennings MR (2002). Spatial tests of the pesticide drift, habitat destruction, UV-B, and climate-change hypotheses for California amphibian declines. Conserv Biol 16：1588–1601.

Day T (2001). Parasite transmission modes and the evolution of virulence. Evolution 55：2389–2400.

Day T (2002). Virulence evolution via host exploitation and toxin production in spore-producing pathogens. Ecol Lett 5：471–476.

Day T (2003). Virulence evolution and the timing of disease life-history events. Trends Ecol Evol 18：113–118.

Day T, Proulx SR (2004). A general theory for the evolutionary dynamics of virulence. Am Nat 163：E40–E63.

de Castro F, Bolker B (2005). Mechanisms of disease-induced extinction. Ecol Lett 8：117–126.

de Roode JC, Pansini R, Cheesman SJ, Helinski MEH, Huijben S, Wargo AR, Bell AS, Chan BHK, Walliker D, Read AF (2005). Virulence and competitive ability in genetically diverse malaria infections. Proc Natl Acad Sci U S A 102：7624–7628.

DeBenedictis PA (1974). Interspecific competition between tadpoles of *Rana pipiens* and *Rana sylvatica*：an experimental field study. Ecol Monogr 44：129–151.

Deng GC, Li SJ, Xie J, Bai JJ, Chen KC, Ma DM, Jiang XY, Lao HH, Yu LY (2011). Characterization of a ranavirus isolated from cultured largemouth bass (*Micropterus salmoides*) in China. Aquaculture 312：198–204.

Dhabhar FS (2009). Enhancing versus suppressive effects of stress on immune function：implications for immunoprotection and immunopathology. Neuroimmunomodulation 16：300–317.

Duffus ALJ, Pauli BD, Wozney K, Brunetti CR, Berrill M (2008). Frog virus 3-like infections in aquatic amphibian communities. J Wildl Dis 44：109–120.

Duffus ALJ, Nichols RA, Garner TWJ (2013). Investigations into the life history stages of the common frog (*Rana temporaria*) affected by an amphibian ranavirus in the United Kingdom. Herpetol Rev 44: 260-263.

Duffus ALJ, Marschang RE, Waltzek TB, Stöhr A, Allender MC, Gotesman M, Whittington R, Hick P, Hines M (2015). Distribution and host range of ranaviruses. In: Gray MJ, Chinchar VG (eds) Ranaviruses: lethal pathogens of ectothermic vertebrates. Springer, Secaucus.

Earl JE, Gray MJ (2014). Introduction of ranavirus to isolated wood frog populations could cause local extinction. Ecohealth. doi: 10. 1007/s10393-014-0950-y.

Eaton JG, Scheller RM (1996). Effects of climate warming on fish thermal habitat in streams of the United States. Limnol Oceanogr 41: 1109-1115.

Ebert D (1999). The evolution and expression of parasite virulence. In: Stearns SC (ed) Evolution in health and disease. Oxford University Press, New York.

Ebert D, Mangin KL (1997). The influence of host demography on the evolution of virulence of a microsporidian gut parasite. Evolution 51: 1828-1837.

Echaubard P, Little K, Pauli B, Lesbarrères D (2010). Context-dependent effects of ranaviral infection on Northern leopard frog life history traits. PLoS One 5: e13723.

Echaubard P, Leduc J, Pauli B, Chinchar VG, Robert J, Lesbarrères D (2014). Environmental dependency of amphibian-ranavirus genotypic interactions: evolutionary perspectives on infectious diseases. Evol Appl 7: 723-733.

Farnsworth S, Seigel R (2013). Responses, movements, and survival of relocated box turtles during construction of the intercounty connector highway in Maryland. Transp Res Rec 2362: 1-8.

Forson D, Storfer A (2006a). Effects of atrazine and iridovirus infection on survival and life history characteristics in long-toed salamanders, *Ambystoma macrodactylum*. Environ Toxicol Chem 25: 168-173.

Forson DD, Storfer A (2006b). Atrazine increases ranavirus susceptibility in the tiger salamander, *Ambystoma tigrinum*. Ecol Appl 16: 2325-2332.

Fox S, Greer AL, Torres-Cervantes R, Collins JP (2006). First case of ranavirus-associated morbidity and mortality in natural populations of the South American frog *Atelognathus patagonicus*. Dis Aquat Organ 72: 87-92.

Fraker ME, Hu F, Cuddapah V, McCollum SA, Relyea RA, Hempel J, Denver RJ (2009). Characterization of an alarm pheromone secreted by amphibian tadpoles that induces behavioral inhibition and suppression of the neuroendocrine stress axis. Horm Behav 55: 520–529.

Gahl MK, Calhoun AJK (2008). Landscape setting and risk of ranavirus mortality events. Biol Conserv 141: 2679–2689.

Gahl MK, Calhoun AJK (2010). The role of multiple stressors in ranavirus-caused amphibian mortalities in Acadia national park wetlands. Can J Zool 88: 108–121.

Gandon S, van Baalen M, Jansen VAA (2002). The evolution of parasite virulence, superinfection, and host resistance. Am Nat 159: 658–669.

Gantress J, Maniero GD, Cohen N, Robert J (2003). Development and characterization of a model system to study amphibian immune responses to iridoviruses. Virology 311: 254–262.

Gobbo F, Cappellozza E, Pastore MR, Bovo G (2010). Susceptibility of black bullhead *Ameiurus melas* to a panel of ranavirus isolates. Dis Aquat Organ 90: 167–174.

Goldberg TL (2002). Largemouth bass virus: an emerging problem for warmwater fisheries? In: Philipp DP, Ridgway MS (eds) American Fisheries Society symposium 31, Bethesda.

Goodman RM, Miller DL, Ararso YT (2013). Prevalence of ranavirus in Virginia turtles as detected by tail-clip sampling versus oral-cloacal swabbing. Northeast Nat 20: 325–332.

Gosner KL (1960). A simplified table for staging anuran embryos and larvae with notes on identification. Herpetologica 16: 183–190.

Granoff A, Came PE, Breeze DC (1966). Viruses and renal carcinoma of *Rana pipiens* I. The isolation and properties of virus from normal and tumor tissue. Virology 29: 133–148.

Grant EC, Philipp DP, Inendino KR, Goldberg TL (2003). Effects of temperature on the susceptibility of largemouth bass to largemouth bass virus. J Aquat Anim Health 15: 215–220.

Grant EC, Inendino KR, Love WJ, Philipp DP, Goldberg TL (2005). Effects of practices related to catch-and-release angling on mortality and viral transmission in juvenile largemouth bass infected with largemouth bass virus. J Aquat Anim Health 17: 315–322.

Gray MJ, Miller DL, Schmutzer AC, Baldwin CA (2007). *Frog virus 3* prevalence in tadpole popu lations inhabiting cattle-access and non-access wetlands in

Tennessee, USA. Dis Aquat Organ 77: 97–103.

Gray MJ, Miller DL, Hoverman JT (2009a). Ecology and pathology of amphibian ranaviruses. Dis Aquat Organ 87: 243–266.

Gray MJ, Miller DL, Hoverman JT (2009b). First report of ranavirus infecting lungless salamanders. Herpetol Rev 40: 316–319.

Gray MJ, Brunner JL, Earl JE, Ariel E (2015). Design and analysis of ranavirus studies: surveillance and assessing risk. In: Gray MJ, Chinchar VG (eds) Ranaviruses: lethal pathogens of ectothermic vertebrates. Springer, Secaucus.

Green DE, Converse KA (2005). Diseases of frogs and toads. In: Majumdar SK, Huffman JE, Brenner FJ, Panah AI (eds) Wildlife diseases: landscape epidemiology, spatial distribution and utilization of remote sensing technology. Easton, Pennsylvania.

Green DE, Converse KA, Schrader AK (2002). Epizootiology of sixty-four amphibian morbidity and mortality events in the USA, 1996-2001. Ann N Y Acad Sci 969: 323–339.

Greer AL, Collins JP (2008). Habitat fragmentation as a result of biotic and abiotic factors controls pathogen transmission throughout a host population. J Anim Ecol 77: 364–369.

Greer AL, Berrill M, Wilson PJ (2005). Five amphibian mortality events associated with ranavirus infection in south central Ontario, Canada. Dis Aquat Organ 67: 9–14.

Greer AL, Briggs CJ, Collins JP (2008). Testing a key assumption of host-pathogen theory: density and disease transmission. Oikos 117: 1667–1673.

Greer AL, Brunner JL, Collins JP (2009). Spatial and temporal patterns of *Ambystoma tigrinum* virus (ATV) prevalence in tiger salamanders (*Ambystoma tigrinum nebulosum*). Dis Aquat Organ 85: 1–6.

Grizzle JM, Brunner CJ (2003). Review of largemouth bass virus. Fisheries 28: 10–14.

Grizzle JM, Altinok I, Fraser WA, Francis-Floyd R (2002). First isolation of largemouth bass virus. Dis Aquat Organ 50: 233–235.

Groocock GH, Grimmett SG, Getchell RG, Wooster GA, Bowser PR (2008). A survey to determine the presence and distribution of largemouth bass virus in wild freshwater bass in New York State. J Aquat Anim Health 20: 158–164.

Gruia-Gray J, Desser SS (1992). Cytopathological observations and epizootiology of frog erythrocytic virus in bullfrogs (*Rana catasbeiana*). J Wildl Dis 28: 34-41.

Gruia-Gray J, Petric M, Desser S (1989). Ultrastructural, biochemical, and biophysical properties of an erythrocytic virus of frogs from Ontario, Canada. J Wildl Dis 25: 497-506.

Haddad JJ, Saadé NE, Safieh-Garabedian B (2002). Cytokines and neuro-immune-endocrine interactions: a role for the hypothalamic-pituitary-adrenal revolving axis. J Neuroimmunol 133: 1-19.

Haislip NA, Gray MJ, Hoverman JT, Miller DL (2011). Development and disease: how susceptibility to an emerging pathogen changes through anuran development. PLoS One 6: e22307.

Haislip NA, Hoverman JT, Miller DL, Gray MJ (2012). Natural stressors and disease risk: does the threat of predation increase amphibian susceptibility to ranavirus? Can J Zool 90: 893-902.

Hanson LA, Petrie-Hanson L, Meals KO, Chinchar VG, Rudis M (2001). Persistence of largemouth bass virus infection in a northern Mississippi reservoir after a die-off. J Aquat Anim Health 13: 27-34.

Harikrishnan R, Balasundaram C, Heo M-S (2010). Molecular studies, disease status and prophylactic measures in grouper aquaculture: economic importance, diseases and immunology. Aquaculture 309: 1-14.

Harp EM, Petranka JW (2006). Ranavirus in wood frogs (*Rana sylvatica*): potential sources of transmission within and between ponds. J Wildl Dis 42: 307-318.

Haydon DT, Cleaveland S, Taylor LH, Laurenson MK (2002). Identifying reservoirs of infection: a conceptual and practical challenge. Emerg Infect Dis 8: 1468-1473.

Homan RN, Bartling JR, Stenger RJ, Brunner JL (2013). Detection of *Ranavirus* in Ohio, USA. Herpetol Rev 44: 615-618.

Hoverman JT, Gray MJ, Miller DL (2010). Anuran susceptibilities to ranaviruses: role of species identity, exposure route, and a novel virus isolate. Dis Aquat Organ 89: 97-107.

Hoverman JT, Gray MJ, Haislip NA, Miller DL (2011). Phylogeny, life history, and ecology contribute to differences in amphibian susceptibility to ranaviruses. Ecohealth 8: 301-319.

Hoverman JT, Gray MJ, Miller DL, Haislip NA (2012). Widespread occurrence of ranavirus in pond-breeding amphibian populations. Ecohealth 9: 36–48.

Huang SM, Tu C, Tseng CH, Huang CC, Chou CC, Kuo HC, Chang SK (2011). Genetic analysis of fish iridoviruses isolated in Taiwan during 2001-2009. National Taiwan University, Graduate Institute of Veterinary Medicine, Taipei.

Jancovich JK, Jacobs BL (2011). Innate immune evasion mediated by the *Ambystoma tigrinum* virus eukaryotic translation initiation factor 2α homologue. J Virol 85: 5061–5069.

Jancovich JK, Davidson EW, Morado JF, Jacobs BL, Collins JP (1997). Isolation of a lethal virus from the endangered tiger salamander *Ambystoma tigrinum stebbinsi*. Dis Aquat Organ 31: 161–167.

Jancovich JK, Davidson EW, Seiler A, Jacobs BL, Collins JP (2001). Transmission of the *Ambystoma tigrinum* virus to alternative hosts. Dis Aquat Organ 46: 159–163.

Jancovich J, Davidson EW, Parameswaran N, Mao J, Chinchar VG, Collins JP, Jacobs BL, Storfer A (2005). Evidence for emergence of an amphibian iridoviral disease because of human enhanced spread. Mol Ecol 14: 213–224.

Johnson AF, Brunner JL (2014). Persistence of an amphibian ranavirus in aquatic communities. Dis Aquat Organ 111: 129–138.

Johnson AJ, Pessier AP, Jacobson ER (2007). Experimental transmission and induction of ranaviral disease in western ornate box turtles (*Terrapene ornata ornata*) and red-eared sliders (*Trachemys scripta elegans*). Vet Pathol 44: 285–297.

Johnson AJ, Pessier AP, Wellehan JFX, Childress A, Norton TM, Stedman NL, Bloom DC, Belzer W, Titus VR, Wagner R, Brooks JW, Spratt J, Jacobson ER (2008). Ranavirus infection of free-ranging and captive box turtles and tortoises in the United States. J Wildl Dis 44: 851–863.

Johnson AJ, Wendland L, Norton TM, Belzer B, Jacobson ER (2010). Development and use of an indirect enzyme-linked immunosorbent assay for detection of iridovirus exposure in gopher tortoises (*Gopherus polyphemus*) and eastern box turtles (*Terrapene carolina carolina*). Vet Microbiol 142: 160–167.

Johnson PT, Rohr JR, Hoverman JT, Kellermanns E, Bowerman J, Lunde KB (2012). Living fast and dying of infection: host life history drives interspecific variation in infection and disease risk. Ecol Lett 15: 235–242.

Keeling MJ, Rohani P (2008). Modeling infectious diseases in humans and animals.

Princeton University Press, Princeton.

Kerby JL, Storfer A (2009). Combined effects of atrazine and chlorpyrifos on susceptibility of the tiger salamander to *Ambystoma tigrinum* virus. Ecohealth 6: 91–98.

Kerby JL, Hart AJ, Storfer A (2011). Combined effects of virus, pesticide, and predator cue on the larval tiger salamander (*Ambystoma tigrinum*). Ecohealth 8: 46–54.

Kik M, Martel A, Sluijs AS, Pasmans F, Wohlsein P, Grone A, Rijks JM (2011). Ranavirus-associated mass mortality in wild amphibians, the Netherlands, 2010: a first report. Vet J 190: 284–286.

Kimble SJ, Karna AK, Johnson AJ, Hoverman JT, Williams RN (2014). Mosquitoes as a potential vector of ranavirus transmission in terrestrial turtles. Ecohealth http://dx. doi. org/10. 1007/ s10393-014-0974-3.

Langdon JS (1989). Experimental transmission and pathogenicity of epizootic haematopoietic necrosis virus (EHNV) in redfin perch, *Perca fluviatilis* L., and 11 other teleosts. J Fish Dis 12: 295–310.

Langdon JS, Humphrey JD (1987). Epizootic haematopoietic necrosis, a new viral disease in redfin perch, *Perca fluviatilis* L., in Australia. J Fish Dis 10: 289–297.

Langdon JS, Humphrey JD, Williams LM, Hyatt AD, Westbury HA (1986). First virus isolation from Australian fish: an iridovirus-like pathogen from redfin perch, *Perca fluviatilis* L. J Fish Dis 9: 263–268.

Langdon JS, Humphrey JD, Williams LM (1988). Outbreaks of an EHNV-like iridovirus in cultured rainbow trout, *Salmo gairdneri* Richardson, in Australia. J Fish Dis 11: 93–96.

Lenski RE, May RM (1994). The evolution of virulence in parasites and pathogens: reconciliation between two competing hypotheses. J Theor Biol 169: 253–265.

Maceina MJ, Grizzle JM (2006). The relation of largemouth bass virus to largemouth bass population metrics in five Alabama reservoirs. Trans Am Fish Soc 135: 545–555.

Majji S, LaPatra S, Long SM, Sample R, Bryan L, Sinning A, Chinchar VG (2006). *Rana catesbeiana virus* z (RCV-Z): a novel pathogenic ranavirus. Dis Aquat Organ 73: 1–11.

Mao J, Green DE, Fellers G, Chinchar VG (1999). Molecular characterization of iridoviruses isolated from sympatric amphibians and fish. Virus Res 63: 45-52.

Marcogliese DJ, Pietrock M (2011). Combined effects of parasites and contaminants on animal health: parasites do matter. Trends Parasitol 27: 123-130.

Martin LB (2009). Stress and immunity in wild vertebrates: timing is everything. Gen Comp Endocrinol 163: 70-76.

May R, Nowak M (1994). Coinfection and the evolution of parasite virulence. Proc R Soc Lond B Biol Sci 255: 81-89.

Miller D, Gray M, Storfer A (2011). Ecopathology of ranaviruses infecting amphibians. Viruses 3: 2351-2373.

Moody NJG, Owens L (1994). Experimental demonstration of the pathogenicity of a frog virus, *Bohle iridovirus*, for a fish species, barramundi *Lates calcarifer*. Dis Aquat Organ 18: 95-102.

Morales HD, Abramowitz L, Gertz J, Sowa J, Vogel A, Robert J (2010). Innate immune responses and permissiveness to ranavirus infection of peritoneal leukocytes in the frog *Xenopus laevis*. J Virol 84: 4912-4922.

Muths E, Gallant AL, Campbell Grant EH, Battaglin WA, Green DE, Staiger JS, Walls SC, Gunzburger MS, Kearney RF (2006). The Amphibian Research and Monitoring Initiative (ARMI): 5-year report. Scientific investigations report 2006-5224. U. S. Geological Survey, Reston.

Nagasawa K, Cruz-Lacierda ER (eds) (2004). Diseases of cultured groupers. Southeast Asian Fisheries Development Center, Aquaculture Department, Iloilo.

Nazir J, Spengler M, Marschang RE (2012). Environmental persistence of amphibian and reptilian ranaviruses. Dis Aquat Organ 98: 177-184.

Parris MJ, Storfer A, Collins JP, Davidson EW (2005). Life-history responses to pathogens in tiger salamander (*Ambystoma tigrinum*) larvae. J Herpetol 39: 366-372.

Paull SH, Song S, McClure KM, Sackett LC, Kilpatrick AM, Johnson PT (2012). From superspreaders to disease hotspots: linking transmission across hosts and space. Front Ecol Environ 10: 75-82.

Pearman PB, Garner TWJ (2005). Susceptibility of Italian agile frog populations to an emerging strain of *Ranavirus* parallels population genetic diversity. Ecol Lett 8: 401-408.

Pearman PB, Garner TWJ, Straub M, Greber UF (2004). Response of the Italian agile frog (*Rana latastei*) to a ranavirus, frog virus 3: a model for viral emergence in naive populations. J Wildl Dis 40: 660–669.

Petranka JW, Murray SS, Kennedy CA (2003). Responses of amphibians to restoration of a southern Appalachian wetland: perturbations confound post-restoration assessment. Wetlands 23: 278–290.

Petranka JW, Harp EM, Holbrook CT, Hamel JA (2007). Long-term persistence of amphibian populations in a restored wetland complex. Biol Conserv 138: 371–380.

Pfennig DW, Loeb MLG, Collins JP (1991). Pathogens as a factor limiting the spread of cannibalism in tiger salamanders. Oecologia 88: 161–166.

Picco AM, Collins JP (2008). Amphibian commerce as a likely source of pathogen pollution. Conserv Biol 22: 1582–1589.

Picco AM, Karam AP, Collins JP (2010). Pathogen host switching in commercial trade with management recommendations. Ecohealth 7: 252–256.

Plumb JA, Zilberg D (1999a). Survival of largemouth bass iridovirus in frozen fish. J Aquat Anim Health 11: 94–96.

Plumb JA, Zilberg D (1999b). The lethal dose of largemouth bass virus in juvenile largemouth bass and the comparative susceptibility of striped bass. J Aquat Anim Health 11: 246–252.

Plumb JA, Grizzle JM, Young HE, Noyes AD, Lamprecht S (1996). An iridovirus isolated from wild largemouth bass. J Aquat Anim Health 8: 265–270.

Polis GA, Myers CA (1985). A survey of intraspecific predation among reptiles and amphibians. J Herpetol 19: 99–107.

Prasankok P, Chutmongkonkul M, Kanchanakhan S (2005). Characterisation of iridovirus isolated from diseased marbled sleepy goby, *Oxyeleotris marmoratus*. In: Walker P, Lester R, Bondad Reantaso MG (eds) Diseases in Asian aquaculture V. Fish Health Section, Asian Fisheries Society, Manila.

Price S, Garner T, Nichols R, Balloux F, Ayres C, Mora-Cabello de Alba A, Bosch J (2014). Collapse of amphibian communities due to an introduced Ranavirus. Curr Biol 24: 2586–2591.

Qin QW, Chang SF, Ngoh-Lim GH, Gibson-Kueh S, Shi C, Lam TJ (2003). Characterization of a novel ranavirus isolated from grouper *Epinephelus tauvina*. Dis Aquat Organ 53: 1–9.

Raffel TR, Rohr JR, Kiesecker JM, Hudson PJ (2006). Negative effects of changing temperature on amphibian immunity under field conditions. Funct Ecol 20:

819–828.

Reddacliff LA, Whittington RJ (1996). Pathology of epizootic haematopoietic necrosis virus (EHNV) infection in rainbow trout (*Oncorhynchus mykiss* Walbaum) and redfin perch (*Perca fluviatilis* L). J Comp Pathol 115: 103–115.

Reeve BC, Crespi EJ, Whipps CM, Brunner JL (2013). Natural stressors and ranavirus susceptibility in larval wood frogs (*Rana sylvatica*). Ecohealth 10: 190–200.

Reilly SM, Lauder GV, Collins JP (1992). Performance consequences of a trophic polymorphism: feeding behavior in typical and cannibal phenotypes of *Ambystoma tigrinum*. Copeia (3).: 672–679.

Restif O, Koella JC (2003). Shared control of epidemiological traits in a coevolutionary model of host-parasite interactions. Am Nat 161: 827–836.

Ridenhour BJ, Storfer AT (2008). Geographically variable selection in *Ambystoma tigrinum* virus (Iridoviridae) throughout the western USA. J Evol Biol 21: 1151–1159.

Robert J, Abramowitz L, Gantress J, Morales HD (2007). *Xenopus laevis*: a possible vector of ranavirus infection? J Wildl Dis 43: 645–652.

Robert J, George E, De Jesús Andino F, Chen G (2011). Waterborne infectivity of the Ranavirus frog virus 3 in *Xenopus laevis*. Virology 417: 410–417.

Rojas S, Richards K, Jancovich JK, Davidson EW (2005). Influence of temperature on ranavirus infection in larval salamanders, *Ambystoma tigrinum*. Dis Aquat Organ 63: 95–100.

Rollins-Smith LA (1998). Metamorphosis and the amphibian immune system. Immunol Rev 166: 221–230.

Rollins-Smith LA (2009). The role of amphibian antimicrobial peptides in protection of amphibians from pathogens linked to global amphibian declines. Biochim Biophys Acta 1788: 1593–1599.

Roy BA, Kirchner JW (2000). Evolutionary dynamics of pathogen resistance and tolerance. Evolution 54: 51–63.

Ryder JJ, Miller MR, White A, Knell RJ, Boots M (2007). Host-parasite population dynamics under combined frequency- and density-dependent transmission. Oikos 116: 2017–2026.

Sabelis MW, Metz JAJ (2002). Taking stock: relating theory to experiment. In:

Dieckmann U, Metz J, Sabelis M, Sigmund K (eds) Adaptive dynamics of infectious diseases: in pursuit of virulence management. Cambridge University Press, New York.

Schloegel LM, Picco AM, Kilpatrick AM, Davies AJ (2009). Magnitude of the U. S. trade in amphibians and presence of *Batrachochytrium dendrobatidis* and ranavirus infection in imported North American bullfrogs (*Rana catesbeiana*). Biol Conserv 142: 1420–1426.

Schloegel LM, Daszak P, Cunningham AA, Speare R, Hill B (2010). Two amphibian diseases, chytridiomycosis and ranaviral disease, are now globally notifiable to the world organization for animal health (OIE): an assessment. Dis Aquat Organ 92: 101–108.

Schock DM, Bollinger TK, Chinchar VG, Jancovich JK, Collins JP (2008). Experimental evidence that amphibian ranaviruses are multi-host pathogens. Copeia 1: 133–143.

Sheafor B, Davidson EW, Parr L, Rollins-Smith L (2008). Antimicrobial peptide defenses in the salamander, *Ambystoma tigrinum*, against emerging amphibian pathogens. J Wildl Dis 44: 226–236.

Sheridan JF, Dobbs C, Brown D, Zwilling B (1994). Psychoneuroimmunology: stress effects on pathogenesis and immunity during infection. Clin Microbiol Rev 7: 200–212.

Southard GM, Fries LT, Terre DR (2009). Largemouth bass virus in Texas: distribution and management issues. J Aquat Anim Health 21: 36–42.

Speare R, Smith JR (1992). An iridovirus-like agent isolated from the ornate burrowing frog *Limnodynastes ornatus* in northern Australia. Dis Aquat Organ 14: 51–57.

St-Amour V, Wong WM, Garner TW, Lesbarrères D (2008). Anthropogenic influence on prevalence of 2 amphibian pathogens. Emerg Infect Dis 14: 1175–1176.

Stark T, Laurijssens C, Weterings M, Spitzen-van der Sluijs A, Martel A, Pasmans F (2014). Death in the clouds: ranavirus associated mortality in assemblage of cloud forest amphibians in Nicaragua. Acta Herpetol 9: 125–127.

Storfer A, Alfaro ME, Ridenhour BJ, Jancovich JK, Mech SG, Parris MJ, Collins JP (2007). Phylogenetic concordance analysis shows an emerging pathogen is novel and endemic. Ecol Lett 10: 1075–1083.

Sutton W, Gray M, Hoverman JT, Secrist R, Super P, Hardman R, Tucker J, Miller D (2014). Trends in ranavirus prevalence among plethodontid salamanders in the Great Smoky Mountains National Park. EcoHealth http://dx. doi. org/10. 1007/s10393-014-0994-z.

Teacher AGF, Garner TWJ, Nichols RA (2009). Population genetic patterns suggest a behavioural change in wild common frogs (*Rana temporaria*) following disease outbreaks (*Ranavirus*). Mol Ecol 18: 3163-3172.

Teacher AGF, Cunningham AA, Garner TWJ (2010). Assessing the long term impact of ranavirus infection in wild common frog populations. Anim Conserv 13: 514-522.

Terrell KA, Quintero RP, Murray S, Kleopfer JD, Murphy JB, Evans MJ, Nissen BD, Gratwicke B (2013). Cryptic impacts of temperature variability on amphibian immune function. J Exp Biol 216: 4204-4211.

Titus VR, Green TM (2013). Presence of *Ranavirus* in green frogs and eastern tiger salamanders on Long Island, New York. Herpetol Rev 44: 266-267.

Todd-Thompson M (2010). Seasonality, variation in species prevalence, and localized disease for *Ranavirus* in Cades Cove (Great Smoky Mountains National Park) amphibians. Dissertation, University of Tennesse, Knoxville.

Tollrian R, Harvell D (1999). The ecology and evolution of inducible defenses. Princeton University Press, Princeton.

Torrence SM, Green DE, Benson CJ, Ip HS, Smith LM, McMurry ST (2010). A new ranavirus isolated from *Pseudacris clarkii* tadpoles in playa wetlands in the southern high plains, Texas. J Aquat Anim Health 22: 65-72.

Une Y, Sakuma A, Matsueda H, Nakai K, Murakami M (2009). Ranavirus outbreak in North American bullfrogs (*Rana catesbeiana*), Japan, 2008. Emerg Infect Dis 15(7).: 1146-1147.

Waltzek TB, Miller DL, Gray MJ, Drecktrah B, Briggler JT, MacConnell B, Hudson C, Hopper L, Friary J, Yun SC, Malm KV, Weber ES, Hedrick RP (2014). New disease records for hatchery-reared sturgeon. I. Expansion of frog virus 3 host range into Scaphirhynchus albus. Dis Aquat Org 111: 219-227.

Warne RW, Crespi EJ, Brunner JL (2011). Escape from the pond: stress and developmental responses to ranavirus infection in wood frog tadpoles. Funct Ecol 25: 139-146.

Wheelwright NT, Gray MJ, Hill RD, Miller DL (2014). Sudden mass die-off of a large population of wood frog (*Lithobates sylvaticus*) tadpoles in Maine, USA, likely due to ranavirus. Herpetol Rev 45: 240–242.

Whittington RJ, Reddacliff GL (1995). Influence of environmental temperature on experimental infection of redfin perch (*Perca fluviatilis*) and rainbow trout (*Oncorhynchus mykiss*) with epizootic haematopoietic necrosis virus, an Australian iridovirus. Aust Vet J 72: 421–424.

Whittington RJ, Philbey A, Reddacliff GL, Macgown AR (1994). Epidemiology of epizootic haematopoietic necrosis virus (EHNV) infection in farmed rainbow trout, *Oncorhynchus mykiss* (Walbaum): findings based on virus isolation, antigen capture elisa and serology. J Fish Dis 17: 205–218.

Whittington RJ, Reddacliff LA, Marsh I, Kearns C, Zupanovic Z, Callinan RB (1999). Further observations on the epidemiology and spread of epizootic haematopoietic necrosis virus (EHNV) in farmed rainbow trout *Oncorhynchus mykiss* in southeastern Australia and a recommended sampling strategy for surveillance. Dis Aquat Organ 35: 125–130.

Whittington RJ, Becker JA, Dennis MM (2010). Iridovirus infections in finfish – critical review with emphasis on ranaviruses. J Fish Dis 33: 95–122.

Williams PD, Day T (2001). Interactions between sources of mortality and the evolution of parasite virulence. Proc R Soc Lond B Biol Sci 268: 2331–2337.

Williams T, Barbosa-Solomieu V, Chinchar VG (2005). A decade of advances in iridovirus research. Adv Virus Res 65: 173–248.

Woodland JE, Brunner CJ, Noyes AD, Grizzle JM (2002a). Experimental oral transmission of largemouth bass virus. J Fish Dis 25: 669–672.

Woodland JE, Noyes AD, Grizzle JM (2002b). A survey to detect largemouth bass virus among fish from hatcheries in the southeastern USA. Trans Am Fish Soc 131: 308–311.

第四章 蛙病毒复制：分子、细胞及免疫学事件

詹姆斯·K. 贾柯维奇①，秦启伟②，张奇亚③和 V. 格雷戈里·钦察尔④

1 引言

105 自 50 年前发现蛙病毒以来，我们对蛙病毒生物学的认识经历了两个显著不同的阶段。第一阶段，蛙病毒的研究得到了艾伦·格拉诺夫（Allan Granoff）及其合作者以及欧洲和美国的科学工作者研究工作的推动，这个时期发生在 1965 年至 1985 年。在这一初始阶段，蛙病毒在细胞中的复制事件通过 FV3（Frog virus 3）的研究得以阐明，该病毒隶属虹彩病毒科（Iridoviridae），是蛙病毒属（*Ranavirus*）的代表性病毒种。蛙病毒研究的第二个浪潮始于 20 世纪 90 年代，并持续到现在。美国及亚洲和欧洲越来越多的研究都集中在蛙病毒基因和基因组，以及运用一系列不同的现代分子生物学方法来确定病毒复制中特定基因的作用。此外，目前的研究工作已扩展到超出 FV3 以外的其他病毒，并利用其他蛙病毒种以及来自其他属的虹彩病毒，特别是细胞肿大病毒属（*Megalocytivirus*）的病毒来进行研究。最近的一些研究已鉴定一些病毒基因，这些基因不仅在蛙病毒复制中直接起着结构和酶的作用，而且有可能在特定细

①J. K. Jancovich/Department of Biological Sciences, California State University, San Marcos, CA 92096, USA.

②Q. W. Qin/Key Laboratory of Tropical Marine Bio~Resources and Ecology, South China Sea Institute of Oceanology, Chinese Academy of Sciences, Guangzhou, China.

③Q. Y. Zhang/State Key Laboratory of Freshwater Ecology and Biotechnology, Institute of Hydrobiology, Chinese Academy of Sciences, Wuhan, Hubei, China.

④V. G. Chinchar/Department of Microbiology, University of Mississippi Medical Center, Jackson, MS 39216, USA/e-mail: vchinchar@umc.edu.

胞和宿主环境中促进病毒的复制，逃避宿主的免疫应答，导致毒力增强。目 106
前，正在进行的研究涉及病毒基因敲除、基因功能敲低以及重组蛙病毒蛋白的分析，从而为蛙病毒基因的功能提供一个更全面的认识。此外，还进行了在病毒毒力中起着关键作用的基因鉴定。这些研究对低等脊椎动物的先天性和获得性、保护性免疫反应，提供了更好的认识，并且促进了有效抗蛙病毒疫苗的开发。本章重点除集中在生物化学和遗传学研究以外，还用大量的信息来描述蛙病毒和其他脊椎动物虹彩病毒对野生和养殖变温脊椎动物的不良影响（Duffus 等人，2015；Brunner 等人，2015）。为了给蛙病毒病分子机制的认识提供基础，我们将在下面描述病毒复制的突出事件以及在这一过程中具体病毒基因的作用。虽然侧重点主要集中于病毒复制策略以及确定病毒基因功能的经典和现代的方法，但是我们也同样简单地接触到病毒的分类和抗病毒的免疫反应，这两个主题在本书的其他章节也涵盖了一定的篇幅（Jancovich 等人，2015；Grayfer 等人，2015）。

2　蛙病毒的分类及基因组

蛙病毒属是虹彩病毒科内 5 个属之一（见表 1）。组成虹彩病毒科的 5 个属包括只感染无脊椎动物的 2 个属，即虹彩病毒属（*Iridovirus*）和绿虹彩病毒属（*Chloriridovirus*），以及以冷血脊椎动物为靶宿主的 3 个属，即蛙病毒属、细胞肿大病毒属和淋巴细胞囊肿病毒属（*Lymphocystivirus*）（Jancovich 等人，2012），但细胞肿大病毒属和淋巴细胞囊肿病毒属的病毒只感染鱼类。蛙病毒尽管与它们齐名，但其靶宿主不仅有鱼类，还有两栖类和爬行类。此外，它们广泛的宿主范围说明，一些蛙病毒可能感染来自不同脊椎动物类别的宿主。例如，博乐虹彩病毒既能感染两栖类，也能感染鱼类（Moody 和 Owens，1994）。当蛙病毒感染多种脊椎动物细胞时（包括哺乳动物），病毒的离体混杂性特别明显。虹彩病毒是虹彩病毒科所有成员的总称，这个科的病毒具有一个二十面体的病毒衣壳，该衣壳包含一个双链 DNA（dsDNA）基因组。虹彩病毒基因组的特征（包括物种的缩写和分离株的命名）具体见表 1，如表所示，基因组依据具体的病毒的不同，而有大小上的变化，并且包括 92 ~ 211 个不同的推导开放阅读框（open reading frame，ORF）（Jancovich等人，2012）。虹彩病毒科所有病毒成员保守的一组 26 个基因的系统发生分析，支持将该科分成 4 个不同的类群，即蛙病毒群（Ranavirus）、细胞肿大病毒群（Megalocytivirus）、淋巴细胞囊肿病毒群（Lymphocystivirus）

以及虹彩病毒群/绿虹彩病毒群（Iridovirus/Chloriridovirus）（Eaton 等人，2007）。虽然在系统发生上没有显著区别，但是虹彩病毒和绿虹彩病毒在鸟嘌呤与胞嘧啶（G+C）残基的百分比、病毒粒子的大小以及宿主范围方面，都表现出明显的不同，这些特征是否为目前分成两个属提供了足够的根据，有待进一步确定。

早期对蛙病毒基因组的研究都集中在 FV3。该病毒具有一个线性、双链 DNA 基因组，该基因组循环置换（circularly permutated）且末端冗长。这是
107 虹彩病毒科中所有病毒基因的一个典型特征（Goorha 和 Murti，1982）。除此之外，FV3 基因组以及其他所有脊椎动物的虹彩病毒，除新加坡石斑鱼虹彩病毒（SGIV）以外，显示出在每个 CpG 甲基化的双核苷酸内部，每个胞嘧啶都高度甲基化（Willis 和 Granoff，1980）。这些早期的研究虽然对蛙病毒基因组的整个结构作了深入观察，但是关于蛙病毒基因组的遗传组成、编码能力（coding capacity）以及蛙病毒之间的基因组变异，则知之甚少。蛙病毒研究第二个浪潮标志着又一个里程碑的树立，从而使我们对蛙病毒基因组的认识很快得到了扩展。从 1997 年对淋巴细胞囊肿病毒基因组测序开始（Tidona 和 Darai，1997），许多虹彩病毒的全基因组序列也相继被测定，其中包括多个蛙病毒种以及感染一系列不同宿主的病毒分离株（见表 1）。现
108 有的全基因组序列信息允许分析整个基因组的组织结构、蛋白序列变异以及在蛙病毒之间的多态区。

表 1　虹彩病毒分类：病毒属及种

属名	病毒种[a]	大小（bp）	数量（ORFs）	(G+C)%	基因库登录号
虹彩病毒属（*Iridovirus*）	***IIV-9***	206 791	191	31	GQ918152
	IIV-6	212 482	211	29	AF303741
绿虹彩病毒属（*Chloriridovirus*）	***IIV-3***	191 132	126	48	DQ643392
淋巴细胞囊肿病毒属（*Lymphocystivirus*）	***LCDV-1***	102 653	108	29	L63545
	LCDV-C	186 250	178	27	AY380826

续表

属名	病毒种[a]	大小（bp）	数量（ORFs）	(G+C)%	基因库登录号
蛙病毒属（*Ranavirus*）	***ATV***	106 332	92	54	AY150217
	FV3	105 903	97	55	AY548484
	EHNV	127 011	100	54	FJ433873
	ADRV	106 734	101	55	KC865735
	STIV	105 890	103	55	EU627010
	CMTV	106 878	104	55	JQ231222
	TFV	105 057	105	55	AF389451
	RGV	105 791	106	55	JQ654586
	ESV	127 732	136	54	JQ724856
	SGIV	140 131	139	49	AY521625
	GIV	139 793	139	49	AY666015
细胞肿大病毒属（*Megalocytivirus*）	***ISKNV***	111 362	117	55	AF371960
	RBIV	112 080	116	53	AY532606
	RSIV	112 414	93	53	BD143114
	OSGIV	112 636	116	54	AY894343
	TRBIV	110 104	115	55	GQ273492
	LYCIV	111 760	ND	ND	AY779031

[a] IIV-9 表示无脊椎动物虹彩病毒 9 型，IIV-6 表示无脊椎动物虹彩病毒 6 型（螟虫虹彩病毒，Chilo iridovirus），IIV-3 表示无脊椎动物虹彩病毒 3 型，LCDV-1 表示淋巴囊肿病毒 1 型，LCDV-C 表示淋巴细胞囊肿病毒-中国型，TFV 表示虎纹蛙病毒，ATV 表示虎纹钝口螈病毒，FV3 表示蛙病毒 3，RGV 表示蛙虹彩病毒，CMTV 表示普通产婆蟾病毒，STIV 表示中华鳖虹彩病毒，ADRV 表示中国大鲵蛙病毒，EHNV 表示动物流行性造血器官坏死病毒，ESV 表示欧洲六须鲶病毒，SGIV 表示新加坡石斑鱼虹彩病毒，GIV 表示石斑鱼虹彩病毒，ISKNV 表示传染性脾肾坏死病毒，RBIV 表示石鲷虹彩病毒，RSIV 表示真鲷虹彩病毒，OSGIV 表示斜带石斑鱼虹彩病毒，TRBIV 表示大菱鲆红体病虹彩病毒，LYCIV 表示大黄鱼虹彩病毒。表中粗斜体病毒名表示已经得到国际病毒分类委员会的认定；标准体病毒名既表示暂定种，也表示认定种的分离株，ND 表示还没有确定。

蛙病毒基因组大小为 105~140 kbp，并显示 G+C 含量为 49%~55%，预测能编码 92~139 个病毒蛋白（见表 1）。目前，4 个独特的基因组结构可分为两个类群，并通过蛙病毒基因组全序列的 dot-plot 和系统发生分析得到了确认

(Chen 等人，2013)。石斑鱼虹彩病毒（GIV）样蛙病毒包括 GIV 和 SGIV，组成一个类群；而两栖类样蛙病毒（ALRV）组成第二个类群。GIV 样蛙病毒与属中的其他病毒成员比较时，只有很短的片段具有基因组的共线性，而 ALRV 类群的蛙病毒成员包括 ATV 样蛙病毒、CMTV 样蛙病毒以及 FV3 样蛙病毒，具有较长的共线性区域（Jancovich 等人，2010；Chen 等人，2013；Mavian 等人，2012)。然而，在 ALRV 的三个亚类之间，存在着转位、缺失以及添加，从而有利于互相之间进行区分。

蛙病毒基因组编码 92~139 个推导的基因产物，这些产物已经通过长于 50

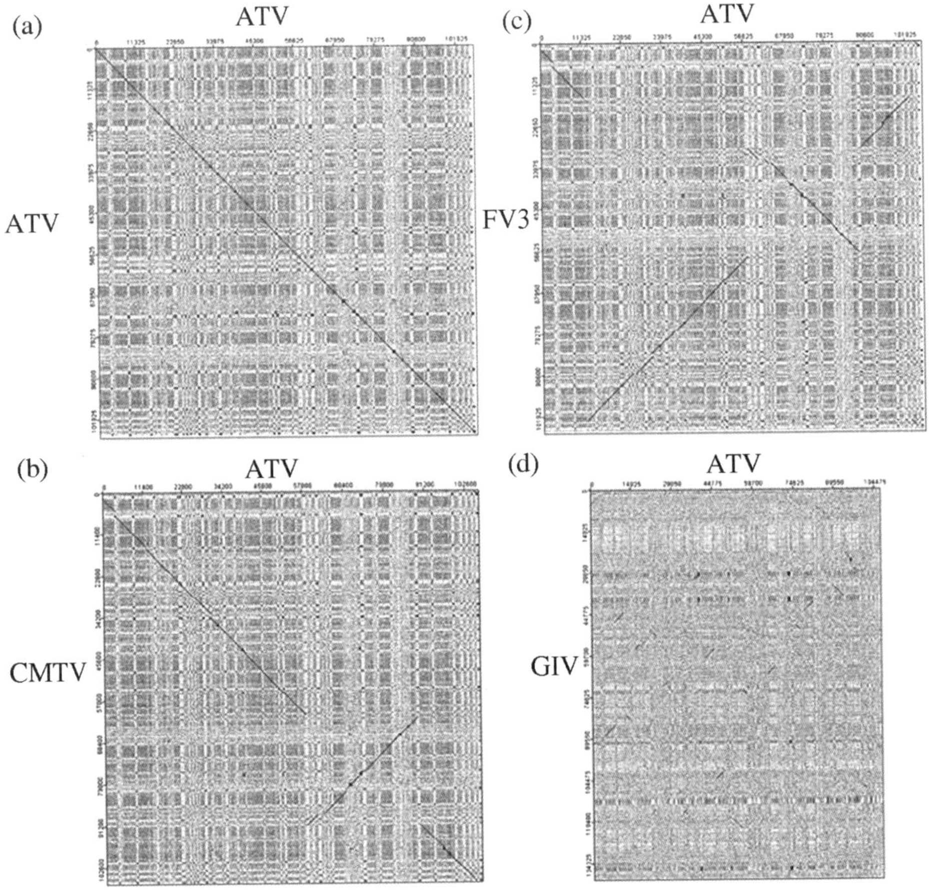

图 1 蛙病毒基因组结构：蛙病毒属代表性病毒成员的 dot-plot 分析。ATV 病毒基因组序列与（a）ATV，（b）CMTV，（c）FV3，（d）GIV 的比较。dot-plot 通过运用 JDotter 生成（Brodie 等人，2004)。图中的线条表示序列相似性/共线性区域。

个氨基酸 ORF 缺失、聚丙烯酰胺凝胶分析纯化后的病毒粒子和病毒感染的细 109
胞、病毒转录本的微阵列分析以及病毒子的蛋白组分析予以确定（Eaton 等人，2007；Majji 等人，2009；Song 等人，2006）。这些基因大约三分之一的功能，是通过与其他已知蛋白或进行基因类比推导出来的。尽管剩余基因的功能还不知道，但是大多数基因与虹彩病毒科的基因同源，这说明这些基因在病毒生物起源中起着很重要的作用。所有蛙病毒都含有 26 个虹彩病毒的核心基因，同时还含有 72 个其他基因，这 72 个基因是蛙病毒属所有病毒成员共有的（Eaton 等人，2007）。由于后者仅见于蛙病毒之中，因而我们推测，通过确定这些基因在病毒复制中的各种作用，可以鉴定出在独特宿主环境中行使功能并造成不同宿主发病的基因。

除编码区以外，蛙病毒基因组还含有回纹结构（palindromes）、微卫星（microsatellites）、重复序列区（repeat regions）以及基因间和基因内的可变区（areas of inter- and intragenic variation）（Eaton 等人，2010；Jancovich 等人，2003；Lei 等人，2012b；Mavian 等人，2012；Morrison 等人，2014；Tan 等人，2004）。重复序列区和可变区有可能作为促进重组或调控基因表达的位点，而在病毒信息 3′端的回纹序列有可能作为转录终止信号。此外，亲缘关系较近的 FV3 样病毒之间相比较，其毒力变化各异，从而提示，基因内差异以及重复序列变化可能影响病毒的致病性（Morrison 等人，2014）。

越来越多的证据显示，蛙病毒编码微小 RNA（miRNA），这些微小 RNA 调控宿主和病毒的基因表达，并在逃避宿主抗病毒免疫中起着重要作用。例如，通过 Illumina/Solexa 深度测序所鉴定的 16 个新 SGIV 编码的 miRNA 中，当用茎环定量 RT-PCR 和荧光素酶报告分析时（Yan 等人，2011），有 11 个 miRNA 存在于 SGIV 感染的石斑鱼细胞中，并具有相应的功能。有一种 miRNA 即 SGIV 编码的 miR-homoHSV 可以致弱 SGIV，诱导了细胞凋亡（apoptosis），进而促进病毒的复制（Guo 等人，2013）。总之，这些数据提示，蛙病毒之间的序列差异可显著影响蛙病毒的宿主范围及致病性。为达这一目的，我们对蛙病毒基因组的认识将随着其他蛙病毒基因组被测序而得以拓展。

3　蛙病毒的复制策略

这一部分，我们将讨论在传染性病毒粒子生成中起着直接作用的病毒编码事件。之后，我们将讨论病毒感染对宿主细胞的影响以及病毒与宿主免疫系统之间的相互作用。在大多数情况下，蛙病毒复制都以 FV3 作为模型进行阐述，除此之外，也适当地运用一些其他蛙病毒甚至是虹彩病毒科的其他属的病毒来进行讨论。蛙病毒复制的关键事件如图 2 所示。在大多数情况下，图 2 所描述的事件以 FV3 的研究工作为依据。似乎所有脊椎动物虹彩病毒的复制基本上都是采用相同的策略。蛙病毒、细胞肿大病毒以及淋巴细胞囊肿病毒之间的差
110 别可能就在于，在细胞与免疫学水平，这些病毒与宿主是如何相互作用的。其他关于蛙病毒复制的资料可见于几篇综述之中（Chinchar 等人，2009，2011；Williams，1996；Williams 等人，2005；Willis 等人，1985）。

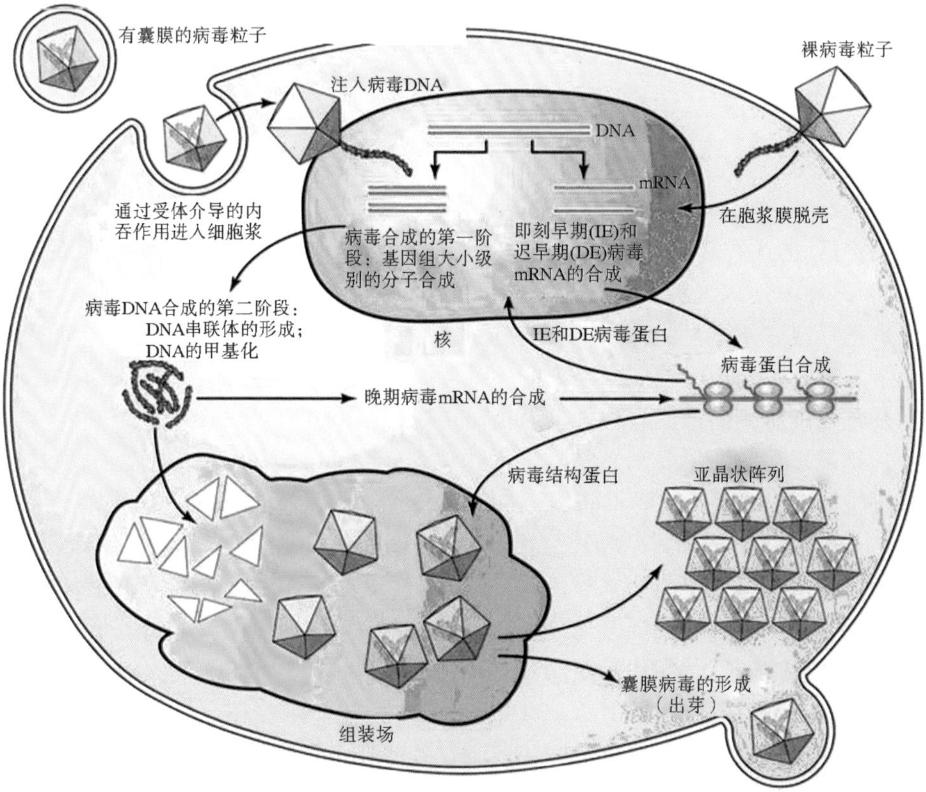

图 2 蛙病毒复制示意图。病毒粒子可通过两个途径进入细胞，病毒复制的启动事件发生在核中（早期病毒的转录以及单位长度基因组的合成）。病毒基因组随后被转运到细胞质，它们在这里进行甲基化，并作为模板形成串联体（concatemer）。病毒组装场所包括病毒 DNA 和许多病毒编码的蛋白质，并作为病毒粒子的形成位点。新合成的病毒粒子可游离于细胞质之中，或见于亚晶状阵列之内。至少在离体条件下，少数病毒粒子会从质膜以出芽的方式排出来，在这一过程中，通过出芽获得囊膜。

3.1　病毒的进入

蛙病毒粒子由复杂的多层结构组成，从里到外，有一个核心，核心由病毒双链 DAN（dsDNA）基因组构成。基因组连有一个或多个病毒编码的蛋白质；一个内脂质膜，膜内含有好几个蛋白质；一个二十面体的衣壳（icosahedral capsid），这个衣壳构成了几乎整个 48 kDa 的主衣壳蛋白（MCP）；以出芽方式释放的病毒粒子还含有一个病毒囊膜（viral envelope），该病毒囊膜是由上述胞质膜衍化而来，并含有一个或多个病毒编码的蛋白质（Darcy-Tripier 等 111
人，1984）。在外观上，蛙病毒粒子与非洲猪瘟病毒以及藻类 DNA 病毒科（Phycodnaviridae）的病毒成员非常类似（Tulman 等人，2009；Wilson 等人，2009）。然而，与其他病毒科的大多数成员相比，其他病毒科成员的感染形式要么是有囊膜的病毒粒子，要么是无囊膜的病毒粒子。而蛙病毒，无论是无囊膜的蛙病毒粒子，还是有囊膜的蛙病毒粒子，都具有传染性。有囊膜的病毒粒子的传染性似乎要大得多，它们通过受体介导的内吞作用进入细胞（Braunwald 等人，1979；Ma 等人，2014）。虽然蛙病毒无论是离体，还是在体，都具有广泛的宿主范围，但是病毒和细胞受体蛋白的鉴定还未曾知晓。无囊膜的病毒粒子与细胞质膜相互作用，随后病毒脱衣壳并直接将病毒 DNA 核心释放到细胞质中。相反，有囊膜的病毒粒子通过受体介导的内吞作用进入细胞，接下来便将无囊膜的病毒粒子释放到细胞质中。随后，病毒粒子被转运到核膜，紧接着，病毒 DNA 被注入宿主细胞核中（Braunwald 等人，1985；Gendrault 等人，1981）。有囊膜的病毒粒子传染性增强，有可能反映出病毒囊膜蛋白与细胞受体能更有效地结合或者受体介导的内吞作用能更有效地摄入和释放病毒粒子。除这些机制以外，病毒进入宿主细胞可能还涉及病毒粒子和细胞膜穴样内陷之间的相互作用（Guo 等人，2011b，2012；Jia 等人，2013；Wang 等人，2014）。

3.2　细胞核复制事件

病毒核心进入细胞核以后，便拉开了蛙病毒复制的序幕，即进行早期病毒转录本的合成以及病毒基因组单位长度拷贝的生成。与其他 DNA 病毒一样，如疱疹病毒（Herpesviruses），蛙病毒利用宿主 RNA 聚合酶 II 来转录病毒信息（Goorha，1981）。然而，与疱疹病毒相反的是，蛙病毒的转录需要一个或多个病毒子关联蛋白的存在，因而脱蛋白的病毒基因组 DNA 不能进行转录，也没有传染性（Willis 等人，1990；Willis 和 Granoff，1985；Willis 和 Thompson，1986）。第一个合成的病毒转录本被称作“即刻早期”（“immediate-early”，

IE）转录本。在它们的基因产物之间，有一个或多个蛋白是第二类早期转录本合成所需要的，第二类早期转录本被命名为“迟早期”（“delayed early”，DE）转录本（Willis 和 Granoff，1978）。总体来讲，IE 和 DE 转录本有可能编码调控蛋白和毒力蛋白以及关键催化蛋白如 RNA 聚合酶 II（vPOL-II）大小亚单位以及病毒 DNA 聚合酶（Majji 等人，2009）。紧随 FV3 基因表达的微阵列分析之后，对应于接近 FV3 编码区的一半序列的 33 IE 和 22 DE 转录本也被鉴定（Majji 等人，2009）。IE 和 DE 基因产物的类似含量水平也见于其他蛙病毒中（Chen 等人，2006；Teng 等人，2008）。宿主 RNA 聚合酶 II（Host POL-II）负责病毒 IE mRNA 的转录，也许还负责 DE mRNA 的转录。然而，病毒 RNA 聚
112 合酶 II（vPOL-II）指导后期病毒信息的转录后文还要述及。如同宿主转录本一样，病毒转录本也存在加帽和甲基化，但不像细胞信使 RNA，蛙病毒 mRNA 缺乏多聚 A 尾巴和内含子（introns）。

3.3 细胞质复制事件

当病毒 DNA 被转运到细胞质以后，就会被一个病毒编码的胞嘧啶特异性 DNA 甲基转移酶（DNA methyltransferase，DMTase）甲基化（Willis 等人，1984；Willis 和 Granoff，1980）。CpG 双核苷酸中的每个胞嘧啶是甲基化的目标。胞嘧啶的甲基化可达到 20%~25%，是脊椎动物病毒中所见到的最高水平的 DNA 甲基化。虽然存在着这一超出平常水平的甲基化，但其在病毒生活史中的精确作用，目前还不清楚。有人提出，甲基化是为了保护病毒基因组 DNA 免受病毒编码的限制性修饰酶的攻击，病毒编码的限制性修饰酶的作用目标是未甲基化的宿主 DNA（Kaur 等人，1995）。另外一种观点就是，甲基化被认为是阻止病毒基因组 DNA 通过模式识别受体（pattern recognition receptors），如 TLR-9 的识别，从而阻断免疫反应的激活（Hoelzer 等人，2008；Krug 等人，2001，2004）。在 5-氮杂胞苷（5′-azacytidine，azaC）和甲基化抑制子存在的条件下，FV3 感染并不影响病毒的转录或翻译，但是会导致 DNA 合成的降低以及病毒产量的明显下降（Goorha 等人，1984）。为了支持 DNA 甲基化具有保护作用这一观点，有学者进行了梯度分析，检测在有 5-氮杂胞苷存在的条件下所合成的病毒 DNA 的单链 DNA 断裂。这些断裂被认为是阻断 DNA 的组装以及传染性病毒粒子生成的原因。

在细胞质内，以单位长度基因组作为模板进行第二阶段病毒 DNA 的合成，从而形成包括十个或更多病毒基因组交错拷贝的大串联体（large concatemers；Goorha，1982）。两个互补类群（涉及第一阶段和第二阶段 DNA 的合成）运用一批温度敏感的突变体与病毒 DNA 合成有着紧密的关联（Chinchar 和 Granoff，

1986；Goorha 和 Dixit，1984；Goorha 等人，1981）。由于序列分析只能鉴定单一病毒 DNA 多聚酶基因，很可能一个互补类群编码一个病毒 DNA 聚合酶，该酶在这两个事件中发挥着作用。而另一个互补类群有可能编码将病毒 DNA 从胞核转运到胞质中所需要的一个蛋白或者肩负着串联体形成的相关功能。

3.4　病毒的组装

病毒粒子的形成发生在胞质形态学不同的区域，这一区域被称为病毒组装场或病毒工厂。组装场是胞质电子透明区域，缺乏细胞器（Murti 等人，1985，1988；Zhang 和 Gui，2012）。不像自噬体（autophagosomes），蛙病毒组装场没
有封闭在膜内，而是被中间丝包裹，周围环绕着线粒体和核糖体（图 3）。无 113
论是早期病毒蛋白，还是晚期病毒蛋白，都与组装场相连，同时还与推测的串联病毒 DNA 相连。病毒组装场还检测出反义吗啉代寡核苷酸（antisense morpholino oligonucleotide，asMO），该寡核苷酸能阻断晚期基因的表达（Sample，2010；Sample 等人，2007）。这些结果提示，组装场的产生不需要晚期病毒蛋白的表达，从而巩固了早期的研究，表明早期蛋白足以满足组装场的形成（Chinchar 等人，1984；Sample，2010）。

对于形成传染性病毒粒子所需要的具体步骤，目前仍知之甚少。*ts* 突变体 114
（*ts* mutants）的遗传分析鉴定出了 12 个互补类群。这些类群在表面合成所有病毒蛋白和病毒 DNA，但不能产生感染性的病毒粒子（Chinchar 和 Granoff，1986）。用透射电子显微镜（TEM）对这些互补类群进行分析，鉴定了好几个 *ts* 突变体。在这些突变体中，病毒粒子的结构没有形成。然而，还见到了其他有完整结构的病毒粒子，但不具传染性（Purifoy 等人，1973；Sample，2010）。很明显，多个结构和催化蛋白必须参与传染性病毒粒子的形成。通过与 ASFV 进行类比分析（Rouiller 等人，1998；Tulman 等人，2009），结果发现，病毒粒子的组装首先是一个豆蔻酰化病毒蛋白（例如，FV3 ORF 53R）结合到内质网的片段上，随后在反面结合主衣壳蛋白（MCP）（Whitley 等人，2010），然后持续不断地加入 53R 和 MCP 膜片段，最后导致月牙状结构的形成，这一结构最终与病毒 DNA 相连，从而形成病毒粒子。与这一模型相一致的是，沿着这一通路形成的病毒粒子的中间体可以通过透射电子显微镜检测出来（见图 3）。

病毒被衣壳包裹的过程还没有确定，虽然病毒 DNA 的一小部分衣壳化被解释为是由于循环排列和末端冗长 DNA 基因组而存在（Goorha 和 Murti，1982）。但还不知道蛙病毒 DNA 是否如同在一些双链 DNA 病毒中所见到的一样，进入一个独特病毒粒子的入口，或者通过不断增大的二十面体病毒 DNA 的吞食（Cardone 等人，2007；Chang 等人，2007）而完成衣壳包裹。此外，

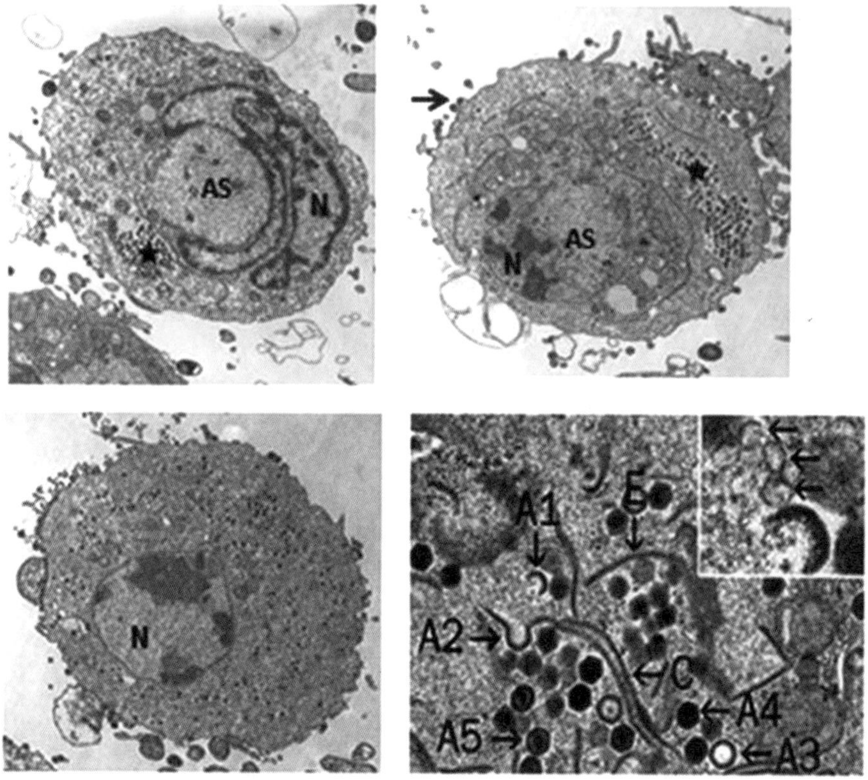

图 3 FV3 感染 FHM 细胞的透射电子显微照片。左上图和右上图分别显示典型病毒感染的细胞具有核（N），并显示染色质的凝聚、界限清晰的病毒组装场（AS）；胞质内亚晶状阵列（星号）以及病毒粒子从质膜出芽释放（箭头）。左下图显示在一个感染病毒的细胞中，病毒散布在整个胞质之中。右下图是倍数放大后的一个病毒组装场，显示处于各个组装阶段的病毒粒子，同时还显示出完整的病毒粒子（A4 和 A5）和空的病毒粒子（A3）以及两个病毒粒子中间体（A1 和 A2）和两个异常病毒粒子形式（C 和 E），右上角的插图表示质膜（箭头），这些质膜可能起源于内质网，该膜对病毒粒子的形态发生起着很重要的作用。

虽然串联体 DNA 被认为是最终包装成病毒粒子的 DNA 来源，但是病毒基因组完整拷贝是如何从串联体 DNA 中解脱出来的，目前还不清楚。完整的病毒粒子存在于病毒组装场、细胞质亚晶状阵列（paracrystalline arrays），同时还可见到与质膜相连（见图 3）。在培养的细胞当中，大多数病毒粒子仍保持与细胞相连，并且在细胞溶解以后以裸病毒粒子的形式释放。然而，有数量不定的病毒粒子从质膜以出芽的方式释放，并在这一过程中获得一个囊膜。决定一个特定病毒粒子是保持与细胞相连，还是形成囊膜的因素，目前还不清楚。冷冻断裂电镜观察表明，成熟的 FV3 病毒粒子显示出像一个俄罗斯娃娃一样的结构，具有多瘤的、由病毒 DNA 组成的球状核心与多个蛋白相连，封闭在一个 48 kDa 的主衣壳蛋白（MCP）衣壳之内（Darcy-Tripier 等人，1984；Devauchelle 等人，1985）。对隶属于虹彩病毒属的螟虫虹彩病毒（Chilo iridescent virus）的电子显微镜观察研究，鉴定了 3 个小蛋白，分别命名为指蛋白、拉链蛋白和锚蛋白，均与病毒粒子相连，但还没有找出它们与特异性病毒 ORF 的联系（Yan 等人，2009）。除大小结构蛋白以外，有人认为其他病毒蛋白（如负责关闭翻译以及 IE 转录的反式激活蛋白）同样也与成熟的病毒粒子相连。如同格雷费尔（Grayfer）等人（2015）在文献中描述的一样，后面的这些蛋白有可能与毒力有关。

早期病毒信息的转录启动是通过宿主 RNA 聚合酶 II（POL-II）在核中进行转录来实现的，与这一转录相反，晚期病毒信息如那些编码 MCP 和其他病毒关联蛋白是通过病毒 RNA 聚合酶 II（vPOL-II）在胞质中进行转录的，vPOL-II 是一种病毒编码的 RNA 聚合酶，该聚合酶的两个大亚基与细胞 RNA 聚合酶 II 对应的亚基是同源的（Tan 等人，2004；Sample 等人，2007）。下面将讨论用 asMO 靶向结合 vPOL-IIα 进行基因敲低研究，结果导致所有晚期蛋白的合成明显降低。直到此时，还不清楚 vPOL-IIα 和 vPOL-IIβ 是否作为一个最 115
小的双亚基 RNA 聚合酶发挥功能作用，或者这两个病毒亚单位是否与其他细胞或病毒蛋白结合形成功能酶。除需要功能性的 vPOL-II 以外，晚期转录同样依赖于病毒 DNA 的持续合成。当病毒 DNA 合成中 *ts* 突变体发生缺失或对 FV3 感染细胞用 DNA 合成抑制剂如磷酰基乙酸（phosphonoacetic acid，PAA）或胞嘧啶阿拉伯糖苷（cytosine arabinoside，araC）进行处理，同样也会抑制晚期基因的表达（Chinchar 和 Granoff，1984，1986）。在细胞培养中，FV3 的复制发生在 12~24 小时，早期病毒信息的转录本可以在前 4 小时之内被检测出来，晚期病毒信息的转录本和蛋白如 MCP 在感染 8 小时之后才能被检测出来，而病毒粒子需要 8~12 小时才能被检测出来。然而，病毒复制的动力学受到宿主细胞感染、温度和代谢多重因素的影响。如上所述，尽管所有脊椎动物虹彩病毒被认为是运用相同的策略进行复制，它们之间的差别有可能是在细胞和免疫学水平影响着它们如何与其宿主相互作用以及如何在体（*in vivo*）影响病毒的复制。

4 病毒感染对宿主细胞的影响

4.1 细胞死亡：坏死、细胞凋亡以及拟凋亡

蛙病毒感染以后，宿主 DNA、RNA 和蛋白质合成受到快速抑制，最后以细胞快速死亡而告终（Goorha 和 Granoff，1979；Raghow 和 Granoff，1979）。有趣的是，无论是有传染性的病毒粒子，还是无传染性的病毒粒子，如用热或紫外线使病毒失活，都能引发细胞转录和翻译的关闭，表明“关闭蛋白”（shut-off protein）是与病毒粒子相关联的（Cordier 等人，1981；Raghow 和 Granoff，1979）。病毒感染的细胞出现了凋亡，表现为染色质的浓缩和细胞 DNA 的破碎（Chinchar 等人，2003）。细胞凋亡既可出现在未灭活的病毒粒子感染以后，也可出现在灭活的病毒粒子感染以后，还可由原先宿主细胞大分子合成的抑制或蛋白激酶 R（protein kinase R，PKR）的激活引起。蛙病毒诱导的细胞凋亡依赖于细胞凋亡蛋白酶（又称半胱天冬酶）的激活，并且可能通过抑制细胞凋亡蛋白酶而被阻止（Chinchar 等人，2003）。

除细胞凋亡以外，蛙病毒属中进化关系最远的病毒即 SGIV，依据细胞类型不同，还可引发不同形式的细胞程序化死亡。在培养的石斑鱼（SGIV 的一种天然宿主）细胞当中，SGIV 的感染引发了一个非凋亡（nonapoptotic）形式的细胞程序化死亡，称作拟凋亡（parapoptosis）。拟凋亡的典型特征为细胞浆中出现空泡（vacuoles）、内质网膨胀以及缺乏 DNA 片段化和凋亡小体，并缺乏细胞凋亡蛋白酶的激活。在非宿主鱼的细胞培养中（如黑头呆鱼，FHM），
116 SGIV 诱发了典型的细胞凋亡，并以细胞凋亡蛋白酶的激活和 DNA 片段化为特征。此外，在石斑鱼细胞中，并没有检测出线粒体跨膜电位的破坏和磷脂酰丝氨酸（phosphatidylserine，PS）的内化。这种现象还见于 SGIV 感染 FHM 细胞之后（Huang 等人，2011a），在 GIV 感染的石斑鱼肾细胞中也获得了类似的结果。然而，所有的蛙病毒调控宿主细胞和非宿主细胞的存活是否都存在差异，还需要作进一步的研究（Pham 等人，2012）。在宿主鱼中，有丝分裂原活化蛋白激酶（mitogen-activated protein kinase，MAPK）信号链参与了 SGIV 诱导的拟凋亡，包括胞外信号调节激酶（extracellular signal-regulated kinase，ERK）、p38 有丝分裂原活化蛋白激酶（p38 MAPK）以及 c-Jun 氨基端激酶（c-Jun N-terminal kinase，JNK）信号传导（Huang 等人，2011a，b）。此外，石斑鱼免疫基因如干扰素调节因子 1（interferon regulatory factor 1，IRF1）、白

细胞介素-8（interleukin-8，IL-8）以及肿瘤坏死因子 α（tumor necrosis factor alpha，TNF-α）的转录都受到 JNK 的调控，但只有 TNF-α 受到 p38 MAPK 调节。因此，有人提出，JNK 信号通路，在病毒感染期间，对 SGIV 的复制以及调控炎症反应是较为重要的（Huang 等人，2011b）。有趣的是，在 FHM 细胞感染 FV3 以后，也见到了类似基因的激活（Cheng 等人，2014）。

4.2　宿主对病毒基因产物的关闭及选择性表达

虽然宿主大分子的合成受到快速抑制，但是病毒 DNA 复制、转录以及翻译仍未受到影响，病毒进入宿主细胞之后会在 24 小时之内产生传染性病毒粒子的高丰度水平（Willis 等人，1985）。面临细胞翻译明显的抑制，病毒蛋白的合成却持续维持，这有可能是几个因素共同作用的结果。首先，真核翻译启动因子 2α（eukaryotic translational initiation factor 2α，vIF-2α）的病毒同源物作为一个伪底物（pseudosubstrate）与 PKR 结合，从而阻止磷酸化和随后的细胞 eIF-2α 失活（Jancovich 和 Jacobs，2011）。此外，高效病毒信使的丰度水平超出了宿主的转录本水平，从而进入细胞翻译机器中（Chinchar 和 Yu，1990a，b）。相反，还不清楚，蛙病毒是如何选择性地抑制宿主转录的。有可能蛙病毒在传染过程中，以宿主的 RNA 聚合酶 II 作为靶向目标，并依靠病毒 RNA 聚合酶 II（vPOL-II）在胞浆中合成病毒的转录本。假如这一模型是正确的，那么这一模型对于确定被合成的早期基因，在晚些时候是否利用 vPOL-II 继续合成，则具有指导意义。随着病毒基因的转录和翻译的进行，宿主 DNA 合成会被阻断。宿主 DNA 合成的抑制被认为是细胞蛋白和 RNA 合成提前被阻断的结果。此外，一个病毒编码的核酸内切酶（endonuclease）有可能是限制性修饰系统的组成部分，同时，还降解宿主 DNA，这不仅对宿主转录产生负面影响，而且所形成的核苷酸可以再次用于病毒 DNA 的合成（Feighny 等人，1981；Kaur 等人，1995）。

5　宿主与病毒的相互作用决定生物水平上的感染结果

毒力是好几个因素的函数，其中有两个因素可能是最重要的，一个因素涉 117
及增强病毒的复制，另一个因素涉及阻断宿主的免疫防御。大 DNA 病毒（如疱疹病毒和痘病毒）编码许多基因，从而显现毒力。对蛙病毒的一些初步发现表明，它们之间没有差别（Eaton 等人，2007）。就增强病毒复制的基因而言，主要例子就是，核糖核苷酸还原酶（ribonucleotide reductase，RR）大小亚基的病毒同源物。无论是牛痘病毒（Vaccinia virus，VACV），还是单纯性疱疹病毒1（Herpes simplex virus 1，HSV1），都编码 RR 基因。VACV RR2 亚单位与宿主 RR1 形成功能性的复合物，从而提供足够的 dNTP 用于病毒的复制（Gammon 等人，2010）。同样地，HSV 和伪狂犬病病毒（Pseudorabies virus）编码的 RR 能将核糖核苷二磷酸转换成对应的脱氧核糖核苷酸，并通过维持 dNTP 池的大小，在病毒 DNA 合成中起着关键性的作用（Conner 等人，1994a，b；Daikoku 等人，1991）。与病毒生物起源的关键作用相一致，缺失 RR 表达的突变体，在体是无毒的（dc Wind 等人，1993）。然而，除在 dNTP 合成中的作用以外，HSV1 和 HSV2 RR1 亚单位同样可以保护细胞免受凋亡（Chabaud 等人，2007；Langelier 等人，2002）。

与 α 疱疹病毒和 β 疱疹病毒相比，小鼠巨细胞病毒（Murine cytomegalovirus）的 RR1 亚单位是催化性失活的，并在增加 dNTP 池的大小上不起任何作用。而且该病毒还进化出了一种新的功能，就是抑制一种细胞接头蛋白（cllular adaptor protein）即 RIP1 的产生，并能阻断涉及先天性免疫和炎症的信息传递通路（Lembo 和 Brune 2009）。其他有助于提高复制的病毒基因包括编码病毒同源物的脱氧尿苷焦磷酸酶（dUTPase）和胸苷激酶（thymidine kinase）的基因（Eaton 等人，2007）。由 FV3 编码的 RR1、RR2、dUTPase 和胸苷激酶同源物在病毒复制中有可能起着相似的关键性作用。

除能够增强病毒复制的基因以外，病毒同样还编码抑制或对抗宿主免疫反应的基因。痘病毒（Poxviruses）含有超过 24 个抑制或对抗宿主免疫反应的基因，其唯一的功能就是阻断先天性或获得性免疫的一个方面或多个方面（Seet 等人，2003）。并且有人估计，由人类巨细胞病毒（Human cytomegalovirus）所编码的基因，可能有一半参与了免疫逃避（Eberhardt 等人，2013）。例如，痘病毒编码 IFNα/β、IFN-γ、IL-1β、IL-18、各种趋化因子和补体的诱骗受体（decoy receptors）。这些受体分别与其同源靶分子结合，从而下行性调节着效应因子的功能（Johnston 等人，2005；Johnston 和 McFadden，2003，2004；Seet 等人，2003）。此外，病毒编码的 miRNA 可能在阻断宿主的免疫反应或控制病毒复制方面起着关键性的作用（Babu 等人，2014；Hook 等人，2014；Pavelin 等人，2013）。

对 FV3 和其他脊椎动物虹彩病毒基因组进行序列分析，已鉴定出好几个推导的基因。这些基因在削弱宿主免疫力方面起着很重要的作用。这些推导的免疫逃逸蛋白包括上述 eIF-2α 的病毒同源物（vIF-2α），还有一个 RNAse III 病毒同源物、一个病毒编码的含 CARD 结构域即半胱天冬酶募集结构域的蛋白、一个 β 羟基类固醇脱氢酶（β-HSD）病毒同源物、一个病毒 Bcl-2 同源物

以及一个或多个 TNFR 同源物（Huang 等人，2013b；Lin 等人，2008；Tan 等 118
人，2004）。除这些推导的免疫效应因子以外，还有其他一些病毒基因对特定的病毒或特定的病毒属而言是独特的，并且代表着相应的蛋白质，仅仅作用于特异性的宿主物种、组织或细胞。例如，有十几个开放阅读框仅见于蛙病毒中，并在特定的变温脊椎动物中，编码增强病毒复制或损伤免疫力的独特多肽（Eaton 等人，2007）。然而，通过相似性的搜索（similarity searches）来鉴定毒力蛋白，其挑战是相当大的。这是因为，即使是在哺乳动物系统中，病毒与细胞同源物之间的相似水平，也相当之低。鉴定一种病毒蛋白是否是一种特定功能的潜在抑制因子，难度相当之大，除非一些关键的基序（motifs）是保守的。对于一个特定的病毒蛋白在毒力当中的具体作用，要获得结论性证据还需要用遗传和生物化学的方法，这将在以后展开讨论。

6 抗病毒免疫

由于格雷费尔等人（2015）对宿主抗病毒免疫已作了详细的论述，在这里，我们只提出一个简单的总结。运用一个模型将分子水平最好鉴定的蛙病毒 FV3 与已做过最完整免疫系统鉴定的非洲爪蟾（*Xenopus laevis*）进行配对，来对蛙病毒感染引起的宿主免疫反应进行富有成效的研究（Robert，2010）。罗伯特及其同事所做的研究显示，具有免疫力的成年蛙的感染只限于肾脏并且在几周之后就会消退，而且死亡率极低（Gantress 等人，2003）。相反，自然缺乏 MHC I 表达的蝌蚪和免疫抑制的成年动物极易感染，并且展示出相当高的发病率和死亡率（Tweedell 和 Granoff，1968；Gantress 等人，2003；Robert 等人，2005）。在免疫缺陷或免疫抑制的非洲爪蟾中，感染始于肾脏，然后发展成全身性感染，扩散到多个器官，包括肝脏、胃肠道和皮肤（Gantress 等人，2003）。与此观察结果相一致的是，鱼类通过接种灭活的病毒粒子或 DNA 疫苗得到免疫保护，不发生蛙病毒和细胞肿大病毒诱导的疾病（Caipang 等人，2006a，b；Ou-yang 等人，2012）。有研究显示，抗体反应在 FV3 感染中起着保护性作用（Maniero 等人，2006）。细胞介导的免疫反应，在保护宿主免受 FV3 感染中起着重要作用（Morales 和 Robert，2007）。最近的研究显示，巨噬细胞容易遭受 FV3 的感染（Morales 等人，2010）。虽然巨噬细胞有可能在免疫中起着关键性的作用，但是它们的感染可能存在着两个方面的不良作用：一是消除巨噬细胞处理抗原和提供病毒抗原的能力；二是提供持续感染细胞的来源，从而促使病毒在种群中长期存在。

虽然它们在存活中的作用还没有被确定，但是可能有多个细胞基因，在抗病毒免疫和病毒复制中起着各种不同的作用，这些基因在细胞被蛙病毒感染后得到了诱导。与这一观点相一致的是，程光（Cheng Kwang）等人（2014）最
119 近用野生型的 FV3 和进行基因敲除缺乏 vIF-2α 的病毒突变体，研究考查了 FHM 细胞对感染的反应，并观察到在转录水平诱导了多个免疫反应基因的表达，其中包括 IFN、IL-8、GILT 和 IRF-3。同样，石斑鱼接种灭活的 SGIV，结果诱导了多个免疫相关基因的表达，其中包括 Mx1、ISG15、IL-8、IL-1β 以及 MHC I/II。从而表明免疫反应在鱼类的不同种当中是保守的，类似的情况也见于哺乳动物（Ou-yang 等人，2012）。

最近格雷费尔等人（2014）报道了非洲爪蟾干扰素（Xl-IFN）的克隆与测序。它们的研究显示，重组 Xl-IFN 保护了两栖类细胞 A6 免受 FV3 的感染，并暂时保护了蝌蚪。不出所料，感染 FV3 不久的成年动物，合成了 Xl-IFN，并且其数量要高于蝌蚪。令人奇怪的是，虽然成年动物存在着更强烈的免疫反应，但在成年动物中，病毒的含量要高于蝌蚪。然而，正如上面报道的一样，病毒在成年动物长期存活的时间也要长一些。此外，非洲爪蟾干扰素虽然能明显地损害病毒复制，但是不能阻止病毒引起的死亡。重组载体处理（vector-treated）的蝌蚪对照组，平均存活了 26 天，在处理后第 32 天，死亡率为 90%。而 rXl-IFN 处理的成年动物显示，平均死亡时间为 36 天，40 天后死亡率才达到 90%。这些结果提示，FV3 对蝌蚪的致病性，比原来所想象的要高得多，即使是低滴度的病毒，也能给内脏器官造成广泛的损害，最终导致死亡。为了支持 IFN 具有保护性作用这一观点，有人将表达龟 IFN-γ 的载体转染给培养的 STA 细胞，结果导致病毒基因的表达降低接近 90%，并使中华鳖虹彩病毒的产量降低至原来的 1/10（Fu 等人，2014）。依据上面所显示的结果，被蛙病毒感染以后所见到的宿主免疫反应，非常类似于鱼类感染其他病毒以后所见到的（Verrier 等人，2011）。

7 病毒基因功能的确定

随着病毒复制大概轮廓为人们所熟知，目前的挑战就在于，阐明特定病毒基因的功能，从而为有效的化学治疗提供靶向目标，或者为疫苗的研发提供帮助。在最近 10 年，随着各种强大的分子方法逐渐被人们接受，蛙病毒基因功能的确定已明显加速，对这些现代方法以及经典方法，我们将在下面予以讨论。

7.1　生物化学和遗传学方法

在鉴定蛙病毒基因功能过程中，最早的尝试是依赖于遗传学和生物化学相结合的方法。在采用后者即生物化学方法的过程中，我们使用过各种抑制因子，试图阻断病毒复制的特定方面。例如，运用环己酰亚胺（cycloheximide，CHX）阻断整个蛋白的合成，并限制病毒转录成 IE mRNA。而氟苯丙氨酸（flurophenylalanine，FPA）、磷酰乙酸（phosphonoacetic acid，PAA）以及胞嘧 120
啶阿拉伯糖苷（cytosine arabinoside，araC），只允许 IE 和 DE 病毒基因产物的表达（Goorha 和 Granoff，1979；Willis 等人，1985）。有学者认为，FPA 可以结合到一个或多个晚期基因表达所需要的病毒蛋白之中（例如 RNA 聚合酶 II 的 2 个大亚基的病毒同源物），并通过改变蛋白质的构象，而产生功能抑制。PAA 和 araC 可以通过特异性的（如 PAA）或非特异性的（如 araC）途径抑制病毒 DNA 聚合酶的活性，从而阻断病毒 DNA 的合成。所有晚期基因表达要求不断地进行 DNA 复制，这可能反映出既需要 DNA 模板的增加，也需要 DNA 模板构象上的改变。

宿主 RNA 聚合酶 II 在早期而不是晚期病毒 mRNA 转录过程中的作用，已运用宿主 RNA 聚合酶 II 的抑制剂 α-鹅膏蕈碱（α-amanitin）予以展示（Goorha，1981）。在感染前或感染时加入 α-鹅膏蕈碱，可以阻断所有病毒基因的表达，而在晚期加入时，则对病毒功能没有丝毫影响。从而提示，宿主 RNA 聚合酶 II 只在感染的早期才需要（Goorha，1981）。有学者用甲基化抑制物 5-氮杂胞苷（5-azacytidine，azaC）探讨了 DNA 甲基化在病毒复制过程中的重要作用，在 azaC 存在时，病毒 RNA 和蛋白质的合成没有受到抑制，而病毒 DNA 的合成也只受到轻度的影响（Goorha 等人，1984）。然而，在 azaC 存在时，新合成的病毒 DNA 缺乏甲基化的胞嘧啶，而且会产生单链 DNA 的断裂，造成病毒产量减少到原来的百分之一左右。

上述研究成功地阐明了几个病毒编码蛋白的作用，这是因为抑制剂只以有限几个病毒基因产物如病毒 DNA 聚合酶、病毒 DNA 甲基转移酶为靶向目标。但是所能鉴定的只是推导的接近 100 个 FV3 开放阅读框中少数几个病毒编码蛋白的作用。有学者运用传统遗传学的方法，侧重于耐药（PAA^{R} 和 azaC^{R}）突变体和温度敏感突变体的产生与鉴定，来尝试鉴定病毒基因及其功能。有学者对连有一个针对去甲基化酶（DMTase）活性的 26 kDa 蛋白的抗 azaC^{R} 突变体进行了鉴定（Essani 等人，1987）。用抗 PAA 突变体的研究确认，药物的靶向目标是 DNA 聚合酶（Chinchar 和 Granoff，1984）。通过温度突变体互补分析，鉴定了病毒复制所必需的 19 个基因（Chinchar 和 Granoff，1986；Naegele 和

Granoff，1971；Purifoy 等人，1973），这些基因包括在病毒 RNA 合成（5 个互补类群）和病毒 DNA 合成（2 个互补类群）缺失的突变体。两个 DNA 缺失互补类群的分析，支持两个阶段 DNA 合成的学说，即在胞核中单位大小基因组合成以及在胞浆中大串联体的形成（Goorha 和 Dixit，1984；Goorha 等人，1981）。在晚期病毒基因表达中，至少存在 5 个互补类群表现出的缺失，有可能反映出突变的靶向目标是病毒 RNA 聚合酶的大小亚基以及晚期 mRNA 合成所需要的其他基因产物，如转录延伸因子 SII（transcription elongation factor SII）。

121 大多数温度敏感突变体 *ts*（12 个互补类群）既合成早期蛋白和晚期蛋白，也合成病毒 DNA，在不适宜的温度条件下，不能形成有传染性的病毒粒子（Chinchar 和 Granoff，1986）。这一发现表明，蛙病毒粒子的组装不是一个简单浓缩的相关事件，在这个事件中，MCP 即使不是唯一成分，也是主要成分。电子显微镜观察显示，随着有些突变体逐渐向外成熟，有些无传染性的病毒粒子也同时形成，并伴随着其他病毒粒子形成的夭折（Sample，2010）。最近的研究表明，多个病毒蛋白组成成熟的虹彩病毒粒子，这有可能反映出，这些缺失既可以是关键结构蛋白如 MEP、53R、锚蛋白、指蛋白、拉链蛋白和病毒组装所需的推导支架蛋白（putative scaffold proteins）的缺失，也可以是病毒复制启动所需的与病毒粒子相连的一个或多个蛋白，例如推导的 IE 转录反式激活因子（putative transactivator of IE transcription）的缺失。

人们尝试用作图的方法来确定温度敏感突变体在基因组当中的位置，这一想法正面临着挑战，但是，依据重组频率（recombination frequencies），许多温度突变体的相对位置已被确定（Chinchar 和 Granoff，1986）。如同在疱疹病毒和痘病毒中所做的一样，对界定的限制性片段进行特异突变的自然分配（physical assignment）还没有取得成功。携带蛙病毒的质粒（plasmids）不能拯救温度突变体，这一结果被认为是由一个病毒编码的内切酶（endonuclease）对新输入质粒降解引起的。该内切酶以未甲基化的质粒 DNA 为靶向目标而导致降解。幸运的是，随着 ATV、FV3 及其他蛙病毒全基因组的测序，确定基因功能的其他方法也成为现实。

7.2 基因敲低策略：反义吗啉代低聚核苷和 RNA 干扰

FV3 和其他蛙病毒基因组完整核苷酸序列的确定，开启了通过对特异性病毒基因进行“打靶”（targeting）来直接确定基因功能的大门。无论是这一部分内容所描述的基因敲低（knock down，KD），还是下一部分内容所描述的基因敲除（knock out，KO），都已得到成功运用。在基因敲低中，运用反义吗啉代低聚核苷（asMO）或小干扰 RNA（siRNA）来抑制 FV3 基因的功能（Sample 等人，2007；Whitley 等人，2010，2011）。asMO 是启动密码子 AUG 下游区域的互补寡核苷酸（长度大约为 25 个核苷）片段，它们通过阻止核糖体的运动（又称“扫描”）和抑制蛋白质的合成而下行性调控基因的表达（Hudziak 等人，2000；Summerton 和 Weller，1997；Summerton，2007）。小干扰 RNA 是小的双链分子（大约 22 个核苷），在它们结合到 RNA 诱导沉默复合体（RNA-induced silencing complex）以后，互补靶信使的链会结合到靶信使链上，既可导致 RNA 的降解，也可导致翻译的抑制（Hannon 2002）。与 asMO 相反，siRNA 既可结合到编码区，也可结合到非编码区。虽然存在着计算来预测究竟是哪段序列能使小干扰 RNA 有效，但它们是不确定的，因而需要进行潜在 siRNA 的验证实验，从而保证成功的抑制。

反义吗啉代低聚核苷已成功地用于几个 FV3 基因的打靶，其中包括编码 122
MCP、18K 即刻早期蛋白、病毒 RNA 聚合酶 IIα（vPOL-IIα）、53R、一个推导的十八烷基化膜蛋白以及一个 SGIV 编码的组蛋白结合蛋白（Sample 等人，2007；Tran 等人，2011；Whitley 等人，2010，2011）。基因敲低通过 SDS-PAGE 或免疫印迹实验之后，目标蛋白缺乏或明显减少而得以验证，其对病毒复制的影响通过透射电子显微镜进行观察并测定病毒产量。MCP 和 53R 的基因敲低导致靶蛋白丰度的明显减少，对非靶蛋白影响很小，甚至完全没有影响。实验表明，对 MCP 和 53R 合成的抑制使病毒产量减少大约 90%，并分别在病毒组装场出现非典型元件（MCP）和颗粒（53R）外观，这有可能是重要结构元件的明显减少，导致了异常结构的积累。与此相反，vPOL IIα 的基因敲低，导致了晚期蛋白合成的整体减少，并在病毒组装场缺乏所有结构元件。18K IE 蛋白的基因敲低只影响靶蛋白，而对病毒的产量没有产生不良影响。这些结果表明，MCP、53R 以及 vPOL IIα 是离体复制的必需蛋白，而 18K 即刻早期蛋白则是在培养的 FHM 细胞中，是复制非必需的（Sample 等人，2007；Whitley 等人，2010）。有学者试图把这些研究扩展到其他病毒基因产物，由于通过 SDS-PAGE 或者免疫印迹实验（immunoblotting）不能检测到靶蛋白的丢失而陷入困境。然而，我们观察到 FV3 接触到 asMO 以后，其产量范围为 ORF 41R 对照组的 8% 到 ORF 95R 对照组的 43%，asMO 的靶目标分别是 ORF 2L（推导的膜蛋白）、ORF 9L（NTP 酶/DEAD/H 螺旋酶）、ORF 41R（一种未知的蛋白，分子量为 129 kDa）、ORF 32R（神经丝三联 H1 蛋白，neurofilament triplet H1 protein）、ORF 38R（RRα）、ORF 57R（丝氨酸/苏氨酸激酶，Ser/Thr kinase）、ORF 80L（核糖核酸酶 III 样蛋白，ribonuclease III-like protein）、ORF 91R（46kDa，即刻早期蛋白）以及 ORF 95R（DNA 修复蛋白，

RAD2）（Whitley 等人，2011；VGC，未发表的研究资料）。基因片段（partial）产量的减少可能反映出靶病毒蛋白补充了现存宿主的功能蛋白（如核苷酸还原酶），以及在培养细胞中，靶蛋白有可能对于复制不是绝对需要的。还有一种解释就是病毒基因片段的减少，有可能是因为不完整的基因敲低。此外，不能通过 SDA-PAGE 确定的基因敲低有可能是由于靶蛋白与丰度高的同等分子量大小蛋白的共迁移（此种情况见于 53R，Whitley 等人，2010）或者靶蛋白在感染细胞中只有很低的含量水平（Whitley 等人，2010）。尽管免疫印迹实验对于鉴定共迁移或者低丰度的蛋白有可能是非常有效的，但是用于检测特异性病毒蛋白的抗体却不容易获得。总体来讲，KD 研究对于鉴定三类蛋白是有潜在优势的：（1）“必需”蛋白（essential proteins）。这类蛋白对于病毒粒子生成是绝对需要的，如 MCP 和 vPOL-IIα。（2）“效率”蛋白（efficiency proteins）。这类蛋白在特定环境中能促进病毒的复制，如病毒的核苷酸还原酶同源物，但对病毒复制不是绝对需要的。（3）“免疫逃避”蛋白（immune evasion proteins）。这类蛋白以宿主抗病毒免疫反应的先天性和获得性成分为靶目标，如真核翻译启动因子 2α（eukaryotic translational initiation factor 2α，vIF-2α）病毒同源物或者 β 羟基类固醇脱氢酶（β-hydroxysteroid dehydrogenase，β-HSD）。

123 除用 asMO 进行 KD 研究以外，还有数量有限的运用 siRNA 使病毒基因的表达沉默的研究（Dang 等人，2008；Kim 等人，2010；Whitley，2011；Xie 等人，2005）。siRNA 介导的 FV3 MCP、vPOL-IIα 以及病毒编码的 DNA 甲基转移酶（virus-encoded DNA methyltransferase）的 KD 使得同源信使水平减少达 90%或导致病毒产量大大减少。电子显微镜的证据显示，病毒粒子形成很少，或是根本没有形成（Whitley 等人，2011；Xie 等人，2005）。然而，与通过 siRNA 介导的 KD 相反，asMO 介导的 KD，在病毒感染以后，可以检测出病毒感染复制数（或称转染倍数，multiplicities of infection，MOI）为 10 PFU/细胞或更高。而 siRNA 介导的 KD，当使用0. 01~0. 1MOI 时就可以见到。转染倍数较高时（例如 MOI 为 1~10），病毒的产量并不能通过 siRNA 的处理而减少。虽然其中的原因还不清楚，但是有可能蛙病毒就像其他病毒一样，能够编码一种基因产物，这一基因产物通过与双链 DNA 结合，阻止 RNA 沉默复合物（RNA silencing complexes）的形成，从而阻断 RNA 的干扰（RNA interference，RNAi）。

总的来讲，运用 siRNA 和 asMO 抑制特异性病毒基因表达的研究，对阐明病毒基因的功能以及确定一个特定基因是否是离体复制所“必需的”极为有用。然而对影响宿主体液免疫和细胞免疫方面的病毒基因来说，siRNA 和 asMO 方法是不适用的，这是因为靶基因的功能有可能只是在体所需要的。基于这一原因，对于推导性的病毒编码的免疫逃避基因的研究，可能用下面讨论的基因敲除方法会更好一些。

7.3 基因敲除的突变体

直到最近，对蛙病毒的研究由于质粒导入感染病毒基因组时不能获得重组而受到限制。这一困难妨碍了通过标记救援获得温度敏感突变体的物理图谱（physical mapping）及重组病毒的产生。最近，运用有效筛选技术开发出了分离稀少病毒重组子的方法。例如，构建了重组 SGIV 来表达增强型绿色荧光蛋白 EGFP（enhanced green fluorescent protein）与囊膜蛋白 VP55（envelope protein VP55）的融合蛋白，并用于病毒复制的动态学评估（Huang 等人，2011c）。同样，携带抗药基因的蛙病毒已引入到 ATV、蛙虹彩病毒以及 FV3（见下述内容）。因此，虽然最初产生重组蛙病毒很困难，但是依据荧光标记技术或者抗药性病毒的筛选技术，可以对它们进行有效的分离。

在 BIV 中，运用同源重组的方法，将 ICP18 启动子（ICP18 promoter）控制下的抗新霉素基因（neomycin resistance gene）以及在病毒 MCP 启动子控制下的成年澳州蟾蜍（*Bufo marinus*）球蛋白基因引入到 vIF-2α 位点（vIF-2α locus），产生了第一个重组蛙病毒（Pallister 等人，2007）。然而，其研究的重点是鉴定 vIF-2α 基因的功能，在这一研究中所开发的技术，促进了重组蛙病毒实验方案的开发，从而确定了 vIF-2α 基因产物的非必需性质。基于这项研究，基因敲除和基因敲入突变体（KO and KI mutants）已在 ATV、FV3 以及 124
RGV 当中产生（Chen 等人，2011；He 等人，2012b；Huang 等人，2011c；Jancovich 和 Jacobs，2011）。进行蛙病毒基因敲除的第一步涉及构建一个质粒，这个质粒中目标基因序列的上下游两侧都带有可选择的标记物。用野生型的病毒感染细胞，随后用构建的重组子进行转染。在可选标记物两侧附近进行同源重组，导致了编码可选择标记物的基因替代靶基因。重组病毒在特异性抑制剂如新霉素或者嘌呤霉素存在的条件下，能够复制后进行分离，或在引入绿色荧光蛋白（GFP）基因的情况下，通过鉴定荧光噬斑（fluorescent plaques）而进行分离。由于病毒和细胞生长对新霉素和嘌呤霉素比较敏感，因而具有抗这些抗生素的基因已被成功地用于筛选基因敲除的蛙病毒（Jancovich 和 Jacobs，2011；Chen 等人，2011）。此外，GFP 已被用于可选标记物（Huang 等人，2011c），一旦分离，基因敲除突变体的鉴定就需要通过表型的变化来确定基因的功能。例如，vIF-2α 基因敲除突变体的离体复制，就没有明显的改变，说明 vIF-2α 是一个非必需基因（Chen 等人，2011；Jancovich 和 Jacobs，2011；Pallister等人，2007）。然而，当蛙或者蝾螈感染 vIF-2α 基因敲除的突变体时，可以观察到死亡率的下降，从而提示，vIF-2α 在体起着很重要的作用。除 vIF-2α 同源物以外，病毒 18K 即刻早期基因同样也是靶基因（Chen 等人，2011）。FV3 的 18K 基因缺失对病毒的离体复制没有丝毫影响，但是在感染的蝌蚪当中，造成了较低的死亡率，再次提示，这一基因与病毒毒力有关。以 vCARD 和 β- HSD 为靶基因的其他两个 FV3 基因敲除突变体的鉴定提示，这两者都是离体复制非必需的（J Robert 和 G Chen，美国罗彻斯特大学和 VGC，未发表资料）。总体来讲，这些研究显示了基因敲除的突变体在确定“非必需”病毒基因当中的作用。

7.4 条件性致死突变体

由于必需基因的缺失会消除病毒复制的能力，因而基因敲除策略只能针对非必需基因。有两个潜在解决办法可以打破这一僵局：（1）构建转染子（trans）中表达缺失基因产物的补充细胞系（complementing cell lines），从而允许病毒复制；（2）在适当诱导物（inducer）如 IPTG 或四环素（tetracycline）存在的条件下，构建完全表达靶基因的条件性致死突变体，但在缺少诱导物的情况下，靶基因表达很少或完全不表达。在非洲猪瘟病毒（African swine fever virus，ASFV，非洲猪瘟病毒科，*Asfarviridae*）以及牛痘病毒（Vaccinia virus，VACV，痘病毒科，*Poxviridae*）中，已经构建了条件性致死突变体（Garcia-Escudero等人，1998；Nichols 等人，2008）。这两个病毒科与虹彩病毒科（*Iridoviridae*）、囊泡病毒科（*Ascoviridae*）以及藻类 DNA 病毒科（*Phycodnaviri-*
125 *dae*）的成员一起，构成系统发生上亲缘关系较近的一个类群，称作核质巨 DNA 病毒（Nuclear cytoplasmic large DNA viruses）（Colson 等人，2012；Koonin 和 Yutin，2010）。

最近，张奇亚及其同事采用同源重组的方法，构建了条件性致死突变体，靶基因为蛙病毒、RGV 的 53R 和 2L 基因（He 等人，2012b，2013，2014）。简单地说，乳糖操纵子的 *lacI* 调节基因编码乳糖阻遏蛋白（lac repressor protein），在蛙病毒 ICP18 即刻早期基因（ICP18 immediate-early gene）启动子的控制下，通过同源重组，导入 RGV TK 基因，结果产生了一个称为 RGV-lacI 的重组病毒。随后，53R 或者 2L 基因，在杂合启动子（hybrid promoter，p50-lacO）的控制下被置入。在杂合启动子当中，乳糖操纵子 lacO 序列在与 lacI 阻遏蛋白结合时，会引起下游转录的沉默，可以定位到 RGV 基因 ORF50 启动子中 TATA 样盒（TATA-like box）的下游。含有这一构建物的质粒通过同源重组导入到 RGV-lacI，从而产生条件性致死突变体（*cl* mutants），即 i53R-RGV-lacIO 和 i2L-RGV-lacIO。在缺少诱导物的情况下，IPTG 乳糖阻遏物会结合到乳糖操纵子 LacO 序列上，从而抑制下游基因的转录。相反，在有诱导物存在的情况下，IPTG 则与阻遏物结合，从而消除转录阻断，允许病毒基因全部表达。就 53R 和 2L 条件性致死突变体来说，靶转录本和靶蛋白的表达明显减少，但在缺少 IPTG 的情况下，表达没有完全消除。因此，在感染后 72 小时，病毒产量减少超过 90%，并伴随着 53R 转录本水平降低达 75%。类似的结果也见于 2L 突变体。总体说来，这些结果表明，53R 和 2L 对于病毒的在体复制都是必需的，同时还表明，和 ASFV 和 VACV 一样，条件性致死突变体可用来探明病毒必需基因。但是，这些突变体广泛运用的一个潜在缺陷就是靶基因的表达常常不是完整的。然而，如同这两个突变体所显示的一样，除靶基因低水平的表达可能满足整个复制的情况以外，所获得的表达抑制（就 53R 而言，转录抑制水平达 75%）足以确定基因产物的必需性质。

7.5　重组病毒蛋白的异位表达

除基因敲低及运用 *ts*、*cl* 和 KO 突变体的研究以外，另外一个探讨病毒基因功能的有效方法就是异位表达重组病毒蛋白并监测它们的活性。运用这一方法所进行的研究已用来确定许多 ALRV 蛋白（如 vIF-2α、β-HSD、dUTPase、ERV1、50L、2L 和 53R）的定位及功能，其总结见表 2。重组病毒蛋白的离体表达，在确定催化蛋白的作用中显得特别有用，这些催化蛋白影响细胞或免疫功能，但在确定病毒结构蛋白或者必须与其他病毒蛋白相互作用后才能行使正常功能的那些蛋白中，则并非如此。

病毒翻译启动因子 2α（vIF-2α）： vIF-2α 是一种真核生物翻译启动因子 2α（eukaryotic translational initiation factor 2）亚基的病毒同源分子（homolog）。
该同源分子被认为，在双链 RNA 激活的蛋白激酶 R（dsRNA-activated PKR） 127
引起整个翻译阻断时，对持续维持病毒蛋白的合成起着关键性的作用。有人推测 vIF-2α 是作为伪底物（pseudosubstrate），类似于 VACV K3L 与 PKR 结合，来阻止磷酸化和 eIF-2α 的失活（Beattie 等人，1991；Langland 和 Jacobs，2002）。

为了阐明 vIF-2α 在维持感染细胞病毒蛋白复制中的作用，罗滕堡（Rothenburg）等人（2011）使用了异源酵母系统（heterologous yeast system），结果显示，来自蛙病毒中的牛蛙病毒-Z（*Rana catesbeiana* virus-Z，RCV-Z）vIF-2α 的异位表达阻断了人类和斑马鱼 PKR 的生长抑制效应。有研究显示，RCV-Z vIF-2α 可作为人类和斑马鱼 PKR 的抑制物，阻断 PKR 介导 eIF-2α 的磷酸化，而 VACV K3L 则显示出宿主特异性，只能阻断人类 PKR 的活性。此外，构建 vIF-2α 缺失的实验显示，N 末端和螺旋结构域足以造成 PKR 的抑制。而碳末端结构域则是可有可无的。vIF-2α 不能替代 eIF-2α 抑制 PKR 的功能。因此，作者建议将其重新命名为蛙病毒蛋白激酶抑制物 R（ranavirus inhibitor of protein kinase R，RIPR）。虽然这一研究为当宿主翻译被关闭时，vIF-2α/RIPR 维持病毒蛋白合成起着关键作用这一观点提供了强有力的支持。但是应该指出的是，无论是 FV3，还是中华鳖虹彩病毒（STIV）编码的 vIF-2α 平末端分子（truncated versions），在 N 端丢失了全长产物的四分之三（Huang 等人，2009；Tan 等人，2004）。FV3 和 STIV 都能关闭宿主翻译，并维持高水平病毒蛋白的合成，这一结果显示，虽然 vIF-2α 在宿主翻译被关闭的情况下，对于维持病毒翻译很重要，但其他蛋白有可能在这一过程中也起着作用，像 VACV 的 E3L/K3L 系统能提供高水平的翻译就是如此（Langland 等人，2006；Langland 和 Jacobs，2002）。

β-羟基类固醇脱氢酶（β-HSD）：3β 羟基类固醇脱氢酶在细胞类固醇合成中起着关键性的作用，其对应的 VACV 同源分子被认为在钝化宿主的抗病毒反应中起着重要的作用（Reading 等人，2003）。有学者克隆了一个 1 068 bp/355 aaβ-HSD 的 RGV 同源分子，并显示出是一个即刻早期基因产物（Sun 等人，
128 2006）。共聚焦显微镜观察显示，在 EPC 细胞中，βHSD-EGFP 隔合蛋白的异位表达与带有线粒体特有标记物的 pDsRed2-Mito 存在共定位。而且在 EPC 细胞中，βHSD-EGFP 的过表达抑制了 RGV 诱导的细胞致病效应（cytopathic effect，CPE）。鉴于 β-HSD 在类固醇合成中的推导作用，这一蛋白在病毒感染中有可能损害宿主的免疫反应。β-HSD 离体实验中是否存在着其他作用来抑制 CPE，还有待进一步确定。如同所见到的缺乏 β-HSD（Reading 等人，2003）的 VACV 突变体一样，我们推测，以 β-HSD 基因为靶目标的突变体完全像 *wt* 病毒（*wt* virus）一样，不能抑制宿主的免疫反应，并且在体实验显示出是一个致弱的表型。

脱氧尿苷焦磷酸酶（dUTPase，dUTP pyrophosphatase）：脱氧尿苷焦磷酸酶是一种普遍存在并负责调控 dUTP 浓度的酶，dUTP 与 DNA 结合以后，会对病毒复制产生有害作用，该酶通过补救通路（salvage pathway）提高 dTTP 的水平（Kato 等人，2014；Oliveros 等人，1999）。脱氧尿苷焦磷酸酶催化 dUTP 转换成 dUMP。GGV 编码的脱氧尿苷焦磷酸酶是一个含有 164 个氨基酸的蛋白质，通过反转录 PRC（RT-PCR）和蛋白印迹技术分析，该蛋白质被确定为一种早期基因产物。dUTPase-EGFP 融合蛋白的异位表达见于细胞浆当中，并且通过免疫荧光实验确定了在病毒感染细胞浆中的位置（Zhao 等人，2007）。脱氧尿苷焦磷酸酶的过表达，对于 RGV 的复制，没有产生可检测出的影响，从而提示，该酶对病毒的离体复制没有负面或正面的影响。因此，在 EPC 细胞

当中，dUTPase 的过表达，不能促进病毒的复制，因为 *wt* 病毒所提供的表达水平足以满足整个病毒的复制。此外，SGIV dUTPase 同源分子的研究，鉴定了一个核输出信号，该信号对于 SGIV dUTPase 从核中转运到胞浆当中是至关重要的（Gong 等人，2010）。最后要说明的是，尽管 BLAST 分析提示，RGV dUTPase 在调节 UTP/dTTP 水平中发挥着功能作用，但是在疱疹病毒中发现的 dUTPase 基因还涉及新的功能，其中包括免疫功能的失调。

表 2　用重组蛋白对蛙病毒基因功能的评估

基因（病毒和 ORF）[a]	表型	参考文献
vIF-2α（RCV）	通过阻断 eIF-2α 的磷酸化来维持病毒的翻译，见于大多数但不是全部蛙病毒中	Rothenburg 等人（2011）
ICP46（SGIV ORF 162L）	促进 GP 细胞的生长，作为核衣壳的一个结构性蛋白增强 SGIV 的复制	Xia 等人（2010）
ICP18（SGIV ORF 086R）	促进 GP 细胞的生长，作为病毒的一个非囊膜蛋白增强 SGIV 的复制	Xia 等人（2009）
LITAF（SGIV ORF 136R）	通过诱导细胞凋亡在细胞死亡中起着关键性的作用	Huang 等人（2008）
LITAF（FV3 ORF 75L）	ORF 75L C 端的一半明显地类似于细胞 LITAF、75L 以及病毒感染细胞中相关联的 LITAF	Eaton 等人（2013）
TNFR（SGIV ORF 096R）	通过调控宿主细胞凋亡反应，促进病毒的复制	Huang 等人（2013b）
dUTPase（SGIV ORF 049R）	包含一个核输出信号	Gong 等人（2010）
dUTPase（RGV ORF 67R）	调控 dUTP 的水平，促进 dTTP 的合成和病毒的复制	Zhao 等人（2007）
H3 结合蛋白（SGIV ORF 158L）	促进病毒的复制，作为组蛋白 H3 的分子伴侣蛋白，行使着调控细胞基因表达和病毒复制的功能	Tran 等人（2011b）
病毒胰岛素样生长因子［IGF］（SGIV ORF 062R）	通过促进 G1/S 的转换，刺激细胞生长和病毒的复制，在非宿主细胞中过表达会导致细胞凋亡的增加	Yan 等人（2013）

续表

基因（病毒和 ORF）[a]	表型	参考文献
病毒轴突导向因子（SGIV ORF 155R）	调节宿主细胞的骨架结构和免疫反应，促进病毒的复制	Yan 等人（2014）
VP088（SGIV ORF 088）	作为一种病毒囊膜蛋白，在病毒进入细胞中起着重要作用	Zhou 等人（2011）
VP019（SGIV ORF 019）	病毒囊膜蛋白	Huang 等人（2013a）
VP18（SGIV ORF 018R）	一种推导的丝/苏氨酸激酶，在病毒粒子组装和晚期基因表达中起着关键性的作用	Wang 等人（2008a）
ORF 38R（SGIV）	具有一个 RGD 基序的病毒蛋白，可能在病毒进入细胞过程起作用	Wan 等人（2010b）
ORF 20R（TFV）	具有一个 RGD 基序的病毒蛋白，可能在病毒进入细胞过程起作用，与 SGIV ORF 38R 类似	Wang 等人（2008b）
ORF 97R（FV3）	ORF 97R 定位于内质网（ER），诱导 ER 的内陷（invagination）和外层核膜进入核中	Ring 等人（2013）
β-HSD（RGV ORF 52L）	在宿主类固醇合成中起着关键性的作用；在 EPC 中过表达会抑制 CPE	Sun 等人（2006）
ERV1/ALR（RGV ORF 88R）	一种巯基氧化酶，通过与 ASFV 同源分子类比研究，ERV1/ALR 被认为在病毒粒子组装中起着关键性的作用	Ke 等人（2009）
53R（RGV ORF 53R）	推导的十八烷基化膜蛋白，在病毒粒子形成中起着关键性的作用	Kim 等人（2010），Zhao 等人（2008）
50L（RGV ORF 50L）	与病毒粒子相连的蛋白，在病毒组装和病毒基因表达中起作用	Lei 等人（2012a）
2L（RGV ORF 2L）	推导的膜蛋白，在病毒复制中起重要作用	He 等人（2014）

a 病毒基因产物的鉴定既依据它们推导的功能（如 dUTPase），也依据参考它们 FV3 的同源物（如 ICP 46）。此外，发现有基因产物的病毒和特异性开放阅读框也作了标明。假如功能不清楚或者 FV3 的同源分子不存在，基因则通过其开放阅读框的名称进行简单的鉴定。

复制和存活必需蛋白 1（essential for replication and viability，ERV1）：酵母蛋白 ERV1 以及哺乳动物同源分子 ALR（肝再生增强因子，augmenter of liver regeneration）是巯基氧化酶，在蛋白折叠中起着关键性的作用（Thorpe 等人，2002）。与这一功能作用相一致的是，非洲猪瘟病毒（SFV）含有一个 ERV1/ALR 分子即 9GL，该分子发现于病毒组装场之中，并且在病毒粒子的成熟过程中起着关键性的作用（Lewis 等人，2000）。一个 9GL 缺失的突变体显示，病毒产量以 2 为底的对数级下降，所形成的病毒粒子当中，有 90%~99% 包含无核心的核结构（acentric nucleoid structures）。与非洲猪瘟病毒一样，RGV ERV1 同源分子（88R）是一种晚期蛋白（Ke 等人，2009），该同源分子包含高度保守的 ERV1 基序（ERV1 motif）Cys-X-X-Cys，并且在核和胞浆中都能检测到。然而，不像 ASFV ERV1/ALR KO 突变体，通过 RNA 干扰 88R 表达的基因敲低（knock down of 88R expression by RNAi）不能导致病毒产量的下降。从而提示，ERV1 在病毒复制中起着其他作用，敲低不完全，或者是宿主细胞保留了足够的巯基氧化酶，能够补偿病毒酶的缺失。

蛙虹彩病毒 50L（RGV 50L）：蛙虹彩病毒 50L 编码一个长度为 499 个氨基酸的蛋白，预测分子量为 55.5 kDa。其同源分子的全长为 499（STIV）—541（EHNV）个氨基酸，该分子存在于各种不同的蛙病毒之中，而且还在 FV3 中发现了一个更短的、只包含蛋白质 C 端一半长度的、含有 249 个氨基酸的同 129
源分子。RGV 50L 已被克隆，并在大肠杆菌（*E. coli*）中得到表达，其表达产物已用于小鼠抗 50L 抗体的制备。此外，RGV 50L 已被克隆到真核表达载体 pcDNA3.1 之中，从而产生 50L-pcDNA3.1（Lei 等人，2012a）。免疫荧光染色检测显示，RGV 50L 主要存在于胞浆内、病毒组装场以及核中。此外，RGV 50L 在核内的存在依赖于中心区域蛋白的一个核定位信号。用转染的 50L-pcDNA3.1重组载体感染细胞，显示存在更高水平的 53R mRNA，从而提示，RGV 50L 的表达可能影响 RGV 基因的表达。

脂多糖诱导的肿瘤坏死因子（LITAF）：SGIV 和 FV3 共同编码的蛋白质，与名叫 LITAF 即脂多糖诱导的肿瘤坏死因子 α（LPS-induced TNFα factor）的细胞蛋白同源。LPS 是单核细胞和巨噬细胞的一个有效刺激因子，能触发 TNFα 和前炎症细胞因子（proinflammatory cytokines）的分泌。LITAF 被鉴定为新型转录因子，调控 TNFα 的表达，并在调控炎症细胞因子中起着很重要的作用（Tang 等人，2006）。SGIV 中的 ORF136 编码一个早期病毒基因产物，并与 LITAF 同源。用表达 ORF136 的载体转染石斑鱼细胞，结果发现，该分子与线粒体有着明显的联系。SGIV LITAF 离体的过表达诱导了细胞凋亡，其表现是细胞凋亡小体增加、线粒体膜电位的去极化及半胱氨酸天冬氨酸蛋白酶 3（caspase-3）的激活，从而提示，SGIV LITAF 可能在 SGIV 诱导的细胞死亡中起着关键性的作用（Huang 等人，2008）。与 SGIV ORF 136 相类似，FV3 ORF

75L 编码的一个蛋白与细胞 LITAF C 端所发现的一个保守结构域有较高的序列相似度（Eaton 等人，2013）。在共转染或病毒感染以后，结果发现细胞 LITAF 和 FV3 75L，在乳绿猴细胞（baby green monkey cells，BGMK）和两栖类细胞 A6 中，都共定位在晚胞内体/溶酶体（late endosomes/lysosomes）中。有趣的是，SGIV ORF136 复查确定，SGIV ORF136 定位在 BGMK 细胞的线粒体中，并且显示，如同 FV3 75L 一样，SGIV ORF136 定位到两栖类细胞 A6 的胞内体/溶酶体（endosomes/lysosomes）中。迄今为止，病毒 LITAF 同源分子的确切功能和暂时性的类别仍未确定。有人提出，病毒 LITAF 的作用是以一种显性位点失活突变的方式来阻断细胞 LITAF 的功能（Eaton 等人，2013）。假如细胞 LITAF 确实具有抗病毒的功能，那么 RNA 干扰或 asMO 基因敲低后的细胞 LITAF 水平，应该离体产生更高滴度的 FV3，从而提示其病毒同源物的功能作用。另外，用 LITAF 基因敲除后的突变体进行在体感染，结果导致了弱毒感染和较低的病毒滴度。

含有 RGD 基序的蛋白（RGD motif-containing proteins）： 含有 RGD 基序的蛋白质存在于虹彩病毒科的所有属中，同时还包括 SGIV VP38（Wan 等人，2010）、虎纹蛙病毒 ORF 20R（Wang 等人，2008b）以及大黄鱼虹彩病毒（Yellow croaker iridovirus，YCIV，属于细胞肿大病毒属）037L（Ao 和 Chen，2006）。尽管在大小上存在着显著不同，但是这三个基因都包含一个 RGD 基序。有学者报道，该基序在病毒附着和进入细胞过程中起着很重要的作用。在这三个研究中，运用重组蛋白产生特异性的抗血清，来鉴定病毒组装场的蛋白质和病毒囊膜。总体来说，这些研究表明，含有 RGD 的蛋白质见于病毒组装
130 场以及病毒囊膜中，并有可能在随后病毒感染进入细胞中起着重要作用。然而，有研究暗示，含有 RGD 基序的蛋白质还有其他作用，一个 YCIV 编码的硫氧化还原蛋白 037L 融合蛋白（YCIV-encoded thioredoxin-037L fusion protein）在转染到 BF-2 细胞单层之后，会引起细胞变圆、脱落以及集聚。下面将简要描述用重组蛋白已经探明功能的其他蛙病毒 ORF，并在表 2 中进行了总结。

其他重组 SGIV 和 ISKNV 编码的蛋白质（additional recombinant SGIV- and ISKNV-encoded proteins）： 到目前为止，分别构建了 SGIV 即刻早期（IE）基因 ORF 162L 和 ORF 086C 及编码 FV3 ICP46 和 ICP18 同源分子的表达载体。两种基因产物的分布在胞浆中占优势，并且它们的过表达能促进石斑鱼胚胎细胞的生长，同时促进 SGIV 病毒的复制。SGIV ORF 096（VP96）编码一种 TNFR 的同源分子，该分子包含两个胞外富含半胱氨酸的结构域（cysteine-rich domains，CRDs），但缺乏 C 端跨膜结构域。ORF 096 离体的过表达能够增强细胞增殖，提高细胞的存活，从而提示，SGIV 可能利用一个 TNFR 同源分子来

调控病毒复制时细胞的凋亡反应（Huang 等人，2013b）。如同 RGV 一样，SGIV 编码一种 dUTPase 同源分子（ORF 049R），在其 C 端具有富含亮氨酸的核输出信号（leucine-rich nuclear export signal，NES）；SGIV dUTPase 是一个胞浆蛋白质，其 NES 对于 dUTPase 从核中转移到胞浆中至关重要（Gong 等人，2010）。SGIV ORF 158L 见于核和病毒组装场，在石斑鱼胚胎细胞中将其进行基因敲低，可以导致病毒产量的显著减少。进一步分析揭示，ORF 158L 可能充当一个组蛋白的分子伴侣（chaperon），能够调控宿主细胞基因的表达，并促进病毒的复制（Tran 等人，2011）。目前，还没有发现 ORF 158L 普遍存在于蛙病毒之中，但是其同源分子显示与 SGIV 产物有 33% 的相似性，并见于 EHNV、ADRV 和 CMTV 之中。SGIV ORF 062R 编码一个新型胰岛素样生长因子，该因子能促进石斑鱼 GP 细胞的生长，同时还通过促进 G1/S 期的转换而增强病毒的复制。此外，在 SGIV 感染的非宿主 FHM 细胞中，ORF 062R 的过表达，会引起细胞凋亡略有增加（Yan 等人，2013）。SGIV ORF 155R 编码一个轴突导向因子同源分子，该分子能离体促进病毒的复制，并能弱化细胞免疫反应。有研究显示，ORF 155R 的异位表达能够改变鱼类的细胞骨架结构，这一改变的特征表现为形成核周围的一个微管周向环（circumferential ring）和一个断裂的微丝组织（Yan 等人，2014）。一个丰度比较高的蛋白 ORF 018R 已被鉴定，该蛋白在丝氨酸/苏氨酸的磷酸化（serine/threonine phosphorylation）和病毒粒子的组装中起着关键性的作用（Wang 等人，2008a）。两个晚期基因即 SGIV ORF 088 和 ORF 019 编码病毒的囊膜蛋白（viral envelope proteins）。此外，rVP88 显示出能与一个 94 kDa 的宿主细胞膜蛋白结合，从而提示，rVP88 可能充当一个附着蛋白，在病毒进入细胞时起作用（Huang 等人，2013a；Zhou 等人，2011）。最后要提的是，类似的方法已应用到 ISKNV（属于细胞肿大病毒属），用来鉴定病毒 TRAF 蛋白（He 等人，2012a）、介导模拟基底膜形成并为淋巴内皮细胞提供附着位点的蛋白（Xu 等人，2010）、能抑制 TNFα 诱导的 NF-κB 信号传递的病毒锚蛋白重复序列蛋白（Guo 等人，2011a）以及病毒编码的血管内皮生长因子（Wang 等人，2008c）。总体来说，重组病 131
毒蛋白的产生和表达为确定细胞定位以及病毒编码蛋白质的功能，提供了一个强有力的方法学。

8　最后的几个见解

学者们运用现代分子技术，正逐步阐明蛙病毒离体复制以及在体引起疾病的特定步骤。本研究工作得益于对疱疹病毒、痘病毒以及非洲猪瘟病毒的开拓性研究。这些研究使人们对病毒复制和疾病进展过程中起重要作用的病毒基因有了更深入的认识。比较乐观的是，导致病毒复制增强以及逃避宿主免疫反应等蛙病毒基因的鉴定，有助于我们构建疫苗，从而有效地保护濒危两栖类、鱼类和爬行类。鉴于世界各地的两栖类种群数量正在下降，以及越来越依赖于海水养殖和淡水养殖作为人类消费的一种蛋白质来源，因而，在这一领域的成功显得至关重要。此外，搞清楚病毒基因是如何相互作用和如何调控免疫反应的机制，将大大拓展我们对低等脊椎动物免疫和免疫系统的起源的认识。

致谢

我们感谢理查德·康迪特（Richard Condit，佛罗里达州立大学）和雅克·罗伯特（Jacques Robert，罗切斯特大学医学中心）对本稿的仔细评阅。本项工作得到了国家科学基金（基金号：IOS 07-42711）和密西西比大学医学中心（VGC）、国家重大基础研究项目（项目号：2010CB126303）、中国国家自然科学基金（基金号：31270213）以及国家淡水生态与生物技术重点实验室项目（项目号：2011FBZ12）（QZ）、中国国家基础研究项目（973项目，项目号：2012CB114402）、中国国家自然科学基金（基金号：31330082）（QQ）以及国家卫生研究院（NIH）资助项目1R15-AI-101889-01（JKJ）的部分资助。

开放存取出版物的发行得到了田纳西大学（林业、野生动物与渔业系，研究与交流办公室，农业研究所）、华盛顿州立大学图书馆、戈登州立大学（学术事务办公室）、两栖爬行动物兽医协会以及两栖爬行动物保护协会的资金赞助。

开放存取

参考文献

Ao J, Chen X (2006). Identification and characterization of a novel gene encoding an RGD-containing protein in large yellow croaker iridovirus. Virology 355:213-222.

Babu SG, Pandeya A, Verma N, Shukla N, Kumar RV, Saxena S (2014). Role of HCMV miR-UL70-3p and miR-UL148D in overcoming the cellular apoptosis. Mol Cell Biochem 393:89-98.

Beattie E, Tartaglia J, Paoletti E (1991). Vaccinia virus-encoded eIF-2 alpha homolog abrogates the antiviral effect of interferon. Virology 183:419-422.

Braunwald J, Tripier F, Kirn A (1979). Comparison of the properties of enveloped and naked frog virus 3 (FV3) particles. J Gen Virol 45:673-682.

Braunwald J, Nonnenmacher H, Tripier-Darcy F (1985). Ultrastructural and biochemical study of frog virus 3 uptake by BHK-21 cells. J Gen Virol 66(pt 2): 283-293.

Brodie R, Roper RL, Upton C (2004). JDotter: a JAVA interface to multiple display plots generated by dotter. Bioinformatics 20:279-281.

Brunner JL, Storfer A, Gray MJ, Hoverman JT (2015). Ranavirus ecology and evolution: from epidemiology to extinction. In: Gray MJ, Chinchar VG (eds) Ranaviruses: lethal pathogens of ectothermic vertebrates. Springer, New York.

Caipang CM, Hirono I, Aoki T (2006a). Immunogenicity, retention and protective effects of the protein derivatives of formalin-inactivated red seabream iridovirus (RSIV) vaccine in red seabream, *Pagrus major*. Fish Shellfish Immunol 20: 597-609.

Caipang CM, Takano T, Hirono I, Aoki T (2006b). Genetic vaccines protect red seabream, *Pagrus major*, upon challenge with red seabream iridovirus (RSIV). Fish Shellfish Immunol 21:130-138.

Cardone G, Winkler DC, Trus BL, Cheng N, Heuser JE, Newcomb WW, Brown JC, Steven AC (2007). Visualization of the herpes simplex virus portal in situ by cryo-electron tomography. Virology 361:426-434.

Chabaud S, Sasseville AM, Elahi SM, Caron A, Dufour F, Massie B, Langelier Y (2007). The ribonucleotide reductase domain of the R1 subunit of herpes simplex virus type 2 ribonucleotide reductase is essential for R1 antiapoptotic function. J Gen Virol 88: 384-394.

Chang JT, Schmid MF, Rixon FJ, Chiu W (2007). Electron cryotomography reveals the portal in the herpesvirus capsid. J Virol 81:2065-2068.

Chen LM, Wang F, Song W, Hew CL (2006). Temporal and differential gene expression of Singapore grouper iridovirus. J Gen Virol 87:2907-2915.

Chen G, Ward BM, Yu KH, Chinchar VG, Robert J (2011). Improved knockout methodology reveals that *Frog virus* 3 mutants lacking either the *18K* immediate-early gene or the truncated *vIF-2 alpha* gene are defective for replication *in vivo*. J Virol 85:11131-11138.

Chen Z, Gui J, Gao X, Pei C, Hong Y, Zhang Q (2013). Genome architecture changes and major gene variations of Andrias davidianus ranavirus (ADRV). Vet Res 44:101.

Cheng K, Escalon BL, Robert J, Chinchar VG, Garcia-Reyero N (2014). Differential transcription of fathead minnow immune-related genes following infection with frog virus 3, an emerging pathogen of ectothermic vertebrates. Virology 456-457: 77-86.

Chinchar VG, Granoff A (1984). Isolation and characterization of a frog virus 3 variant resistant to phosphonoacetate: genetic evidence for a virus-specific DNA polymerase. Virology 138:357-361.

Chinchar VG, Granoff A (1986). Temperature-sensitive mutants of frog virus 3: biochemical and genetic characterization. J Virol 58:192-202.

Chinchar VG, Yu W (1990a). Frog virus 3-mediated translational shut-off: frog virus 3 messages are translationally more efficient than host and heterologous viral messages under conditions of increased translational stress. Virus Res 16: 163-174.

Chinchar VG, Yu W (1990b). Translational efficiency: iridovirus early mRNAs outcompete tobacco mosaic virus message in vitro. Biochem Biophys Res Commun 172:1357-1363.

Chinchar VG, Goorha R, Granoff A (1984). Early proteins are required for the formation of frog virus 3 assembly sites. Virology 135:148-156.

Chinchar VG, Bryan L, Wang J, Long S, Chinchar GD (2003). Induction of apoptosis in frog virus 3-infected cells. Virology 306:303-312.

Chinchar VG, Hyatt A, Miyazaki T, Williams T (2009). Family Iridoviridae: poor viral relations no longer. Curr Top Microbiol Immunol 328:123-170.

Chinchar VG, Yu KH, Jancovich JK (2011). The molecular biology of frog virus 3 and other iridoviruses infecting cold-blooded vertebrates. Viruses 3:1959-1985.

Colson P, de Lamballerie X, Fournous G, Raoult D (2012). Reclassification of giant viruses composing a fourth domain of life in the new order Megavirales. Intervirology 55:321-332.

Conner J, Cross A, Murray J, Marsden H (1994a). Identification of structural domains within the large subunit of herpes simplex virus ribonucleotide reductase. J Gen Virol 75(pt 12):3327-3335.

Conner J, Marsden H, Clements BH (1994b). Ribonucleotide reductase of herpesviruses.Rev Med Virol 4:25-34.

Cordier O,Aubertin AM,Lopez C,Tondre L (1981). Inhibitiion de la traduction par le FV3:action des proteines virales de structure solubilisees sur la synthese proteique in vivo et in vitro.Ann Virol (Inst Pasteur) 132 E:25-39.

Daikoku T,Yamamoto N,Maeno K,Nishiyama Y (1991). Role of viral ribonucleotide reductase in the increase of dTTP pool size in herpes simplex virus-infected Vero cells.J Virol 72:1441-1444.

Dang LT,Kondo H,Hirono I,Aoki T (2008). Inhibition of red seabream iridovirus (RSIV) replication by small interfering RNA (siRNA) in a cell culture system. Antiviral Res 77:142-149.

Darcy-Tripier F,Nermut MV,Braunwald J,Williams LD (1984). The organization of frog virus 3 as revealed by freeze-etching.Virology 138:287-299.

Davison AJ, Stow ND (2005). New genes from old: redeployment of dUTPase by herpesviruses.J Virol 79:12880-12892.

de Wind N, Berns A, Gielkens A, Kimman T (1993). Ribonucleotide reductase-deficient mutants of pseudorabies virus are avirulent for pigs and induce partial protective immunity.J Gen Virol 74(pt 3):351-359.

Devauchelle G, Stoltz DB, Darcy-Tripier F (1985). Comparative ultrastructure of iridoviridae.Curr Top Microbiol Immunol 116:1-21.

Duffus ALJ,Waltzek TB,Stöhr AC,Allender MC,Gotesman M,Whittington RJ,Hick P,Hines MK,Marschang RE (2015). Distribution and host range of ranaviruses. In:Gray MJ,Chinchar VG (eds) Ranaviruses:lethal pathogens of ectothermic vertebrates.Springer,New York.

Eaton HE, Metcalf J, Penny E, Tcherepanov V, Upton C, Brunetti CR (2007). Comparative genomic analysis of the family Iridoviridae: re-annotating and defining the core set of iridovirus genes.Virol J 4:11.

Eaton HE,Ring BA,Brunetti CR (2010). The genomic diversity and phylogenetic relationship in the family Iridoviridae.Viruses 2:1458-1475.

Eaton HE, Ferreira Lacerda A, Desrochers G, Metcalf J, Angers A, Brunetti CR (2013). Cellular LITAF interacts with frog virus 3 75L protein and alters its subcellular localization.J Virol 87:716-723.

Eberhardt MK, Deshpande A, Chang WL, Barthold SW, Walter MR, Barry PA (2013). Vaccination against a virus-encoded cytokine significantly restricts viral challenge.J Virol 87:11323-11331.

Essani K, Goorha R, Granoff A (1987). Mutation in a DNA-binding protein reveals an association between DNA-methyltransferase activity and a 26,000-Da polypeptide in frog virus 3-infected cells. Virology 161:211-217.

Feighny RJ, Henry BE II, Pagano JS (1981). Epstein-Barr virus-induced deoxynuclease and the reutilization of host-cell DNA degradation products in viral DNA replication. Virology 115:395-400.

Fu JP, Chen SN, Zou PF, Huang B, Guo Z, Zeng LB, Qin QW, Nie P (2014). IFN-gamma in turtle: conservation in sequence and signalling and role in inhibiting iridovirus replication in Chinese soft-shelled turtle Pelodiscus sinensis. Dev Comp Immunol 43:87-95.

Gammon DB, Gowrsihankar B, Duraffour S, Andrei G, Upton C, Evans DH (2010). Vaccinia virus-encoded ribonucleotide reductase subunits are differentially required for replication and pathogenesis. PLoS Pathog 6:e1000984.

Gantress J, Maniero GD, Cohen N, Robert J (2003). Development and characterization of a model system to study amphibian immune responses to iridoviruses. Virology 311:254-262.

Garcia-Escudero R, Andres G, Almazan F, Vinuela E (1998). Inducible gene expression from African swine fever virus recombinants: analysis of the major capsid protein p72. J Virol 72:3185-3195.

Gendrault JL, Steffan AM, Bingen A, Kirn A (1981). Penetration and uncoating of frog virus 3 (FV3) in cultured rat Kupffer cells. Virology 112:375-384.

Glaser R, Litsky ML, Padgett DA, Baiocchi RA, Yang EV, Chen M, Yeh PE, Green-Church KB, Caligiuri MA, Williams MV (2006). EBV-encoded dUTPase induces immune dysregulation: implications for the pathophysiology of EBV-associated disease. Virology 346:205-218.

Gong J, Huang YH, Huang XH, Zhang R, Qin QW (2010). Nuclear-export-signal-dependent protein translocation of dUTPase encoded by Singapore grouper iridovirus. Arch Virol 155:1069-1076.

Goorha R (1981). Frog virus 3 requires RNA polymerase II for its replication. J Virol 37:496-499.

Goorha R (1982). Frog virus 3 DNA replication occurs in two stages. J Virol 43:519-528.

Goorha R, Dixit P (1984). A temperature-sensitive (TS) mutant of frog virus 3 (FV3) is defective in second-stage DNA replication. Virology 136:186-195.

Goorha R, Granoff A (1979). Icosahedral cytoplasmic deoxyriboviruses. In: Fraenkel-Conrat H, Wagner RR (eds) Comprehensive virology. Plenum Press, New York, pp 347-399.

Goorha R, Murti KG (1982). The genome of frog virus 3, an animal DNA virus, is circularly permuted and terminally redundant. Proc Natl Acad Sci U S A 79: 248-252.

Goorha R, Willis DB, Granoff A, Naegele RF (1981). Characterization of a temperature-sensitive mutant of frog virus 3 defective in DNA replication. Virology 112: 40-48.

Goorha R, Granoff A, Willis DB, Murti KG (1984). The role of DNA methylation in virus replication: inhibition of frog virus 3 replication by 5-azacytidine. Virology 138: 94-102.

Grayfer L, De Jesús Andino F, Robert J (2014). The amphibian (*Xenopus laevis*) type I interferon response to Frog Virus 3: new insight into ranavirus pathogenicity. J Virol 88: 5766-5777.

Grayfer L, Edholm E-S, De Jesús Andino F, Chinchar VG, Robert J (2015). Ranavirus host immunity and immune evasion. In: Gray MJ, Chinchar VG (eds) Rana-viruses: lethal pathogens of ectothermic vertebrates. Springer, New York.

Guo CJ, Chen WJ, Yuan LQ, Yang LS, Weng SP, Yu XQ, He JG (2011a). The viral ankyrin repeat protein (ORF124L) from infectious spleen and kidney necrosis virus attenuates nuclear factor-{kappa} B activation and interacts with I{kappa} B kinase{beta}. J Gen Virol 92: 1561-1570.

Guo CJ, Liu D, Wu YY, Yang XB, Yang LS, Mi S, Huang YX, Luo YW, Jia KT, Liu ZY, Chen WJ, Weng SP, Yu XQ, He JG (2011b). Entry of tiger frog virus (an Iridovirus) into HepG2 cells via a pH-dependent, atypical, caveola-mediated endocytosis pathway. J Virol 85: 6416-6426.

Guo CJ, Wu YY, Yang LS, Yang XB, He J, Mi S, Jia KT, Weng SP, Yu XQ, He JG (2012). Infectious spleen and kidney necrosis virus (a fish iridovirus) enters Mandarin fish fry cells via caveola-dependent endocytosis. J Virol 86: 2621-2631.

Guo C, Yan Y, Cui H, Huang X, Qin Q (2013). miR-homoHSV of Singapore grouper iridovirus (SGIV) inhibits expression of the SGIV pro-apoptotic factor LITAF and attenuates cell death. PLoS One 8: e83027.

Hannon GJ (2002). RNA interference. Nature 418: 244-251.

He BL, Yuan JM, Yang LY, Xie JF, Weng SP, Yu XQ, He JG (2012a). The viral TRAF protein (ORF111L) from infectious spleen and kidney necrosis virus interacts with TRADD and induces caspase 8-mediated apoptosis. PLoS One 7:e37001.

He LB,Ke F,Zhang QY (2012b). Rana grylio virus as a vector for foreign gene expression in fish cells. Virus Res 163:66-73.

He LB,Gao XC,Ke F,Zhang QY (2013). A conditional lethal mutation in Rana grylio virus ORF 53R resulted in a marked reduction in virion formation. Virus Res 177:194-200.

He LB,Ke F,Wang J,Gao XC,Zhang QY (2014). Rana grylio virus (RGV) envelope protein 2L:subcellular localization and essential roles in virus infectivity revealed by conditional lethal mutant. J Gen Virol 95:679-690.

Hoelzer K,Shackelton LA,Parrish CR (2008). Presence and role of cytosine methylation in DNA viruses of animals. Nucleic Acids Res 36:2825-2837.

Hook LM,Grey F,Grabski R,Tirabassi R,Doyle T,Hancock M,Landais I,Jeng S, McWeeney S, Britt W, Nelson JA (2014). Cytomegalovirus miRNAs target secretory pathway genes to facilitate formation of the virion assembly compartment and reduce cytokine secretion. Cell Host Microbe 15:363-373.

Huang X,Huang Y,Gong J,Yan Y,Qin Q (2008). Identification and characterization of a putative lipopolysaccharide-induced TNF-alpha factor (LITAF) homolog from Singapore grouper iridovirus. Biochem Biophys Res Commun 373:140-145.

Huang Y,Huang X,Liu H,Gong J,Ouyang Z,Cui H,Cao J,Zhao Y,Wang X,Jiang Y,Qin Q (2009). Complete sequence determination of a novel reptile iridovirus isolated from soft-shelled turtle and evolutionary analysis of Iridoviridae. BMC Genomics 10:224.

Huang X, Huang Y, Ouyang Z, Cai J, Yan Y, Qin Q (2011a). Roles of stress-activated protein kinases in the replication of Singapore grouper iridovirus and regulation of the inflammatory responses in grouper cells. J Gen Virol 92:1292-1301.

Huang X,Huang Y,Ouyang Z,Xu L,Yan Y,Cui H,Han X,Qin Q (2011b). Singapore grouper iridovirus,a large DNA virus,induces nonapoptotic cell death by a cell type dependent fashion and evokes ERK signaling. Apoptosis 16:831-845.

Huang Y,Huang X,Cai J,Ye F,Guan L,Liu H,Qin Q (2011c). Construction of green fluorescent protein-tagged recombinant iridovirus to assess viral replication. Virus Res 160:221-229.

Huang X, Gong J, Huang Y, Ouyang Z, Wang S, Chen X, Qin Q (2013a).

Characterization of an envelope gene VP19 from Singapore grouper iridovirus. Virol J 10:354.

Huang X, Huang Y, Cai J, Wei S, Gao R, Qin Q (2013b). Identification and characterization of a tumor necrosis factor receptor like protein encoded by Singapore grouper iridovirus. Virus Res 178:340-348.

Hudziak RM, Summerton J, Weller DD, Iversen PL (2000). Antiproliferative effects of steric blocking phosphorodiamidate morpholino antisense agents directed against c-myc. Antisense Nucleic Acid Drug Dev 10:163-176.

Jancovich JK, Bremont M, Touchman JW, Jacobs BL (2010). Evidence for multiple recent host species shifts among the ranaviruses (Family Iridoviridae). J Virol 84:2636-2647.

Jancovich JK, Jacobs BL (2011). Innate immune evasion mediated by the Ambystoma tigrinum virus eukaryotic translation initiation factor 2 alpha homologue. J Virol 85:5061-5069.

Jancovich JK, Mao J, Chinchar VG, Wyatt C, Case ST, Kumar S, Valente G, Subramanian S, Davidson EW, Collins JP, Jacobs BL (2003). Genomic sequence of a ranavirus (family Iridoviridae) associated with salamander mortalities in North America. Virology 316:90-103.

Jancovich JK, Chinchar VG, Hyatt A, Miyazaki T, Williams T, Zhang QY (2012). Family Iridoviridae. In: King AMQ, Adams MJ, Carstens EB, Lefkowitz EJ (eds) Virus taxonomy: classification and nomenclature of viruses. Ninth report of the International Committee on Taxonomy of Viruses. Elsevier, Amsterdam, pp 193-210.

Jancovich JK, Steckler N, Waltzek TB (2015). Ranavirus taxonomy and phylogeny. In: Gray MJ, Chinchar VG (eds) Ranaviruses: lethal pathogens of ectothermic vertebrates. Springer, New York.

Jia KT, Wu YY, Liu ZY, Mi S, Zheng YW, He J, Weng SP, Li SC, He JG, Guo CJ (2013). Mandarin fish caveolin 1 interaction with major capsid protein of infectious spleen and kidney necrosis virus and its role in early stages of infection. J Virol 87:3027-3038.

Johnston JB, McFadden G (2003). Poxvirus immunomodulatory strategies: current perspectives. J Virol 77:6093-6100.

Johnston JB, McFadden G (2004). Technical knockout: understanding poxvirus pathogenesis by selectively deleting viral immunomodulatory genes. Cell Microbiol

6:695-705.

Johnston JB, Barrett JW, Nazarian SH, Goodwin M, Ricciuto D, Wang G, McFadden G (2005). A poxvirus-encoded pyrin domain protein interacts with ASC-1 to inhibit host inflammatory and apoptotic responses to infection. Immunity 23: 587-598.

Kato A, Hirohata Y, Arii J, Kawaguchi Y (2014). Phosphorylation of herpes simplex virus 1 dUTPase up-regulated viral dUTPase activity to compensate for low cellular dUTPase activity for efficient viral replication. J Virol 88:7776-7785.

Kaur K, Rohozinski J, Goorha R (1995). Identification and characterization of the frog virus 3 DNA methyltransferase gene. J Gen Virol 76(pt 8):1937-1943.

Ke F, Zhao L, Zhang QY (2009). Cloning, expression and subcellular distribution of a *Rana grylio* virus late gene encoding ERV1 homologue. Mol Biol Report 36: 1651-1659.

Kim YS, Ke F, Lei XY, Zhu R, Zhang QY (2010). Viral envelope protein 53R gene highly specific silencing and iridovirus resistance in fish cells by AmiRNA. PLoS One 5:e10308.

Koonin EV, Yutin N (2010). Origin and evolution of eukaryotic large nucleo-cytoplasmic DNA viruses. Intervirology 53:284-292.

Krug A, Towarowski A, Britsch S, Rothenfusser S, Hornung V, Bals R, Giese T, Engelmann H, Endres S, Krieg AM, Hartmann G (2001). Toll-like receptor expression reveals CpG DNA as a unique microbial stimulus for plasmacytoid dendritic cells which synergizes with CD40 ligand to induce high amounts of IL-12. Eur J Immunol 31:3026-3037.

Krug A, Luker GD, Barchet W, Leib DA, Akira S, Colonna M (2004). Herpes simplex virus type 1 activates murine natural interferon-producing cells through toll-like receptor 9. Blood 103:1433-1437.

Langelier Y, Bergeron S, Chabaud S, Lippens J, Guilbault C, Sasseville AM, Denis S, Mosser DD, Massie B (2002). The R1 subunit of herpes simplex virus ribonucleotide reductase protects cells against apoptosis at, or upstream of, caspase-8 activation. J Gen Virol 83:2779-2789.

Langland JO, Jacobs BL (2002). The role of the PKR-inhibitory genes, E3L and K3L, in determining vaccinia virus host range. Virology 299:133-141.

Langland JO, Cameron JM, Heck MC, Jancovich JK, Jacobs BL (2006). Inhibition of PKR by RNA and DNA viruses. Virus Res 119:100-110.

Lei XY, Ou T, Zhang QY (2012a). Rana grylio virus (RGV) 50L is associated with viral matrix and exhibited two distribution patterns. PLoS One 7: e43033.

Lei XY, Ou T, Zhu RL, Zhang QY (2012b). Sequencing and analysis of the complete genome of Rana grylio virus (RGV). Arch Virol 157: 1559-1564.

Lembo D, Brune W (2009). Tinkering with a viral ribonucleotide reductase. Trends Biochem Sci 34: 25-32.

Lewis T, Zsak L, Burrage TG, Lu Z, Kutish GF, Neilan JG, Rock DL (2000). An African swine fever virus ERV1-ALR homologue, 9GL, affects virion maturation and viral growth in macrophages and viral virulence in swine. J Virol 74: 1275-1285.

Lin PW, Huang YJ, John JA, Chang YN, Yuan CH, Chen WY, Yeh CH, Shen ST, Lin FP, Tsui WH, Chang CY (2008). Iridovirus Bcl-2 protein inhibits apoptosis in the early stage of viral infection. Apoptosis 13: 165-176.

Ma J, Zeng L, Zhou Y, Fiang N, Zhang H, Fan Y, Meng Y, Xu J (2014). Ultrastructural morphogenesis of an amphibian iridovirus isolated from Chinese giant salamander (*Andrias davidianus*). J Comp Pathol 150: 325-331.

Majji S, Thodima V, Sample R, Whitley D, Deng Y, Mao J, Chinchar VG (2009). Transcriptome analysis of Frog virus 3, the type species of the genus Ranavirus, family Iridoviridae. Virology 391: 293-303.

Maniero GD, Morales H, Gantress J, Robert J (2006). Generation of a long-lasting, protective, and neutralizing antibody response to the ranavirus FV3 by the frog *Xenopus*. Dev Comp Immunol 30: 649-657.

Mavian C, Lopez-Bueno A, Balseiro A, Casais R, Alcami A, Alejo A (2012). The genome sequence of the emerging common midwife toad virus identifies an evolutionary intermediate within ranaviruses. J Virol 86: 3617-3625.

Moody NJG, Owens L (1994). Experimental demonstration of pathogenicity of a frog virus, *Bohle iridovirus*, for a fish species, barramundi *Lates Calcarifer*. Dis Aquat Organ 18: 95-102.

Morales HD, Robert J (2007). Characterization of primary and memory CD8 T-cell responses against ranavirus (FV3) in *Xenopus laevis*. J Virol 81: 2240-2248.

Morales HD, Abramowitz L, Gertz J, Sowa J, Vogel A, Robert J (2010). Innate immune responses and permissiveness to ranavirus infection of peritoneal leukocytes in the frog *Xenopus laevis*. J Virol 84: 4912-4922.

Morrison EA, Garner S, Echaubard P, Lesbarreres D, Kyle CJ, Brunetti CR (2014).

Complete genome analysis of a frog virus 3 (FV3) isolate and sequence comparison with isolates of differing levels of virulence. Virol J 11:46.

Murti KG, Goorha R, Chen M (1985). Interaction of frog virus 3 with the cytoskeleton. Curr Top Microbiol Immunol 116:107-131.

Murti KG, Goorha R, Klymkowsky MW (1988). A functional role for intermediate filaments in the formation of frog virus 3 assembly sites. Virology 162:264-269.

Naegele RF, Granoff A (1971). Viruses and renal cancinoma of *Rana pipiens*. XI. Isolation of Frog virus 3 temperature-sensitive mutants; complementation and genetic recombination. Virology 44:286-295.

Nichols RJ, Stanitsa E, Unger B, Traktman P (2008). The vaccinia virus gene I2L encodes a membrane protein with an essential role in virion entry. J Virol 82: 10247-10261.

Oliveros M, Garcia-Escudero R, Alejo A, Vinuela E, Salas ML, Salas J (1999). African swine fever virus dUTPase is a highly specific enzyme required for efficient replication in swine macrophages. J Virol 73:8934-8943.

Ouyang Z, Wang P, Huang X, Cai J, Huang Y, Wei S, Ji H, Wei J, Zhou Y, Qin Q (2012). Immunogenicity and protective effects of inactivated Singapore grouper iridovirus (SGIV) vaccines in orange-spotted grouper, *Epinephelus coioides*. Dev Comp Immunol 38:254-261.

Pallister J, Goldie S, Coupar B, Shiell B, Michalski WP, Siddon N, Hyatt A (2007). *Bohle iridovirus* as a vector for heterologous gene expression. J Virol Methods 146:419-423.

Pavelin J, Reynolds N, Chiweshe S, Wu G, Tiribassi R, Grey F (2013). Systematic microRNA analysis identifies ATP6V0C as an essential host factor for human cytomegalovirus replication. PLoS Pathog 9:e1003820.

Pham PH, Lai YS, Lee FF, Bols NC, Chiou PP (2012). Differential viral propagation and induction of apoptosis by grouper iridovirus (GIV) in cell lines from three non-host species. Virus Res 167:16-25.

Purifoy D, Naegele RF, Granoff A (1973). Viruses and renal carcinoma of Rana pipiens. XIV. Temperature-sensitive mutants of frog virus 3 with defective encapsidation. Virology 54:525-535.

Raghow R, Granoff A (1979). Macromolecular synthesis in cells infected by frog virus 3. X. Inhibition of cellular protein synthesis by heat-inactivated virus. Virology 98:319-327.

Reading PC, Moore JB, Smith GL (2003). Steroid hormone synthesis by vaccinia virus suppresses the inflammatory response to infection. J Exp Med 197: 1269-1278.

Ring BA, Ferreira Lacerda A, Drummond DJ, Wangen C, Eaton HE, Brunetti CR (2013). Frog virus 3 open reading frame 97R localizes to the endoplasmic reticulum and induces nuclear invaginations.J Virol 87:9199-9207.

Robert J (2010). Emerging ranaviral infectious diseases and amphibian decline. Diversity 2:314-330.

Robert J, Morales H, Buck W, Cohen N, Marr S, Gantress J (2005). Adaptive immunity and histopathology in frog virus 3-infected *Xenopus*.Virology 332:667-675.

Rothenburg S, Chinchar VG, Dever TE (2011). Characterization of a ranavirus inhibitor of the antiviral protein kinase PKR.BMC Microbiol 11:56.

Rouiller I, Brookes SM, Hyatt AD, Windsor M, Wileman T (1998). African swine fever virus is wrapped by the endoplasmic reticulum.J Virol 72:2373-2387.

Sample R (2010). Elucidation of Frog Virus 3 gene function and pathways of virion formation. Ph. D. dissertation, University of Mississippi Medical Center, Jackson, MS.

Sample R, Bryan L, Long S, Majji S, Hoskins G, Sinning A, Olivier J, Chinchar VG (2007). Inhibition of iridovirus protein synthesis and virus replication by antisense morpholino oligonucleotides targeted to the major capsid protein, the 18 kDa immediate-early protein, and a viral homolog of RNA polymerase II.Virology 358:311-320.

Seet BT, Johnston JB, Brunetti CR, Barrett JW, Everett H, Cameron C, Sypula J, Nazarian SH, Lucas A, McFadden G (2003). Poxviruses and immune evasion. Annu Rev Immunol 21:377-423.

Song W, Lin Q, Joshi SB, Lim TK, Hew CL (2006). Proteomic studies of the Singapore grouper iridovirus.Mol Cell Proteomics 5:256-264.

Summerton JE (2007). Morpholino, siRNA, and S-DNA compared: impact of structure and mechanism of action on off-target effects and sequence specificity. Curr Top Med Chem 7:651-660.

Summerton J, Weller D (1997). Morpholino antisense oligomers: design, preparation, and properties.Antisense Nucleic Acid Drug Dev 7:187-195.

Sun W, Huang Y, Zhao Z, Gui J, Zhang Q (2006). Characterization of the Rana grylio virus 3beta-hydroxysteroid dehydrogenase and its novel role in suppressing virus-induced cytopathic effect.Biochem Biophys Res Commun 351:44-50.

Tan WG, Barkman TJ, Gregory Chinchar V, Essani K (2004). Comparative genomic analyses of frog virus 3, type species of the genus Ranavirus (family Iridoviridae). Virology 323:70-84.

Tang X, Metzger D, Leeman S, Amar S (2006). LPS-induced TNF-alpha factor (LITAF)-deficient mice express reduced LPS-induced cytokine: evidence for LITAF-dependent LPS signaling pathways. Proc Natl Acad Sci U S A 103: 13777-13782.

Teng Y, Hou Z, Gong J, Liu H, Xie X, Zhang L, Chen X, Qin QW (2008). Whole-genome transcriptional profiles of a novel marine fish iridovirus, Singapore grouper iridovirus (SGIV) in virus-infected grouper spleen cell cultures and in orange-spotted grouper, *Epinephulus coioides*. Virology 377:39-48.

Thorpe C, Hoober KL, Raje S, Glynn NM, Burnside J, Turi GK, Coppock DL (2002). Sulfhydryl oxidases: emerging catalysts of protein disulfide bond formation in eukaryotes. Arch Biochem Biophys 405:1-12.

Tidona CA, Darai G (1997). The complete DNA sequence of lymphocystis disease virus. Virology 230:207-216.

Tran BN, Chen L, Liu Y, Wu J, Velazquez-Campoy A, Sivaraman J, Hew CL (2011). Novel histone H3 binding protein ORF158L from the Singapore grouper iridovirus. J Virol 85:9159-9166.

Tulman ER, Delhon GA, Ku BK, Rock DL (2009). African swine fever virus. Curr Top Microbiol Immunol 328:43-87.

Tweedell K, Granoff A (1968). Viruses and renal carcinoma of Rana pipiens. V. Effect of frog virus 3 on developing frog embryos and larvae. J Natl Cancer Inst 40: 407-410.

Verrier ER, Langevin C, Benmansour A, Boudinot P (2011). Early antiviral response and virus-induced genes in fish. Dev Comp Immunol 35:1204-1214.

Wan QJ, Gong J, Huang XH, Huang YH, Zhou S, Ouyang ZL, Cao JH, Ye LL, Qin QW (2010). Identification and characterization of a novel capsid protein encoded by Singapore grouper iridovirus ORF038L. Arch Virol 155:351-359.

Wang F, Bi X, Chen LM, Hew C-L (2008a). ORF018R, a highly abundant virion protein from Singapore grouper iridovirus, is involved in serine/threonine phosphorylation and virion assembly. J Gen Virol 89:1169-1178.

Wang Q, Luo Y, Xie J, Dong C, Weng S, Ai H, Lu L, Yang X, Yu X, He J (2008b).

Identification of two novel membrane proteins from the Tiger frog virus (TFV). Virus Res 136:35-42.

Wang Z-L, Xu X-P, He B-L, Weng S-P, Xiao J, Wang L, Lin T, Liu X, Wang Q, Yu X-Q, He J-G (2008c). ISKNV ORF48R functions as a new viral vascular endothelial growth factor. J Virol 82:4371-4383.

Wang S, Huang X, Huang Y, Hao X, Xu H, Cai M, Wang H, Qin Q (2014). Entry of a novel marine DNA virus (Singapore grouper iridovirus, SGIV) into host cells occurs via clathrin-mediated endocytosis and macropinocytosis in a pH-dependent manner. J Virol 01744-14 88:13047-13063.

Whitley DJS (2011). Determinations of ranavirus gene function using an antisense morpholino-mediated approach. University of Mississippi Medical Center, Jackson, MS. Ph.D. dissertation.

Whitley DS, Yu K, Sample RC, Sinning A, Henegar J, Norcross E, Chinchar VG (2010). Frog virus 3 ORF 53R, a putative myristoylated membrane protein, is essential for virus replication in vitro. Virology 405:448-456.

Whitley DS, Sample RC, Sinning AR, Henegar J, Chinchar VG (2011). Antisense approaches for elucidating ranavirus gene function in an infected fish cell line. Dev Comp Immunol 35:937-948.

Williams T (1996). The iridoviruses. Adv Virus Res 46:345-412.

Williams T, Barbosa-Solomieu V, Chinchar VG (2005). A decade of advances in iridovirus research. Adv Virus Res 65:173-248.

Willis DB, Granoff A (1978). Macromolecular synthesis in cells infected by frog virus 3. IX. Two temporal classes of early viral RNA. Virology 86:443-453.

Willis DB, Granoff A (1980). Frog virus 3 DNA is heavily methylated at CpG sequences. Virology 107:250-257.

Willis DB, Granoff A (1985). Transactivation of an immediate-early frog virus 3 promoter by a virion protein. J Virol 56:495-501.

Willis DB, Thompson JP (1986). The Iridovirus frog virus 3: a model for transacting proteins. Microbiol Sci 3:59-63.

Willis DB, Goorha R, Granoff A (1984). DNA methyltransferase induced by frog virus 3. J Virol 49:86-91.

Willis DB, Goorha R, Chinchar VG (1985). Macromolecular synthesis in cells infected by frog virus 3. Curr Top Microbiol Immunol 116:77-106.

Willis DB, Essani K, Goorha R, Thompson JP, Granoff A (1990). Transcription of a methylated DNA virus, nucleic acid methylation. Alan R. Liss, Inc., New York, pp 139-151.

Wilson WH, Van Etten JL, Allen MJ (2009). The Phycodnaviridae: the story of how tiny giants rule the world. Curr Top Microbiol Immunol 328: 1-42.

Xia L, Cao J, Huang X, Qin Q (2009). Characterization of Singapore grouper iridovirus (SGIV) ORF 086R, a putative homolog of ICP18 involved in cell growth control and virus replication. Arch. Virol. 154: 1409-1416.

Xia L, Liang H, Huang Y, Ouyang Z, Qin Q (2010). Identification and characterization of Singapore grouper iridovirus (SGIV) ORF162L, an immediate-early gene involved in cell growth control and viral replication. Virus Res 147: 30-39.

Xie J, Lu L, Deng M, Weng S, Zhu J, Wu Y, Gan L, Chan SM, He J (2005). Inhibition of reporter gene and Iridovirus-tiger frog virus in fish cell by RNA interference. Virology 338: 43-52.

Xu X, Weng S, Lin T, Tang J, Huang L, Wang J, Yu X, Lu L, Huang Z, He J (2010). VP23R of infectious spleen and kidney necrosis virus mediates formation of virus-mock basement membrane to provide attaching sites for lymphatic endothelial cells. J Virol 84: 11866-11875.

Yan X, Yu Z, Zhang P, Battisti AJ, Holdaway HA, Chipman PR, Bajaj C, Bergoin M, Rossmann MG, Baker TS (2009). The capsid proteins of a large, icosahedral dsDNA virus. J Mol Biol 385: 1287-1299.

Yan Y, Cui H, Jiang S, Huang Y, Huang X, Wei S, Xu W, Qin Q (2011). Identification of a novel marine fish virus, Singapore grouper iridovirus-encoded microRNAs expressed in grouper cells by Solexa sequencing. PLoS One 6: e19148.

Yan Y, Cui H, Guo C, Li J, Huang X, Wei J, Qin Q (2013). An insulin-like growth factor homologue of Singapore grouper iridovirus modulates cell proliferation, apoptosis and enhances viral replication. J Gen Virol 94: 2759-2770.

Yan Y, Cui H, Guo C, Wei J, Huang Y, Li L, Qin Q (2014). Singapore grouper iridovirus-encoded semaphorin homolog (SGIV-sema) contributes to viral replication, cytoskeleton reorganization and inhibition of cellular immune responses. J Gen Virol 95: 1144-1155.

Zhang QY, Gui JF (2012). Atlas of aquatic viruses and viral diseases. Science Press, Beijing.

Zhao Z, Ke F, Gui J, Zhang Q (2007). Characterization of an early gene encoding for dUTPase in Rana grylio virus. Virus Res 123:128-137.

Zhao Z, Ke F, Huang YH, Zhao JG, Gui JF, Zhang QY (2008). Identification and characterization of a novel envelope protein in Rana grylio virus. J Gen Virol 89: 1866-1872.

Zhou S, Wan Q, Huang Y, Huang X, Cao J, Ye L, Lim TK, Lin Q, Qin Q (2011). Proteomic analysis of Singapore grouper iridovirus envelope proteins and characterization of a novel envelope protein VP088. Proteomics 11:2236-2248.

第五章 蛙病毒的宿主免疫与免疫逃避

利昂·格瑞福，伊娃-斯蒂娜·埃德霍尔姆，弗朗西斯科·德·赫苏斯·安迪诺，
雅克·罗伯特[①]和V·格雷戈里·钦察尔[②]

1 引言

141 由虹彩病毒科蛙病毒属成员引起的变温脊椎动物感染和疾病暴发以及在野生和养殖种群中引起的大规模死亡，最近，正以惊人的速度升级，并引起了极大的关注。然而，鱼类、两栖类以及爬行类单个物种对这些病原的易感性会产生明显的变化。同时，蛙病毒免疫和蛙病毒疾病的决定因素，目前还不清楚。事实上，随着蛙病毒感染流行率以及这些病毒感染新宿主非凡能力的快速增长，蛙病毒如FV3，目前被认为是全球变温脊椎动物的潜在威胁（Gray和Miller，2013）。因此，迫切需要确定一个特定的变温脊椎动物物种的易感性是宿主不能获得一种保护性抗病毒免疫反应，还是病毒具有克服整个免疫屏障的能力。确实如此，蛙病毒具有一连串的免疫逃避和宿主调控机制。因此，在分子和细胞水平，更全面地考查蛙病毒-宿主免疫的相互作用是很有必要的，以便针对这些病毒致病因子制定出全面的预防措施。

与哺乳动物相比较，变温脊椎动物具有一个复杂的免疫系统，但它们的适应性免疫反应相对较弱。一般来说，变温脊椎动物显示出较差的T淋巴细胞扩
142 增、更少的抗体亚型以及比哺乳动物更不发达的免疫记忆反应（Robert和

①L. Grayfer · E. -S. Edholm · F. De Jesús Andino · J. Robert/Department of Microbiology and Immunology, University of Rochester Medical Center, 601 Elmwood Ave, Box 672, Rochester, NY 14642, USA/e-mail: Jacques_ Robert@ urmc. rochester. edu.

②V. G. Chinchar/Department of Microbiology, University of Mississippi Medical Center, 2500 North State Street, Jackson, MS 39216, USA.

Ohta，2009）。因此，这些生物可能更依赖于先天性免疫反应来促进对诸如蛙病毒等病原体的清除。变温脊椎动物的先天性抗病毒防御，有别于哺乳动物中所描述的先天性抗病毒防御，因而这一免疫反应的确切作用和效率需要在蛙病毒感染的背景下予以阐明。

本章总结了我们探索变温脊椎动物先天性免疫反应和适应性免疫反应在消除蛙病毒感染或在感染进程中的作用的最新进展，并概括了这些病原逃避宿主免疫屏障的策略。

2　蛙病毒感染的先天性免疫反应

2.1　蛙病毒感染的抗菌肽反应

抗菌肽是无尾类动物先天性免疫的一个重要元素，能为皮肤和黏膜表面抵抗各种病原提供保护。这些小分子在皮肤颗粒腺中合成和贮存，并在应激反应或受到伤害时分泌到黏液当中（Rollins-Smith，2009；Rollins-Smith 等人，2005）。抗菌肽同样参与蛙病毒的防御。从北美豹蛙（*Rana pipiens*）中分离到的两种抗菌肽，即 Esculentin-2P（E2P）和 Ranatuerin-2P（R2P），在温度低至 0 ℃ 时，可以使 FV3 和斑点叉尾鮰病毒（channel catfish virus，CCV）在几分钟之内失活。这一结果提示，这些分子能与病毒发生直接作用，而不是抑制病毒的复制，从而降低病毒的传染性（Chinchar 等人，2001）。抗菌肽能够在广泛的温度范围内发挥功能作用，有可能反映出宿主的变温特性。特别值得一提的是，50 μM 的 E2P 或 R2P 足以使 99% 的 CCV 失活，而每一种肽的浓度必须高于 10 倍才能使 90% FV3 失活（Chinchar 等人，2001）。据推测，FV3 对于抗菌肽引起的失活有较强的抵抗力，表现出抗菌肽较难对 FV3 衣壳下面内脂质膜（inner lipid membrane）产生靶向作用。很有可能，这一内脂质膜需要破裂才能引起病毒的失活。其他抗菌肽包括 Ranatuerin-2YJ、Dybowskin-YJb、Dybowskin-YJa、Temperin-YJa 以及Temperin-YJb 已被鉴定，并从感染蛙虹彩病毒（RGV）的东北林蛙（*Rana dybowskii*）皮肤中克隆出来（Yang 等人，2012）。有趣的是，所有这些抗菌肽对于 RGV 蚀斑的形成都具有浓度依赖性，并且病毒的清除与编码这些分子的基因表达增加相一致（Yang 等人，2012）。

2.2 先天性和炎症免疫反应

143 蛙病毒感染与宿主的炎症反应有广泛的联系。的确如此，类似于哺乳动物的病毒感染，蛙病毒感染引起的炎症反应就是一把双刃剑，两刃对于清除病毒都是很关键的，但也有可能加剧蛙病毒介导的疾病以及给宿主存活带来不良影响。按照实际情况来说，在一系列的变温宿主（poikilothermic host）物种中，关于蛙病毒感染的先天性免疫反应以及相关的炎症已有大量的文献记载（Carey 等人，1999；Chen 和 Robert，2011；Grayfer 等人，2014；Jancovich 和 Jacobs，2011；Morales 等人，2010）。

2.2.1 无尾两栖类

我们研究小组采用并优化了 FV3 感染非洲爪蟾这一蛙病毒—变温脊椎动物（特别是无尾两栖类）抗病毒免疫的模型。这一模型将 FV3（在分子水平描述最全面）与非洲爪蟾（具有最明显特征的两栖动物免疫系统）进行配对。一般来说，我们的实验方法涉及给非洲爪蟾成体或蝌蚪腹腔（i. p.）注射 FV3，随后评估感染的进展、病毒的复制以及宿主的免疫反应。虽然在非洲爪蟾中，腹腔注射或水浴接种都得到了类似的免疫反应（Robert 等人，2011），但是前者往往更方便，也更符合免疫学研究的要求。通过这种方法，我们能够描绘出非洲爪蟾成体在整个 FV3 感染过程中的先天性和适应性免疫反应的序列推进（Morales 等人，2010）。在非洲爪蟾成体中，组织化学和流式细胞分析显示，在腹腔注射感染后最早 1 天，被激活的单核和多核白细胞被招募而来，并大量聚集在腹腔渗出物（peritoneal exudates）当中（Morales 等人，2010）。随后，在感染 3 天后，我们观察到自然杀伤细胞（natural killer cells，NK）在腹腔聚集和积累。而淋巴细胞（包括 T 细胞）的聚集，在 FV3 感染接种 6 天之后才观察到（Morales 等人，2010）。显而易见，腹腔白细胞（leukocytes）的快速积累，伴随着炎症基因表达剧烈升高。在众多的标志性炎症基因当中，我们观察到，在 FV3 感染接种后第 1 天到第 3 天，非洲爪蟾肿瘤坏死因子 α（*X. laevis* tumor necrosis factor-alpha，TNF-α）的表达显著升高（Morales 等人，2010）。白细胞介素-1β 基因（编码一种由巨噬细胞产生的早期促炎细胞因子）的表达，在 FV3 感染接种后 1~6 天明显升高，而另个一种选择性极化巨噬细胞的标志分子（marker of alternatively polarized macrophages，M2）（Joerink 等人，2006b，c）抗炎精氨酸酶-1（Arg-1）在病毒感染接种后第 1 天升高，随后下降（Morales 等人，2010）。综合这些发现，一个有效而协调良好的免疫反应，表现出先天性和适应性免疫细胞效应因子的有序增加和相应免疫基因的激
144 活。在感染后第 1 天 Arg-1 基因表达水平的升高，反映出原有的而不是新增的

炎症髓细胞群表达的增加。的确如此，继 FV3 腹腔接种以后，我们始终观察到巨噬细胞和颗粒细胞群落刺激因子受体，分别简称为 M-CSFR 和 G-CSFR，其 mRNA 转录本的升高，表明髓细胞浸润的积累（L Grayfer 和 J Robert，美国罗彻斯特大学，未发表资料）。特别值得一提的是，腹腔白细胞 M-CSFR 和 G-CSFR 表达水平的升高，常常伴随着 M1 巨噬细胞标记物和诱导型一氧化氮合酶（inducible nitric oxide synthase，iNOS）表达水平的显著升高，iNOS 催化炎症巨噬细胞抗菌一氧化氮的生成（L Grayfer，F De Jesús Andino 和 J Robert，美国罗彻斯特大学，未发表资料）。这一结果支持了腹腔炎症发生时所观察到的 Arg-1 表达水平的降低，同时表明 Arg-1 和 iNOS 的功能在多个脊椎动物类群中是相反的（Joerink 等人，2006a，b，c；Wiegertjes 和 Forlenza，2010）。

值得强调的是，两栖类对蛙病毒的易感性在不同物种之间、各物种不同发育阶段甚至在同一物种的不同种群之间变化相当大（Miller 等人，2011）。这些差别很可能是多个复杂的决定因素引起的，其中包括宿主和蛙病毒的遗传变异以及各自宿主的免疫状态。对非洲爪蟾所进行的广泛研究揭示，爪蟾蝌蚪具有一个不同于成体的免疫系统，幼体的免疫系统一般来说欠成熟，特别是有关适应性免疫方面，例如不良的 T 细胞和抗体反应。在这方面，完全不奇怪，蝌蚪常常不能完全控制蛙病毒感染并死于这些病原（Bayley 等人，2013；Grayfer 等人，2014；Hoverman 等人，2010；Landsberg 等人，2013；Reeve 等人，2013）。然而，很清楚，在一些两栖类物种中，成年动物也同样死于蛙病毒感染（Sutton 等人，2014）。的确如此，好几个研究报道表明，与特定两栖类物种的幼体相比，其变态体（Brunner 等人，2004；Haislip 等人，2011；Reeve 等人，2013）和成体（Duffus 等人，2013）对蛙病毒的易感性比较高。鉴于蝌蚪的快速发育以及它们剧烈的变态重构，物种特异性免疫发育路径的不同可以解释为蛙病毒易感性的差别。广泛的免疫重构致使变态体在关键发育阶段易感性更高。此外，还存在这样一个事实，就是蛙病毒感染的非洲爪蟾蝌蚪所携带的病毒量要比成蛙低，而且更容易死于 FV3 感染（Grayfer 等人，2014），从而提示，在蝌蚪中检测出病毒要比在成年动物中困难得多，或者说较低的病毒携带量在蝌蚪中能引起更严重的疾病。

为了说明无尾两栖类蝌蚪可能存在先天性免疫反应的无效性，以解释蛙病毒的易感性，我们对 FV3 感染的非洲爪蟾蝌蚪和成体之间免疫基因表达的模式作了广泛的比较（De Jesús Andino 等人，2012），结果表明，与成蛙感染相反，蝌蚪表现出弱的并大大延后的抗 FV3 炎症基因表达反应（De Jesús Andino
等人，2012），在蝌蚪 PLs、脾细胞和肾脏中，TNFα、IL-1β 和 IFN-γ 基因的 145
表达，直到 FV3 感染接种后第 6 天才显著增加，并与感染成体这些基因强劲而快速（1 dpi）的上调（upregulation）形成鲜明的对比（De Jesús Andino 等人，2012）。值得指出的是，用热灭活的大肠杆菌（*Escherichia coli*）对蝌蚪进行刺激，在 24 小时之内，很容易诱导上述基因的表达，从而提示，免疫延迟具有 FV3 特异性（De Jesús Andino 等人，2012）。蝌蚪对 FV3 先天性免疫的这些无效性可能反映出多重非独特性问题（multiple nonexclusive issues），其中包括病毒的免疫逃避（viral immune evasion）、蝌蚪病原前哨受体系统缺失或者为了促进生长和发育放弃能量成本大的炎症反应而进行的生理均衡（physiological tradeoffs）。因此，这些适度和延迟的免疫反应有可能是无尾两栖类蝌蚪对 FV3 易感性更高及发病更严重的成因。

2.2.2 有尾两栖类

比较一致的观点是，宿主对蛙病毒会产生广泛的炎症反应，墨西哥钝口螈（*Ambystoma mexicanum*）感染虎纹钝口螈病毒的一个全面微阵列分析显示，这些动物脾脏和肺脏的大量炎症标志分子和先天性免疫基因都存在着负转录调控（或称上调）（Cotter 等人，2008），这些基因但不限于这些基因，其中包括吞噬受体（phagocytic receptors）和细胞内成分、细胞因子信号分子（cytokine signaling molecules）、补体成分（complement components）、NADPH 氧化酶亚单位（NADPH oxidase subunits）（又称催化活性氧抗菌反应的髓细胞酶）以及髓过氧化物酶（myloperoxidase）（又称催化过氧化氢生成的颗粒细胞酶）（granulocyte enzyme）（Cotter 等人，2008）。与所观察到的非洲爪蟾感染 FV3 的（Morales 等人，2010；Morales 和 Robert，2007）情形相反，在 ATV 感染的炎症反应中，没有观察到淋巴细胞增殖基因的负转录调控（Cotter 等人，2008），该物种缺乏这一有效的适应性反应，有可能是 ATV 对有尾两栖类致死性极强的原因。另外，这些观察可能反映出 ATV 采用的不同感染策略。

2.2.3 硬骨鱼类

有大量文献记载了硬骨鱼类感染蛙病毒后的先天性免疫和相关的炎症反应。用 FV3、欧洲鲶鱼病毒（ECV）、医生鱼病毒（DFV）和动物流行性造血器官坏死病毒（EHNV）四种不同的蛙病毒感染鲤鱼上皮乳头瘤（epithilioma papulosum cyprinid，EPC）硬骨鱼细胞系，导致了不同的炎症基因表达方式（Holopainen 等人，2012）。具体来讲，EHNV 和 FV3 引起了标志性前炎症因子基因 TNFα 和 IL-1β 的表达，而 ECV 和 DFV 诱导了一个常见免疫抑制基因即转移生长因子 β（TGF-β）的瞬时表达（Holopainen 等人，2012）。有趣的是，这四种病毒都引起了细胞凋亡基因和 β2-微球蛋白基因的表达，二者对于表面主要组织相容性复合物 I 类（surface MHC class I）的表达和细胞毒性 T 细胞功
146 能起着关键性的作用，从而提示，硬骨鱼 FV3 的感染，以蝾螈作比较，宿主

所产生的适应性免疫反应可能由蛙病毒感染引起。

最近，一项微阵列研究考查了黑头呆鱼（FHM）细胞在感染野生型 FV3 或缺乏平头 vIF-2α 基因的基因敲除突变体以后的转录反应，结果显示，在野生型 FV3 感染接种 8 小时之后，引起了许多免疫相关基因的上调。这些基因包括 IL-8、IFN、IFN 调节因子-1、IFN 调节因子-2、IFN 调节因子-3 以及 IL-1β，等等。在大多数情况下，类似的基因在感染基因敲除突变体的细胞中被上调，但诱导的程度比较低（Cheng 等人，2014）。

2.3　蛙病毒诱导致病和死亡的决定因素

恰当及时的炎症反应就如同这个反应的过程和进展一样重要，因为长时间的炎症反应会增加组织损伤和宿主死亡的风险（Fullerton 等人，2013）。尽管证据稀少，但仍有证据显示，蛙病毒感染可能加剧炎症反应，这可能是一些所观察到的蛙病毒病理变化的原因。例如，1997 年，在加拿大萨斯喀彻温，从虎纹钝口螈幼体中分离到了一株新型虹彩病毒（Bollinger 等人，1999），这些患病动物表现出严重的发炎、坏死以及在脾脏、肾脏、淋巴和造血组织中见到特征性的由蛙病毒诱导的细胞浆包含体（cytoplasmic inclusions）（Bollinger 等人，1999）。类似地，蛙病毒感染的丽纹攀蜥（*Japalura splendida*）整个种群表现出全身性出血（systemic hemorrhaging）、坏死（necrosis）、肉芽肿、坏死性炎症以及严重的肾脏病变、充血和广泛的肝脏损伤（Behncke 等人，2013），最终大规模死亡。蟒蛇蛙病毒感染表明，炎症可能是蛙病毒病理的一个决定因素（Hyatt 等人，2002）。腹腔注射大嘴鲈病毒（LMBV）感染大嘴鲈引起的死亡，被认为是病毒诱导的炎症和坏死的结果（Zilberg 等人，2000）。与这些炎症症状一致，接种 LMBV 的稚鲈表现出螺旋状游泳和腹胀（Zilberg 等人，2000）。值得一提的是，感染鱼的深层组织未受影响，因而带出一个问题，病毒诱导的损伤究竟是由于靶细胞的易感性，还是由于 LMBV 细胞嗜性（cell tropism）受到限制。后者提示，炎症和坏死引起的死亡可能是在感染起始位置第一损伤的结果。的确如此，以上的观察是对啮齿动物 FV3 感染早期研究的一种回忆（Gut 等人，1981；Kirn 等人，1980，1982）。在啮齿动物 FV3 感染的早期研究中，虽然 FV3 在 37 ℃不能复制（Aubertin 等人，1973），但是最初的病毒接种引起了广泛的炎症、坏死和肝脏损伤。

我们最近对非洲爪蟾的研究，支持 FV3 诱导前炎症反应的观点（Grayfer 147
等人，2014）。请留意这样一个思想，就是非洲爪蟾成体有可能产生有效的抗蛙病毒反应，导致病毒的清除。我们还发现，在急性感染期间，成蟾的 FV3 的病毒载量（viral loads）高出蝌蚪 1～2 个数量级，而一般来说，蝌蚪对 FV3

的易感性更高（Grayfer 等人，2014）。同样，在林蛙蝌蚪中还观察到温度对病毒载量的影响，这些动物在 25 ℃很快死于病毒感染，而且其病毒载量要高于饲养在 15 ℃的那些动物（J Chaney 和 M Gray，美国田纳西大学，未发表资料）。这些结果提示，FV3 的毒力的高低并不完全取决于病毒复制的数量。此外，具有免疫力的蝌蚪相对成体来说，可能更易受到蛙病毒因子和其他环境参数的攻击。为了支持这一观点，我们观察到，重组爪蟾 I 型干扰素（recombinant *X. laevis* type I interferon，rXlIFN）预刺激蝌蚪，其病毒载量要比成体低几个对数级，尽管如此，它们还是死于 FV3 感染（Grayfer 等人，2014）。此外，IFN 处理的幼体虽然具有较低的病毒载量，但是没有广泛的白细胞浸润，最终还是经历了多器官损伤，其中包括由坏死和细胞凋亡引起的组织结构和细胞组织的大量损失（Grayfer 等人，2014）。因此，即使病毒载量显著减少，蛙病毒可能还会在感染早期，给蝌蚪带来不可逆的组织损伤，产生的主要原因是病毒介导的细胞病理而不是病毒复制。在啮齿动物 FV3 模型所见到的也是如此，无论它们在宿主细胞内复制的能力如何，蛙病毒都可能引发毒性和潜在致命的影响（Gendrault 等人，1981）。最近这一观点在石斑鱼虹彩病毒中得到了证实。UV 灭活的复制缺陷型 GIV，在感染的 3 个细胞系中，有 2 个细胞系诱导了细胞凋亡（Pham 等人，2012）。类似地，热灭活和 UV 灭活的 FV3 引起了 FHM 细胞凋亡，并抑制宿主 RNA 和蛋白质的合成（Chinchar 等人，2003；Raghow 和 Granoff，1979）。依据这些发现，我们提出一个假设就是，动物用足够剂量的灭活病毒进行接种将会诱导缺乏病毒复制的毒性作用。如果这个假设适用于其他属和科的病毒成员，那么我们有必要考虑到这些病毒的致病性远比原先想象的要高得多。

3　巨噬细胞系的细胞在蛙病毒感染中的复杂作用

3.1　从啮齿动物 FV3 感染模型中得到的推论

对巨噬细胞系的细胞参与蛙病毒感染的推断，来自 30 年前啮齿类动物作为肝炎模型所进行的最初研究（Gut 等人，1981；Kirn 等人，1980，1982）。这些早期的研究显示，可扑弗氏细胞（Kupffer cells）（又称肝巨噬细胞）是
148 FV3 感染的主要靶细胞，它们的死亡与肝清除病毒能力的丧失有关，并最终导致严重的肝炎和死亡（Gut 等人，1981）。这些研究同样也意味着，炎症作为一个起重要作用的因素介导了 FV3 引起的病理变化，其中包括由肝巨噬细胞

即可扑弗氏细胞所进行的广泛的白细胞三烯释放（Hagmann 等人，1987）。FV3 感染动物对白细胞三烯的抑制，大大减少了病毒引起的肝损伤（Hagmann 等人，1987），从而提示，FV3 的病理，在很大程度上是由炎症反应引起的。

由于 FV3 不是哺乳动物的病原体，除一些选择性早期基因的表达以外（Lopez 等人，1986），不会在 37 ℃进行复制（Aubertin 等人，1973）。尽管如此，这项研究还是支持了目前的假设，由于它们高的巨噬细胞和内吞活性，巨噬细胞系细胞是蛙病毒感染的整体目标。事实上，在 37 ℃不能复制，可以视为研究 FV3 进入细胞机制的一个优点。在培养的老鼠肝巨噬细胞（Kupffer cells）中，病毒粒子在 FV3 感染后迅速出现在吞噬液泡（phagocytic vacuoles）中以及内吞作用的隔间（endocytic compartments）之中（Gendrault 等人，1981）。此外，相当比例附着于细胞之上的 FV3 病毒粒子存在着病毒衣壳-宿主膜融合以及病毒释放核心进入细胞胞浆之中（Gendrault 等人，1981）。这一现象表明，控制蛙病毒侵入细胞的潜在机制是普遍的，并且这种机制会促进病毒进入来自像哺乳动物、鱼类和两栖类这样进化关系比较远的生物细胞之内。依据这一推理，髓细胞系（myeloid lineage）可能充当了蛙病毒的准确攻击靶标，因为它们对细胞外物质的摄取效率很高，并受到大量吞饮或巨噬细胞表面受体（endocytic/phagocytic surface receptors）的促进，其中几个受体能识别并与蛙病毒结合。脊椎动物具有这一特点的特异吞噬细胞，可作为蛙病毒感染的靶标，从而可以解释为什么蛙病毒可以跨宿主物种范围而成功地感染。此外，由于蛙病毒在哺乳动物体温条件下不能进行复制，上述文献意味着，在 FV3 感染啮齿类所见到的病理事件不是整个病毒复制的结果。相反，细胞死亡可能是封装在 FV3 病毒粒子中预先形成的溶解因子所引起的或者是因为早期病毒基因产物的表达（Lopez 等人，1986）。类似于腺病毒粒子的 mRNA（Chung 等人，2003），FV3 在不适宜的温度下，早期基因的表达可能是预包装的毒力因子编码 mRNAs 释放的结果，而不是病毒的重新转录。的确如此，哺乳动物细胞的 FV3 感染能快速诱导细胞 RNA、DNA 和蛋白质合成的抑制（Elharrar 等人，1973）。此外，从 FV3 病毒粒子溶解所释放的因子能导致细胞毒性，并抑制宿主大分子的合成（Aubertin 等人，1973；Kirn 等人，1972）。

3.2　两栖类蛙病毒传播和存活的载体

越来越多来自两栖动物自然感染蛙病毒的证据支持巨噬细胞不仅对抗病毒防御很重要，而且对蛙病毒感染的策略也很重要这一观点。我们利用非洲爪蟾
FV3 感染作为一个平台来研究蛙病毒-宿主之间的免疫联系，我们过去和现在 149
的研究工作都证明了 FV3 与非洲爪蟾巨噬细胞之间的相互作用。我们已经证

实，在分辨出临床上明显发病以后，FV3 在两栖动物宿主中能持续存在几个月（Robert 等人，2007）。此外，FV3 DNA 可以在健康动物体内检出，而这些动物在实验室不能被感染，这表明 FV3 在具有免疫力的宿主中，采用某种静止形式作为一种长期存活的手段。值得注意的是，FV3 在体外和体内都能有效地感染青蛙的腹腔白细胞，并在这些细胞内持续存活，在感染之后，直到第 12 天，出现了非常活跃的病毒转录（Robert 等人，2007）。由于腹腔白细胞主要以巨噬细胞系的细胞占优势，我们的发现不仅证实了蛙病毒-巨噬细胞嗜性（ranavirus-macrophage tropism），而且还揭示，这些最终分化的、长寿命的细胞群是病毒传播理想的载体，或作为“宿主内”（within host）的贮藏宿主。

以上假设已经被我们随后利用透射电子显微镜对 FV3 感染非洲爪蟾腹腔白细胞的分析所证实。在感染的白细胞中，我们在具有巨噬细胞特征的腹腔白细胞中发现了二十面体病毒颗粒（Morales 等人，2010）。这些 FV3 感染的细胞显示了少量的胞内病毒颗粒，这意味着 FV3 可能会把单核巨噬细胞作为传播的贮藏宿主，FV3 -巨噬细胞的相互作用，让人想起了 HIV 与巨噬细胞的关系，病毒粒子集聚在骨髓细胞中作为一种传播机制（Coiras 等人，2009；Goodenow 等人，2003；Gousset 等人，2008；Groot 等人，2008）。有趣的是，这些 FV3 感染的腹腔巨噬细胞不仅包含细胞质病毒粒子，而且它们还将病毒释放到胞外环境当中（Morales 等人，2010），从而确认这些细胞有可能在它们的宿主体内，既作为病毒传播的贮藏宿主，又作为病毒传播的媒介。

总之，用 FV3 腹腔注射感染成年非洲爪蟾，结果导致大量白细胞集聚在感染部位，其中包括大量具有巨噬细胞形态学的细胞（Morales 等人，2010）。这些白细胞的骨髓起源不仅得到巨噬细胞炎症基因如 TNF-α、IL-1β 和精氨酸酶-1（Morales 等人，2010）表达的支持，而且还得到了巨噬细胞系标志分子 M-CSFR 的支持（L Grayfer，F De Jesús Andino 和 J Robert，美国罗彻斯特大学，未发表资料）。有趣的是，直到感染后第 21 天，我们才从感染的非洲爪蟾腹腔中的白细胞中扩增到 FV3 DNA，但 FV3 的早期和晚期转录本，在接种病毒后的第 6 天可以检测到，在第 15 天或者第 21 天则检测不到（Morales 等人，2010）。这说明，病毒基因组在休眠状态时存活在巨噬细胞样细胞之中。在感染后的一段时间内无法检测到腹腔白细胞中的 FV3 基因组，反映出在非洲爪蟾宿主中，这些细胞散播到了宿主体内的远端部位。

在非洲爪蟾中，肾脏是 FV3 复制具有代表性的重点部位。有趣的是，直到感染后第 9 天，活跃的 FV3 基因转录才见于某些 FV3 感染的爪蟾中，但不是所有 FV3 感染的爪蟾中。然而，病毒基因组 DNA，在病毒接种感染 2 周之后，才会从这些感染动物中的一些动物中被扩增到（Morales 等人，2010）。值得注意的是，这一时间间隔要短于 3 周，而在这 3 周时间内，病毒的存在可以在腹腔白细胞群中很可靠地被检测出来。这些差别反映出，FV3 与这些不同靶 150
细胞之间存在着不同的相互作用。据推测，肾脏细胞是最有效的病毒感染器官，并作为 FV3 复制活跃的部位，可见到高病毒滴度和广泛的组织损伤（Grayfer 等人，2014）。相反，巨噬细胞系的细胞是终末分化的，不能再分裂，而且是长寿的细胞，这些细胞很明显是作为蛙病毒休眠（非复制型）的贮藏宿主。

总体来说，这些观察结果意味着，爪蟾巨噬细胞系构成了 FV3 感染的中心目标，作为病毒存活、静止以及传播的靶细胞。的确如此，我们最近观察到，在离体感染培养的腹腔巨噬细胞当中，最初感染的几个月内，可以扩增到 FV3 基因组 DNA，但不能检测出病毒基因的表达（L Grayfer 和 J Robert，美国罗彻斯特大学，未发表资料）。我们认为，进一步揭示蛙病毒-巨噬细胞互相作用及蛙病毒静止的关键取决于离体骨髓细胞培养和相关试剂的开发。

3.3　巨噬细胞贮藏宿主与蛙病毒的再激活

我们最近提供了大量的证据支持这样一个观点，就是两栖动物的巨噬细胞不仅作为蛙病毒在宿主内传播的载体，同时也是疾病重新激活（disease reactivation）的疫源地（foci）。动物感染后第 30 天，从 15 个非洲爪蟾成体中分离腹腔巨噬细胞时，只有一个个体的细胞显示出可检测的 FV3 DNA 水平，并表达了编码病毒 DNA 聚合酶和主衣壳蛋白的转录本（J Robert，L Grayfer 和 F De Jesús Andino，美国罗彻斯特大学，未发表资料）。然而，在腹腔注射热灭活的大肠杆菌以后，同样 15 个动物当中有 9 个表现出了可检测的病毒基因组 DNA 和活跃的病毒基因表达。此外，以 53R、一个 FV3 复制和组装所需要的基因产物以及一种 HAM56（一个非洲爪蟾巨噬细胞标志分子）为靶分子的免疫荧光显微技术观察显示（Nishikawa 等人，1998）腹腔巨噬细胞表现出高效的 FV3 复制（图 1），但是单核巨噬细胞的病毒 DNA 水平较低，而且可以通过炎症重新激活。因此，进一步研究单核巨噬细胞的激活状态（既包括经典的，也包括替代的）对于制定预防性的措施和认识这些复杂病原精确的感染策略显得尤为关键。（Auffray 等人，2007；Nahrendorf 等人，2007；Zhao 等人，2009；Ziegler-Heitbrock，2007）

3.4　其他变温动物细胞中的蛙病毒感染

类似于其他的许多病原体，蛙病毒能突破巨噬细胞的抗菌和抗病毒屏障，在这一点上，这些细胞成了蛙病毒传播和存活的载体。由于蛙病毒属和虹彩病

毒科的其他病毒成员也采用这一宿主渗透和免疫逃避的机制，因而利用巨噬细
151 胞系的细胞作为病毒传播和存活的载体似乎并不局限于 FV3。例如，一种虹彩病毒样病原体感染了六须鲶肾巨噬细胞，并能通过这些细胞离体抑制（down-regulating）豆蔻酸-佛波醇-乙酸酯（phorbol myristate acetate）诱导的活性氧生成（Siwicki 等人，1999）。同样，杂交石斑鱼在感染中国台湾石斑鱼虹彩病毒（Taiwan grouper iridovirus，TGIV）后，检测出了磷酸酶阳性，并出现了高度吞噬性的嗜碱性和嗜酸性单核白细胞的数量增多（Chao 等人，2004）。有趣的是，TGIV 基因组 DNA 在感染的早期仅存在于单核吞噬细胞的核内，而在感染的晚些时候，既见于核中，也见于胞质间隔之中，并且这些单核吞噬细胞失去了吞噬能力（Chao 等人，2004）。很明显，TGIV 已经进化出复杂的和暂时性的调节策略来克服和利用每一种免疫细胞，这些免疫细胞能协调抗病毒的免疫反应。病毒入侵单核巨噬细胞作为一种免疫逃避和传播手段的这一策略，是所有脊椎动物虹彩病毒感染的一个显著特征。

152 来自相关宿主物种的原代巨噬细胞离体培养和感染模型的发展，将有助于人类更深入地认识这些感染策略。

4　蛙病毒感染的抗病毒免疫反应

4.1　变温脊椎动物的抗病毒干扰素

干扰素（IFN）反应为抗病毒免疫作出了重要的贡献，IFN 反应通常是一系列宿主病原体识别受体（pathogen recognition receptors，PRRs）识别病毒产物引起的结果。这些受体包括 Toll 样受体（Toll-like receptors，TLRs）、视黄酸诱导基因 1 样受体（retinoic acid-inducible gene 1-like receptors，RLRs）以及胞质 DNA 感受器（cytosolic DNA sensors）（Baum 与 Garcia-Sastre，2010；Sadler 与 Williams，2008）。这一抗病毒免疫的分支包括三种类型的细胞因子，即 I 型、II 型和 III IFN（Sadler 和 Williams，2008）。IFN-γ 是哺乳动物唯一的 II 型干扰素，值得注意的是，硬骨鱼具有多种 II 型干扰素（Grayfer 等人，2010），起着多重免疫和抗病毒作用，而 IFN-I 型和 IFN-III 型的突出功能是充当抗病毒分子。哺乳动物 IFN-I 型具有广泛的细胞特异性，而 IFN-III 型的靶目标是特异性的细胞亚群（Levraud 等人，2007；Zou 等人，2007）。有趣的是，尽管 IFN-I 型和 IFN-III 型的不同受体系统存在着细胞特异性，但是这两个细胞因子家族都激活同一下游的 JAK 激酶（Janus kinase，JAK）和信号转换器（signal

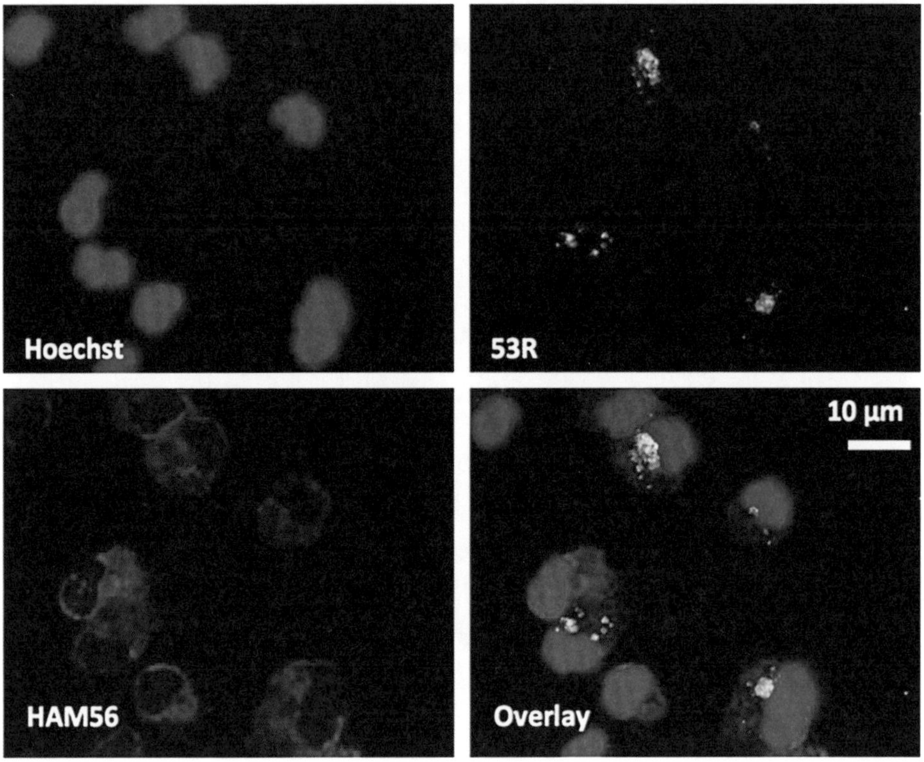

图 1 非洲爪蟾（*Xenopus laevis*）感染 FV3 的 HAM56[+]腹腔巨噬细胞。用感染复数（MOI）为 0. 3 的 FV3 感染非洲爪蟾的腹腔白细胞。巨噬细胞用抗巨噬细胞标记分子抗体进行染色，FV3 用抗 53R 病毒蛋白抗体进行可视化。烟酸己可碱（hoechst）用来对细胞核进行可视化染色。

transducer），并激活转录信号通路激活因子（activator of transcription signaling pathways），最终达到相同的抗病毒结果（Sadler 和 Williams，2008），包括诱导诸如蛋白激 R（protein kinase R，PKR）以及黏液病毒抵抗分子（myxovirus resistance molecules）的表达。

虽然这些 IFN 反应在温血脊椎动物中得到了很好的研究，但是对蛙病毒易感的冷血动物宿主的 IFN 系统，目前还不太清楚。哺乳动物 I 型 IFN 由无内含子的基因编码，由多基因 IFN-α 家族（在人类有 13 个基因）和唯一一个 IFN-β 基因组成（Hervas-Stubbs 等人，2011）。虽然爬行动物和鸟类也存在单一外显子编码的 I 型 IFN（Robertsen，2006；Zou 和 Secombes，2011），但是较低等的脊椎动物包括软骨鱼类和硬骨鱼类以及两栖类都具有 5 个外显子/4 个内含子转录本所编码的 I 型 IFN，并显示出与哺乳动物所对应的编码产物有明显的差异（Chang 等人，2009；Qi 等人，2010；Robertsen，2006；Zou 和 Secombes，2011；Zou 等人，2007）。

目前，仅对硬骨鱼类的 I 型 IFN 系统进行了详细的探索。依据半胱氨酸模式，这些 IFN 被细分成两个类群，即类群 I：2C 和类群 II：4C（Sun 等人，2009；Zou 等人，2007），并且按照系统发生进一步划分为四类即 IFNa 到 IFNd（Chang 等人，2009；Sun 等人，2009）。重要的是，虽然多种不同哺乳动物的 IFN 通过相同的受体复合物而发挥它们的生物学作用（Li 等人，2008；Samuel，
153 2001），但是鱼类类群 I 和类群 II IFN 是通过独特的受体复合物进行信号传递的（Aggad 等人，2009）。有关干扰素的功能性研究主要集中于类群 I 中的鱼类 IFN（Aggad 等人，2009；Altmann 等人，2003；Long 等人，2004；Lopez-Munoz 等人，2009；Robertsen 等人，2003；Zou 等人，2007），研究显示这些 IFN 在建立细胞抗病毒状态方面的能力是不同的（Aggad 等人，2009；Levraud 等人，2007；Li 等人，2010；Lopez-Munoz 等人，2009）。例如，鲑科鱼类（salmonid）IFNa 到 IFNd 具有不同的转录调控方式以及不同的抗病毒功能，因为其中一些细胞因子能够产生有效的抗病毒反应，而另一些则被认为根本就没有抗病毒功能（Svingerud 等人，2012）。两栖类动物和爬行类动物 II 型干扰素系统仍未阐明，而硬骨鱼类的干扰素系统似乎比哺乳动物的要复杂得多（Zou 和 Secombes，2011），在这里不作进一步讨论。

哺乳动物 III 型干扰素由干扰素 λ（IFN-λ）-1、-2 和-3（也称为 IL-28A，IL-28B 和 IL-29）组成。这些分子由含有 5 个外显子/4 个内含子转录本的基因编码，并通过干扰素 λ 受体-1（interferon lambda receptor-1，IFNλR1）和白细胞介素-10 受体-2（interleukin-10 receptor-2，IL-10R2）组成的受体系统进行信号传导（见 Kotenko，2011 的文献综述）。有趣的是，在硬骨鱼类中，真正的 III 型 IFN 要么不存在，要么还没有被鉴定。现在已知两栖类动物具有这两种类型的 I 型 IFN，如同对应的鱼类 I 型 IFN 分子一样，它们都有相同的 5 个外显子/4 个内含子的基因结构，同时两栖类动物还具有真正的 III 型干扰素（Qi 等人，2010）。两栖类动物是鱼类和哺乳动物之间的关键性进化中间体，并且栖息在水域和陆地栖息地，这一点尤其重要。事实上，鱼类和两栖类 I 型 IFN 的一个标志性特征就是具有 5 个外显子/4 个内含子的基因结构，这一点有别于爬行类和哺乳类无内含子 I 型 IFN 的结构（Robertsen，2006；Robertsen 等人，2003；Sun 等人，2009）。到目前为止，关于从鱼类 IFN-I 型到更高等脊椎动物 IFN-I 型和 IFN-III 型之间系统发生的关系仍存在大量的争论。鱼类细胞因子展现出外显子/内含子的基因结构，类似于哺乳动物 III 型 IFN 的基因结构。

同时，还具有更高等脊椎动物I型IFN的标记特征，如具有保守半胱氨酸位点和一个C端CAWE基序（鲶鱼IFN-I型除外）以及一个几乎在所有IFN中都能见到的保守基序（Lutfalla等人，2003；Qi等人，2010；Robertsen，2006；Zou等人，2007）。确定这些分子在鱼类和两栖类动物免疫系统中各自的作用是很有趣的，特别是鱼类似乎只有I型IFN，而蛙类同时具有IFN-I型和IFN-III型（Qi等人，2010）。

4.2 对蛙病毒感染的干扰素反应

如上所述，在干扰素反应过程中，所合成的一种重要抗病毒基因产物是黏病毒抗性蛋白（Myxorirus resistance protein，Mx）（Samuel，2001），黏病毒抗性蛋白被认为对建立由IFN引起的抗病毒状态至关重要（Samuel，2001）。黏病毒抗性蛋白是高分子量GTP酶，属于动力蛋白超家族（dynamin superfamily），现已知该酶能促进细胞内膜的重构以及胞内运输（Kochs等人， 154
2005）。如同哺乳动物一样，硬骨鱼类黏病毒抗性蛋白，其功能是作为抗病毒的介导因子，具有独特抗病毒效果，并有别于其他不同种类黏病毒抗性蛋白的同分异构体。到目前为止，已经证明鱼类当中大多数黏病毒抗性蛋白不能有效地防止虹彩病毒科各病毒成员的感染。例如，日本牙鲆（Japanese flounder）黏病毒抗性蛋白能够抑制2种弹状病毒（*Rhabdovirus*）的复制，但是不能抑制真鲷虹彩病毒（Red seabream iridovirus，RSIV）的复制。真鲷虹彩病毒属于虹彩病毒科肿大细胞病毒属（Caipang等人，2003）。同样，澳州肺鱼黏病毒抗性蛋白能抑制诺达病毒性神经坏死病毒（Nodavirus viral nervous necrosis virus，VNNV）和传染性胰腺坏死病毒（Infectious pancreatic necrosis virus，IPNV）的复制，但没有显示出对中国台湾石斑鱼虹彩病毒（Taiwan grouper iridovirus，TGIV）的抗病毒效果（Wu等人，2012；Wu和Chi，2007）。同样，塞内加尔鳎黏病毒抗性蛋白具有很好的抗传染性胰腺坏死病毒和抗病毒性出血性败血症病毒（VHSV，弹状病毒科，*Rhabdoviridae*）的效果，但没有抗蛙病毒中的欧洲六须鲶病毒（ESV）的效果（Alvarez-Torres等人，2013）。最后，虹鳟Mx1具有抗传染性胰腺坏死病毒、鲑鱼α病毒（Salmonid alpha virus，SAV，属披膜病毒科，*Togaviridae*）以及传染性造血组织坏死病毒（Infectious hematopoetic necrosis virus，IHNV，弹状病毒科）感染的效果，但是不能有效地阻断EHNV的复制（Lester等人，2012；Trobridge等人，1997）。宿主抗病毒反应有可能与当地的蛙病毒分离株之间存在着共进化。因此，在应对外来蛙病毒分离株时，诸如Mx1抗病毒成分的不足，最终会导致全球性威胁，这种威胁由地理上遥远的病毒株通过亚临床感染宿主或因国际贸易进口引入病毒而造成。

在南欧，具有巨大商业价值和水产上最为重要的鱼种就是金头鲷，部分原因是它对大多数病毒病原体具有天然抵抗力（Cano 等人，2006，2009）。事实上，唯一影响商业金头鲷种群的病毒性疾病就是淋巴细胞囊肿病毒（LCDV，淋巴囊肿病毒属 *Lymphocystivirus*，虹彩病毒科 *Iridoviridae*）引起的疾病（Cano 等人，2013）。有趣的是，金头鲷至少含有 3 种 Mx 蛋白，其中一种 Mx 异构体能有效地抑制 VHSV 和 LCDV 的复制；第二种 Mx 分子能有效地抑制欧洲六须鲶病毒和 LCDV 的复制；第三种 Mx 分子能有效地保护宿主免受 VHSV 的感染（Alvarez-Torres 等人，2013；Fernandez-Trujillo 等人，2013）。这是硬骨鱼类 Mx 分子有效抑制 DNA 病毒感染的第一个例子。

有趣的是，LCDV 虽然能引起金头鲷这一物种的疫病，但值得注意的是，与虹彩病毒科许多其他病毒成员造成死亡不同，LCDV 的感染能被金头鲷有效地清除。因而有学者认为，金头鲷可能携带病毒而无症状。因此，硬骨鱼类 IFN/Mx 反应的效力可以很好地说明单一鱼种对诸如虹彩病毒这样强毒力病原体的易感性。值得注意的是，许多鱼类能感染 LCDV，也有许多鱼类能清除 LCDV，因为这些感染涉及鱼类皮肤（Leibovitz，1980）。因此，像 Mx 这样引起的全身性抗病毒反应，有可能对于解决 LCDV 感染显得不太重要。

在另外一个例子当中，日本牙鲆 IFN 诱导的跨膜（IFN-inducible transmembrane，IFITM）蛋白以上调的方式对蛙虹彩病毒感染作出反应（Zhu 等人，2013）。此外，过表达及 siRNA 基因敲低研究显示，牙鲆 IFITM1 在细胞抗
155 RGV 病毒反应中，起着非常重要的作用（Zhu 等人，2013）。IFITM1 通过抑制病毒-宿主细胞的进入，并以高尔基器为靶向目标而发挥着功能作用（Zhu 等人，2013）。

随着对这些单个抗病毒分子所进行的研究越来越多，越来越明显的是，有一些关键的、以前不为人知的因子不仅参与了变温脊椎动物的抗病毒反应，而且参与了所有脊椎动物的抗病毒反应。因此，我们提出，这些 IFN 诱导产生的单个抗病毒成分，极有可能在其他 IFN 调控分子的整个网络上存在着相互依赖的关系。抗病毒的 IFN 反应无论是在细胞水平，还是在整个生物水平，其相对效力都依赖许多细胞和分子成分之间的平衡。现在已知不同的鱼类和两栖动物具有各自非常独特的抗病毒效应分子系统，因此，毫不奇怪，这些不同的生物体对相似的病原体表现出不同的易感性。

感染 ATV 蝾螈的微阵列分析显示，除多层面的炎症基因表达反应以外，同样也显示出这些感染动物上调多个抗病毒干扰素反应基因的表达（Cotter 等人，2008）。蛙病毒感染所诱导的基因有很多，其中包括 Mx1 基因、抗病毒螺旋酶（antiviral helicases）、干扰素调节因子（interferon regulatory actors）、

IFITM 以及核糖核酸酶(Cotter 等人，2008)。然而，编码蝾螈 I 型 IFN 和 II 型 IFN 的基因仍有待鉴定。描绘出抗病毒 IFN 存在于蝾螈基因组中的精确定位，并弄清楚其转录调控以及在抗蛙病毒如 ATV 免疫反应过程中这些分子的功能作用，显得尤为重要。

如上所述，蛙具有 I 型 IFN 和 III 型 IFN 基因，这两个基因在病毒感染以后进行转录的上调（Qi 等人，2010）。然而，还没有两栖类 III 型 IFN 功能研究的报道，我们最近鉴定了一种非洲爪蟾 I 型 IFN，构建了非洲爪蟾 I 型 IFN 重组子（rXlIFN），并在 FV3 感染的背景下鉴定了该分子（Grayfer 等人，2014）。用非洲爪蟾 I 型 IFN 重组子对非洲爪蟾肾上皮培养细胞 A6 进行预处理，结果保护了这些细胞对抗 FV3 的溶细胞作用（Grayfer 等人，2014）。而对照组的培养细胞在 FV3 感染后几乎全部摧毁，并表现出广泛的病毒复制，但是非洲爪蟾 I 型 IFN 重组子预处理的培养细胞几乎没有受到 FV3 感染，并且生长旺盛（Grayfer 等人，2014）。此外，用 rXlIFN 处理细胞 A6 显著上调 Mx1 的表达，提示用这一细胞因子刺激会产生细胞的抗病毒状态（Grayfer 等人，2014）。在 FV3 感染接种后所产生的 I 型 IFN 反应，成年非洲爪蟾比蝌蚪更为强烈（Grayfer 等人，2014）。虽然如此，但是蝌蚪腹腔注射非洲爪蟾 I 型 IFN 重组子表现出脾脏和腹腔白细胞 Mx1 基因的表达显著降低（Grayfer 等人，2014）。此外，FV3 接种后，用 rXlIFN 处理后的蝌蚪出现病毒的复制和转录活性降低（Grayfer 等人，2014）。因此，除延迟先天性与炎症相关的免疫基因表达反应以外，不恰当的 I 型 IFN 反应同样也会导致非洲爪蟾蝌蚪对 FV3 有更高的易感性。

然而，对 FV3 与蝌蚪之间相互作用的复杂性要补充说明的是，虽然 rXlIFN 处理的蝌蚪在 FV3 感染接种以后表现出平均存活时间延长，同时病毒载量也降低好几个对数级，但是这些动物还是发生了广泛的器官损伤，并最终
死于感染（Grayfer 等人，2014）。这一结果与上述描述的观点相一致，该观点 156
认为，依据物种和发育阶段的不同，蛙病毒毒力因子（virulence factors）不依赖于病毒复制。令人吃惊的是，尽管非洲爪蟾 IFN 具有抗病毒效力，但最终 FV3 能导致蝌蚪死亡。阐明两栖类 I 型 IFN 和 III 型 IFN 及其同源受体系统的作用和这些分子在蝌蚪和成蟾蛙病毒感染中各自的作用（以及可能的缺失）将是很有价值的。

5 病毒感染的适应性免疫反应

低等脊椎动物的抗蛙病毒免疫反应是多层面的、复杂的并鲜为人知。然而，很明显，对蛙病毒的清除在很大程度上取决于成功的适应性免疫反应，而到目前为止，这种适应性免疫反应几乎只在非洲爪蟾中进行研究过。

5.1 蛙病毒感染的抗体反应

两栖类动物免疫球蛋白（Ig）重链和轻链基因座的结构与使用，让人想起了哺乳动物的同类对应物包括 V-（D）-J 重排、类别转换重组（class-switch recombination）、体细胞高频突变（somatic hypermutation）以及亲和力成熟（affinity maturation）（Du Pasquier 等人，1989，2000；Hsu，1998）。如同哺乳动物一样，非洲爪蟾 Ig 从 IgM 到 IgY（IgG 的类似物）的类别转换具有胸腺依赖性，并需要 T 细胞 - B 细胞的协作（Blomberg 等人，1980；Turner 和 Manning，1974）。与哺乳动物 IgG 中所见到的 10 000 倍增加相比而言，虽然两栖类 IgY 亲和力成熟只能引起 10 倍的增加，但是已经证明非洲爪蟾的体液免疫是引起抗病毒免疫反应的一个重要因素，尤其是成年爪蟾的体液免疫反应（Maniero 等人，2006）。

成年非洲爪蟾再次感染 FV3 后，会产生大量病毒特异性 IgY，并能在感染一周后被检测出来，感染接种后 3 周左右达到高峰（Gantress 等人，2003）。的确如此，成蟾在第一次感染后 15 个月，再次感染 FV3（缺少免疫佐剂），以胸腺依赖性方式产生抗 FV3 特异性的 IgY 抗体，并在再次免疫后的第 10 天开始直到第 8 周，可检测出该抗体（Maniero 等人，2006）。值得注意的是，FV3 在体外接触这一血清时被有效地中和了（Maniero 等人，2006）。此外，在 FV3 感染之前，给自然易感的非洲爪蟾蝌蚪注射免疫血清，会使蝌蚪产生一定程度但不是显著的抗病毒被动免疫保护（Maniero 等人，2006）。很明显，两栖动物抗体反应能清除一部分蛙病毒感染，这一特定的免疫机制导致成年非洲爪蟾产
157 生的最终抗病毒反应，究竟达到了一个什么样的程度，还有待确定。这些结果与细胞肿大病毒属（*Megalocytivirus*）的真鲷虹彩病毒（Red seabream iridovirus，RSIV）所产生的免疫反应相一致，也就是用灭活的病毒粒子进行疫苗接种，使鱼类免受后来的病毒感染（Caipang 等人，2006；Nakajima 和 Kunita，2005；Nakajima 等人，1999）。

5.2 蛙病毒感染的T细胞反应和免疫记忆

非洲爪蟾蝌蚪不能表达最佳水平的MHC类Ia蛋白（Du Pasquier等人，1989），而它们的脾细胞包括真正的CD8 T细胞（bona fide CD8 T cells）能够表达乏T非洲爪蟾细胞表面标记分子CD5（Jurgens等人，1995），产生完全重排的TCRα/β转录本（Horton等人，1998）。非洲爪蟾蝌蚪不能表达最佳水平的MHC类Ia蛋白，会导致T细胞的分化有别于变态后动物的选择性通路，从而大大地依赖于非多态、非经典MHC类Ib分子。的确如此，在缺乏最优类Ia介导的T细胞选择的情况下，幼体CD8 T细胞可能具有受限制的抗原结合区，从而反映出蝌蚪对蛙病毒的相对易感性。然而，如同在这一节后面部分论述的一样，在蝌蚪中存在着独特的MHC类Ib介导的T细胞选择机制以及T细胞亚群，而T细胞亚群有可能补充常规类Ib限制的CD8 T细胞。

与蝌蚪相反，成年非洲爪蟾显示出常规MHC类Ia限制的CD8细胞毒性T细胞群。虽然缺少有用的抗体，但CD4 T辅助细胞很有可能存在，因为存在着参与CD4 T细胞分化和功能的所有基因、在细胞（CD8-/CD5+）中CD4基因的表达以及由混合淋巴细胞反应所获得的MHC II类依赖增殖反应（Du Pasquier等人，1989）。在成年非洲爪蟾中，清除FV3所需要的T细胞已经通过使用亚致死性照射γ辐射得到了证实，这种辐射可能使起源于胸腺的T细胞衰竭。辐射引起成蟾的T细胞衰竭以后不能控制FV3，并使成蟾最终死于感染（Robert等人，2005）。此外，由于使用抗非洲爪蟾CD8 mAb引起的非洲爪蟾CD8 T细胞的衰减，显著增加了成蟾对FV3感染的易感性（Robert等人，2005）。这些CD8 T细胞衰竭使感染FV3的成蟾经历了严重的水肿和出血。成蟾体内病毒载量的广泛升高，最终死于感染。而对照组的同类则有效地清除了病毒（Robert等人，2005）。因此，细胞毒性CD8 T细胞反应对于清除FV3是至关重要的。有趣的是，使用抗CD8 Ab的蝌蚪既没有造成CD8 T细胞的衰竭，也没有增加对FV3的易感性（Robert等人，2005），再次强调了蝌蚪T细胞群的非常规性质。

再度感染FV3的成蟾显示，病毒的快速清除，伴随着CD5+CD8+脾细胞的早期增殖和肾脏的快速浸润（3与6 dpi）以及中心位置非洲爪蟾FV3的复制（Morales和Robert，2007）。这一结果不仅强调了CD8 T细胞在清除蛙病毒中的重要性，而且还说明成年非洲爪蟾T细胞存在对蛙病毒再度感染的记忆反应。有趣的是，与原发性感染相比，肾浸润在第二次FV3感染接种以后大大 158
加速，但募集的CD8+细胞在第二次免疫反应期间却显著减少（Morales和Robert，2007）。这可能归因于原发性免疫反应当中，更高频度T细胞前身的浸润或者在再度感染过程中，产生了更有效力的CD8+记忆T细胞数量的降低。同时，也不能排除这种温和的继发性反应是进化上原始的两栖动物适应性免疫系统的固有特性。其原因有可能是非洲爪蟾免疫接种后，T细胞扩增程度相对较低，缺乏引流淋巴结和缺乏白髓-红髓脾脏结构（Du Pasquier 等人，1989）。另外，这种温和的继发性 CD8 反应可以说明，在后继抗病毒反应过程中，是其他效应细胞群募集和参与免疫的结果。为了支持这一观点，在继发性抗 FV3 反应期间，存在着快速而强健的 CD8- MHC II+免疫细胞群的募集以及在肾脏的浸润，这些免疫细胞群可能是 B 细胞、CD4 T 细胞或者 CD8-非经典 MHC 类 Ib 限制性不变 T 细胞群。

5.3 非经典 MHC 限制细胞在蛙病毒免疫中的作用

非经典 MHC 类 Ib 分子显示出与类 Ia 具有结构上的相似性，但一般来说，在组织分布上受到限制，并且其多态性显著减少（Flajnik 和 Kasahara，2001）。在哺乳动物中，这些表面糖蛋白中的一些分子涉及不变 T 细胞的不同亚群的分化和功能调控，其中包括 CD1d 限制的 iNKT 细胞和 MR1 限制的黏液相关 iT 细胞（Bendelac 等人，1995，1996，1997；Matsuda 和 Gapin，2005）。这两类淋巴细胞群经历了非常规的分化途径，展示了独特的半不变 T 细胞受体重排，有人认为，这两类细胞参与了抗细菌和抗病毒的免疫反应（Behar 和 Porcelli，2007；Choi 等人，2008；Cohen 等人，2009；Le Bourhis 等人，2010）。

令人吃惊的是，如上所述，当非洲爪蟾幼体产生天然 MHC 类 Ia 缺失时（Du Pasquier 等人，1989），它们会表达许多非经典类 Ia 基因（XNCs），其中有一些基因如 XNC10，表现出偏胸腺性表达（Goyos 等人，2009，2011）。值得一提的是，我们最近鉴定了占优势的非洲爪蟾 iT 免疫细胞亚群，该细胞亚群无论是发育，还是其功能的发挥都需要 XNC10 的参与（Edholm 等人，2013）。运用 XNC10 四聚体以及反向遗传学结合转基因技术和 RNA 干扰，我们确定该 iT 细胞群是 CD8-/CD4-，表达半不变 T 细胞受体，构成一个不变 TCRα（iVα6~Jα1.43），并与限制性 TCRβ 的可变区结合，而且这些细胞群在缺乏 XNC10 或者 XNC10 表达减少的情况下，不能进行发育（Edholm 等人，2013）。值得注意的是，通过有效 RNA 干扰沉默胸腺和脾脏 XNC10 表达的转基因动物，不能使这类 iT 细胞亚群进一步发育。此外，上述转基因动物对 FV3 的易感性显著升高，并且更容易死于 FV3 感染（Edholm 等人，2013）。从而提示，这些细胞在抗蛙病毒防御中是很重要的。值得注意的是，对非洲爪蟾
159 蝌蚪 TCRα 的深度测序分析显示，非洲爪蟾幼体具有几个其他占优势的 iT 细胞群（Edholm 等人，2013），这类细胞群有可能是 XNC 限制性的。同时，极

有可能参与诸如抗蛙病毒的免疫反应。的确如此，我们在成年非洲爪蟾中也鉴定了 XNC10 依赖的 iT 细胞群。因此，有理由推测，在原发性和继发性抗 FV3 反应过程中，这些淋巴细胞亚群有可能存在于上述所讨论的 CD8-肾脏浸润免疫细胞群中（Morales 和 Robert，2007）。表 1 对宿主免疫策略提供了一个全面的总结。

表 1　目前对抗蛙病毒免疫防御的理解

<table>
<tr><th>免疫参数</th><th>物种</th><th>免疫结果</th><th>参考文献</th></tr>
<tr><td colspan="4">细胞免疫</td></tr>
<tr><td rowspan="4">巨噬细胞系细胞</td><td>非洲爪蟾</td><td>FV3 接种引起 Mϕ 增加</td><td>Morales 等人（2010）</td></tr>
<tr><td>非洲爪蟾</td><td>FV3 贮藏宿主</td><td>Robert 等人（2007）</td></tr>
<tr><td>鞍带石斑鱼</td><td>TGIV 贮藏宿主</td><td>Chao 等人（2004）</td></tr>
<tr><td>六须鲶</td><td>TGIV 抑制肾巨噬细胞 ROI</td><td>Siwicki 等人（1999）</td></tr>
<tr><td>先天性免疫</td><td>墨西哥钝口螈</td><td>对 ATV 产生先天性免疫反应</td><td>Cotter 等人（2008）</td></tr>
<tr><td>NK 细胞反应</td><td>非洲爪蟾</td><td>FV3 引起 NK 细胞增加</td><td>Morales 等人（2010）</td></tr>
<tr><td>CD8 反应</td><td>非洲爪蟾</td><td>CD8 增加并清除 FV3</td><td>Morales 等人（2010）</td></tr>
<tr><td>没有淋巴细胞反应</td><td>墨西哥钝口螈</td><td>缺乏与 ATV 易感性相关的适应性免疫</td><td>Cotter 等人（2008）</td></tr>
<tr><td>非经典 MHC Ib 限制 iT 细胞群</td><td>非洲爪蟾</td><td>在蝌蚪和成蛙中对 FV3 的保护不太清楚</td><td>Edholm 等人（2013）</td></tr>
<tr><td colspan="4">体液免疫</td></tr>
<tr><td rowspan="2">抗菌肽</td><td>北美豹蛙</td><td>破坏 FV3 的囊膜</td><td>Chinchar 等人（2001）</td></tr>
<tr><td>东北林蛙</td><td>抑制 RGV 的传染</td><td>Yang 等人（2012）</td></tr>
<tr><td>IgY</td><td>非洲爪蟾</td><td>FV3 的清除；对再感染的记忆反应</td><td>Du Pasquier 等人（1989，2000）；Hsu（1998）</td></tr>
<tr><td colspan="4">炎性细胞因子</td></tr>
</table>

续表

免疫参数	物种	免疫结果	参考文献
TNFα	非洲爪蟾	表达抗 RV 保护的相关物；在蝌蚪中表达轻微和延迟	Morales 等人（2010）
	EPC 细胞系	受 FV3 的诱导	Holopainen 等人（2012）
IL-1β	非洲爪蟾	表达抗 FV3 保护的相关物；在蝌蚪中表达轻微和延迟	Morales 等人（2010）
IFN-γ	非洲爪蟾	表达抗 FV3 保护的相关物；在蝌蚪中表达轻微和延迟	De Jesús Andino 等人（2012）
抗病毒免疫			
Mx	金头鲷	抑制 ESV 的复制	Alvarez-Torres 等人（2013）；Fernandez-Trujillo 等人（2013）
IFN 可诱导基因	墨西哥钝口螈	ATV-引起表达	Cotter 等人（2008）
IFITM1	牙鲆	对 RGV 产生细胞抗病毒反应	Zhu 等人（2013）
I 型 IFN	非洲爪蟾	表达抗 FV3 保护的相关物；在蝌蚪中表达轻微和延迟	Grayfer 等人（2014）

6 蛙病毒逃避宿主抗病毒免疫的策略

如同在痘病毒和其他病毒所见到的一样，蛙病毒编码多个蛋白，这些蛋白的功能就是阻碍宿主的抗病毒反应（Finlay 和 McFadden，2006；Johnston 和 McFadden，2003；Seet 等人，2003）。然而，除真核细胞启动因子大亚基的蛙病毒同源分子（vIF-2α）以外，蛙病毒编码的这些基因产物的功能至今还没有确定。现对 vIF-2α 的功能描述如下，同时对其他潜在的免疫逃避蛋白也作一个简单的描述。

6.1 vIF-2α 阻断 eIF-2α 的磷酸化

蛋白激酶 R（protein kinase R，简称 PKR，又称 EIF2αK2）是一种蛋白激

酶，该酶在对各种环境应激因子包括病毒感染的反应过程中，通过真核细胞翻译启动因子 2α（eukaryotic translation initiation factor 2，简称 eIF-2α）亚基的磷酸化和失活，来调控细胞蛋白质的合成（Proud，1995；Toth 等人，2006）。PKR 在未感染的细胞当中含量很低，是一种无活性的单体在细胞受到刺激后，通过 IFN 诱导生成。然而，在病毒感染过程中，低浓度的病毒双链 DNA 与 PKR 结合，形成二聚体，并进一步通过自身磷酸化（autophosphorylation）而激活（Zhang 等人，2001）。激活的 PKR 随后磷酸化 eIF-2α 亚基，从而导致整个蛋白合成的停止（Panniers 等人，1988；Rowlands 等人，1988）。除 PKR 介导的 eIF-2α 失活对蛋白合成的影响，激活的 PKR 还起着其他作用。激活的 PKR 能使抑制因子 I-κB 磷酸化，并结合到 NF-κB 上，进一步导致后者的释放，随后激活前炎症因子和干扰素基因的表达（Proud，1995）。此外，在病毒感染的细胞中，激活后的 PKR 是触发细胞凋亡的危险信号。因为翻译抑制、NF-κB 激活以及细胞凋亡会给病毒复制带来不利的影响，病毒已经进化出了许多方法来绕过 PKR 介导的抗病毒功能（Diener 等人，1993；Katze，1992）。

为了阻断 eIF-2α 的磷酸化，蛙病毒编码一种 eIF-2α 的伪底物，并命 161
名为 vIF-2α。在大多数蛙病毒中，vIF-2α 作为一种大约 250 个氨基酸长度的蛋白，包含多个蛙病毒、宿主细胞 eIF-2α 以及牛痘病毒 K3L 蛋白共有的一个基序 V［L/I］RVDxxKGY［V/I］D（Majji 等人，2006）。研究显示，K3L 作为 PKR 的一个伪底物，通过阻断 eIF-2α 的磷酸化，使其失活（Beattie 等人，1995）。而 vIF-2α 已显示出以上述相似的方式发挥着作用（参见下述内容）。研究者通过使用缺乏 vIF-2α 的 ATV 基因敲除突变体（Jacobs，2011）显示，该突变体病毒对 IFN 介导的抑制更为敏感，并且不能阻断 eIF-2α 的磷酸化。此外，*wt* ATV 能降解鱼类的 PKZ，PKZ 是一种 IFN 诱导的类似于 PKR 的激酶，而基因敲除的突变体则不能。最后，基因敲除的突变体能在体降低毒力，从而提示，vIF-2α 是一个毒力基因。类似的毒力减弱也见于非洲爪蟾在感染基因敲除的 FV3 突变体之后，该突变体病毒缺乏截短型 vIF-2α（Chen 等人，2011）。截短型 vIF-2α 见于 FV3、中华鳖虹彩病毒和蛙虹彩病毒，其分子结构特点是在原分子中丢失了 N 端的一半。由于这个区域含有上面描述的 VxRVDxxKGYxD 基序，毒力的降低不可能是由 vIF-2α 对 PKR 的影响造成的，而是由蛋白 C 端一半的某些成分造成的。

罗滕堡等人（2011）运用酵母模型进行研究，结果显示，将表达人类 PKR 或表达斑马鱼 PKR 的一个载体转染给酵母细胞，引起了明显的细胞死亡。为了确定 vIF-2α 作为 PKR 拮抗物的作用，有研究者将表达全长牛蛙病毒（RCV）的 vIF-2α 的一个载体与 PKR 一起共转染，结果阻断了人类 PKR 和斑马鱼 PKR 的毒性效应。在提示物种特异性方面，牛痘病毒 K3L 是唯一能够阻断人类 PKR 活性的。虽然上述研究表明，在病毒感染的细胞中 vIF-2α 在维持蛋白合成中起作用，但是在 FV3 和其他亲缘关系较近、含有截短型 vIF-2α 基因并缺少关键性 N 端基序的蛙病毒中的观察表明，vIF-2α 并不是在病毒感染细胞中维持蛋白合成起作用的唯一病毒蛋白。

6.2 核糖核酸酶 III 样蛋白

与痘病毒相似，蛙病毒至少有两个基因阻止 PKR 的激活。牛痘病毒既编码上述的 K3L 基因，也编码一个叫 E3L 的第二基因。这两个基因的编码产物能与双链 DNA 结合，从而阻止二聚体的形成，并激活 PKR（Langland 和 Jacobs，2002；Langland 等人，2006）。虽然蛙病毒没有检测出有蛋白与 E3L 同源，但有一个核糖核酸酶 III 样蛋白（RNAse III-like protein）已被鉴定。RNAse III 以双链 DNA 为靶向目标，蛙病毒 RNAse III 同源物有可能与病毒诱导的双链 DNA 结合，并使之降解，或者阻断其与之相互作用，并激活 PKR，然而，直接验证这一学说的实验还没有被报道。此外，使用反义吗啉代低聚核
162 苷进行 RNAse III 样蛋白表达的基因敲低研究，结果显示病毒产量降低了 40%（Cheng 和 Chinchar，美国密西西比大学，未发表资料），从而提示 RNAse III 样蛋白在蛙病毒复制中起着重要作用。

6.3 β-羟基类固醇脱氢酶与 vCARD

痘病毒包含与 β-羟基类固醇脱氢酶（β-hydroxysteroid dehydrogenase，βHSD）同源的蛋白质。βHSD 在类固醇合成中起着重要的作用。由牛痘病毒表达的一个 βHSD 同源物，能导致免疫抑制，并引起病毒复制的增加（Sroller 等人，1998）。vCARD 是一个分子量为 10 kDa，并由病毒编码的蛋白质，该蛋白含有一个半胱天冬酶激活和募集结构域（caspase activation and recruitment domain，CARD）基序，该基序调控带有相似结构域蛋白之间的相互作用（Kawai 和 Akira，2009，2010）。参与细胞凋亡的蛋白或者诱导 IFN 和前炎症的分子如 RIG-I、MDA5 以及 MAVS 都含有 CARD 基序，因而推测，vCARD 与一个或多个上述炎症信号分子相互作用，从而造成细胞抗病毒免疫的短路（Besch 等人，2009；Meylan 等人，2005）。

6.4　病毒肿瘤坏死因子受体、脱氧尿苷三磷酸酶、DNA 胞嘧啶甲基转移酶

除上述 4 个病毒基因产物以外，蛙病毒同样也含有病毒肿瘤坏死因子受体（virus tumor necrosis factor receptor，vTNFR）和脱氧尿苷三磷酸酶（dUTPase）同源蛋白（Chinchar 等人，2009；Eaton 等人，2007）及一个独特的病毒编码的 DNA 胞嘧啶甲基转移酶（DNA cytosine methyltransferase，DMTase）同源蛋白分子。与它们在痘病毒中对应的蛋白分子相似，蛙病毒的 vTNFR 在功能上能够充当一个诱骗分子，阻断 TNFα 所介导的保护。通常 dUTPase 被认为在病毒 DNA 合成中，通过增加 dTTP 池或者阻断 dUTP 与 DNA 的结合起着重要作用。但是对疱疹病毒 dUTPase 的研究显示，dUTPase 可以阻断抗病毒的免疫反应（Glaser 等人，2006；Oliveros 等人，1999）。最后还要一提的是，蛙病毒的 DMTase 可能在 CpG 基序（CpG motifs）中通过甲基化胞嘧啶残基和阻断 TLR-9 的识别或者阻断胞浆 DNA 感受器及阻止随后的 IFN 和前炎症细胞因子的诱导中起着重要的作用（Krieg，2002；Krug 等人，2004）。

上述所列出的推导性免疫逃避蛋白（putative immune evasion proteins）是基于可以检测到的蛙病毒同源蛋白质，而这些同源蛋白质在其他系统中具有抗病毒作用。而这一点在确定和鉴定蛙病毒免疫逃避基因中，可作为一个非常有用的起点。研究显示，大约 12 个未知功能的其他开放阅读框（ORF）是蛙病毒所独有的。这些 ORF 是否在特异性宿主中编码调控病毒复制蛋白质，或者它们是否编码调控宿主特异性免疫反应的蛋白质，仍有待确定。在未来，有必要用反义吗啉代低聚核苷进行基因敲低实验或者用小 RNA（siRNA）进行干扰及 163
用基因敲除的突变体进行感染，从而弄清楚这些独特的蛙病毒特异性蛋白的功能。

7　结语和未来的方向

从文中所描述的研究中可以看出，抗蛙病毒免疫是多层面的、复杂的以及有可能具有物种特异性和具有发育阶段特异性的。同样明显的是，我们对这些病原体免疫反应的认识，以及对宿主能够产生有效反应，从而抑制和消除这些感染的能力还存在着很多缺陷的认识上还存在很多空白。尤其令人担忧的是，蛙病毒属和虹彩病毒科其他属的病毒在进化中，已经设计出了许多高效率的策略来实施免疫逃避，甚至是利用宿主的免疫成分来达到存活、促进传播以及扩

大宿主范围的目的。很明显，蛙病毒编码大量推导性的基因产物，这些产物既包括潜在的毒力因子，也包括未来干预治疗有发展前景的目标分子。

低等脊椎动物的免疫系统在功能上类似于哺乳动物的免疫系统很容易被忽视，但是越来越多的研究文献表明并非如此。通过对支配这些独特免疫系统的生理和生态压力的基本认识，我们在逐步认识蛙病毒感染的策略和免疫系统，以及这些免疫系统是否有充分的共进化（co-evolved）来阻止它。

对病毒感染和免疫颠覆策略的研究不仅要考虑到定义良好的哺乳动物病原体，还要考虑到蛙病毒有可能代表着独特的病毒病因。与大多数恒温脊椎动物（homeothermic vertebrate）病原体相反，蛙病毒在克服细胞和宿主趋向性屏障（host tropism barriers）方面具有非凡的能力。同时，它们致病的机制至少是部分地依赖于病毒载量。的确如此，变温宿主的免疫系统是不同生理压力和致病压力进化的结果，并受到不同生理压力和致病压力的影响，而这些压力已塑造了哺乳动物的免疫系统。理所当然，蛙病毒作为病原与这些独特的免疫系统发生了共进化。因此，我们必须对二者都有更深刻的认识，才能完全理解。

致谢

我们要感谢路易丝·罗林斯-史密斯（Louise Rollins-Smith，范德堡大学）对早期草稿的评审；利昂·格瑞福（L. Grayfer）得到了加拿大自然科学与工程研究委员会和霍华德·休斯医学研究所生命科学研究基金会博士后奖学金的资助。弗朗西斯科·德·赫苏斯·安迪诺（F. De Jesús Andino）莫里斯基金会的项目资助号：D14ZO-084；雅克·罗伯特（J. Robert）和伊娃-斯蒂娜·埃德霍尔姆（E. S. Edholm）国家卫生研究院（NIH）的研究项目资助号：2R24Al059830－10；雅克·罗伯特（J. Robert）和V·格雷戈里·钦察尔（V. G. Chinchar）国家自然科学（NSF）项目资助号分别为：IOS 0923772 和 IOS 0742711。

开放存取出版物的发行得到了田纳西大学（林业、野生动物与渔业系，研究与交流办公室，农业研究所）、华盛顿州立大学图书馆、戈登州立大学（学术事务办公室）、两栖爬行动物兽医协会以及两栖爬行动物保护协会的资金赞助。

开放存取

参考文献

Aggad D, Mazel M, Boudinot P et al. (2009). The two groups of zebrafish virus-induced interferons signal via distinct receptors with specific and shared chains. J Immunol 183:3924-3931.

Altmann SM, Mellon MT, Distel DL et al. (2003). Molecular and functional analysis of an interferon gene from the zebrafish, *Danio rerio*. J Virol 77:1992-2002.

Alvarez-Torres D, Garcia-Rosado E, Fernandez-Trujillo MA et al. (2013). Antiviral specificity of the *Solea senegalensis* Mx protein constitutively expressed in CHSE-214 cells. Mar Biotechnol (NY) 15:125-132.

Aubertin AM, Hirth C, Travo C et al. (1973). Preparation and properties of an inhibitory extract from frog virus 3 particles. J Virol 11:694-701.

Auffray C, Fogg D, Garfa M et al. (2007). Monitoring of blood vessels and tissues by a population of monocytes with patrolling behavior. Science 317:666-670.

Baum A, Garcia-Sastre A (2010). Induction of type I interferon by RNA viruses: cellular receptors and their substrates. Amino Acids 38:1283-1299.

Bayley AE, Hill BJ, Feist SW (2013). Susceptibility of the European common frog*Rana temporaria* to a panel of ranavirus isolates from fish and amphibian hosts. Dis Aquat Organ 103:171-183.

Beattie E, Paoletti E, Tartaglia J (1995). Distinct patterns of IFN sensitivity observed in cells infected with vaccinia K3L-and E3L-mutant viruses. Virology 210:254-263.

Behar SM, Porcelli SA (2007). CD1-restricted T cells in host defense to infectious diseases. Curr Top Microbiol Immunol 314:215-250.

Behncke H, Stohr AC, Heckers KO et al. (2013). Mass-mortality in green striped tree dragons (*Japalura splendida*) associated with multiple viral infections. Vet Rec 173:248.

Bendelac A, Lantz O, Quimby ME et al. (1995). CD1 recognition by mouse NK1+ T lymphocytes. Science 268:863-865.

Bendelac A, Hunziker RD, Lantz O (1996). Increased interleukin 4 and immunoglobulin E production in transgenic mice overexpressing NK1 T cells. J Exp Med 184:1285-1293.

Bendelac A, Rivera MN, Park SH et al. (1997). Mouse CD1-specific NK1 T cells: development, specificity, and function. Annu Rev Immunol 15:535-562.

Besch R, Poeck H, Hohenauer T et al. (2009). Proapoptotic signaling induced by RIG-I and MDA-5 results in type I interferon-independent apoptosis in human melanoma cells. J Clin Invest 119:2399-2411.

Blomberg B, Bernard CC, Du Pasquier L (1980). In vitro evidence for T-B lymphocyte collaboration in the clawed toad, *Xenopus*. Eur J Immunol 10:869-876.

Bollinger TK, Mao J, Schock D et al. (1999). Pathology, isolation, and preliminary molecular characterization of a novel iridovirus from tiger salamanders in Saskatchewan. J Wildl Dis 35:413-429.

Brunner JL, Schock DM, Davidson EW et al. (2004). Intraspective reservoires: complex life hoistory and the persistance of a lethal ranavirus. Ecology 85:560-566.

Caipang CM, Hirono I, Aoki T (2003). In vitro inhibition of fish rhabdoviruses by Japanese flounder, *Paralichthys olivaceus* Mx. Virology 317:373-382.

Caipang CM, Takano T, Hirono I et al. (2006). Genetic vaccines protect red seabream, *Pagrus major*, upon challenge with red seabream iridovirus (RSIV). Fish Shellfish Immunol 21:130-138.

Cano I, Alonso MC, Garcia-Rosado E et al. (2006). Detection of lymphocystis disease virus (LCDV) in asymptomatic cultured gilt-head seabream (*Sparus aurata*, L.) using an immunoblot technique. Vet Microbiol 113:137-141.

Cano I, Ferro P, Alonso MC et al. (2009). Application of in situ detection techniques to determine the systemic condition of lymphocystis disease virus infection in cultured gilt-head seabream, Sparus aurata L. J Fish Dis 32:143-150.

Carey C, Cohen N, Rollins-Smith L (1999). Amphibian declines: an immunological perspective. Dev Comp Immunol 23:459-472.

Chang M, Nie P, Collet B et al. (2009). Identification of an additional two-cysteine containing type I interferon in rainbow trout *Oncorhynchus mykiss* provides evidence of a major gene duplication event within this gene family in teleosts. Immunogenetics 61:315-325.

Chao CB, Chen CY, Lai YY et al. (2004). Histological, ultrastructural, and in situ hybridization study on enlarged cells in grouper *Epinephelus* hybrids infected by grouper iridovirus in Taiwan (TGIV). Dis Aquat Organ 58:127-142.

Chen G, Robert J (2011). Antiviral immunity in amphibians. Viruses 3:2065-2086.

Chen G, Ward BM, Yu KH et al. (2011). Improved knockout methodology reveals that frog virus 3 mutants lacking either the 18K immediate-early gene or the truncated vIF-2alpha gene are defective for replication and growth in vivo.J Virol 85:11131-11138.

Cheng K, Escalon BL, Robert J et al. (2014). Differential transcription of fathead minnow immune-related genes following infection with frog virus 3, an emerging pathogen of ectothermic vertebrates. Virology 456-457:77-86.

Chinchar VG, Wang J, Murti G et al. (2001). Inactivation of frog virus 3 and channel catfish virus by esculentin-2P and ranatuerin-2P, two antimicrobial peptides isolated from frog skin. Virology 288:351-357.

Chinchar VG, Bryan L, Wang J et al. (2003). Induction of apoptosis in frog virus 3 infected cells. Virology 306:303-312.

Chinchar VG, Hyatt A, Miyazaki T et al. (2009). Family Iridoviridae: poor viral relations no longer. Curr Top Microbiol Immunol 328:123-170.

Choi HJ, Xu H, Geng Y et al. (2008). Bacterial infection alters the kinetics and function of iNKT cell responses. J Leukoc Biol 84:1462-1471.

Chung SW, Arnott JA, Yang Y et al. (2003). Presence of prepackaged mRNA in virions of DNA adenovirus. J Biol Chem 278:50635-50640.

Cohen NR, Garg S, Brenner MB (2009). Antigen presentation by CD1 lipids, T cells, and NKT cells in microbial immunity. Adv Immunol 102:1-94.

Coiras M, Lopez-Huertas MR, Perez-Olmeda M et al. (2009). Understanding HIV-1 latency provides clues for the eradication of long-term reservoirs. Nat Rev Microbiol 7:798-812.

Cotter JD, Storfer A, Page RB et al. (2008). Transcriptional response of Mexican axolotls to *Ambystoma tigrinum virus* (ATV) infection. BMC Genomics 9:493.

De Jesús Andino F, Chen G, Li Z et al. (2012). Susceptibility of *Xenopus laevis* tadpoles to infection by the ranavirus Frog-Virus 3 correlates with a reduced and delayed innate immune response in comparison with adult frogs. Virology 432:435-443.

Diener TO, Hammond RW, Black T et al. (1993). Mechanism of viroid pathogenesis: differential activation of the interferon-induced, double-stranded RNA-activated, M(r) 68,000 protein kinase by viroid strains of varying pathogenicity. Biochimie 75:533-538.

Du Pasquier L, Schwager J, Flajnik MF (1989). The immune system of *Xenopus*. Annu Rev Immunol 7:251-275.

Du Pasquier L, Robert J, Courtet M et al. (2000). B-cell development in the amphibian *Xenopus*. *Immunol Rev* 175:201-213.

Duffus ALJ, Nichols RA, Garner TWJ (2013). Investigations into the life history stages of the common frog (*Rana temporaria*) affected by an amphibian ranavirus in the United Kingdom. Herpetol Rev 44:260-263.

Eaton HE, Metcalf J, Penny E et al. (2007). Comparative genomic analysis of the family *Iridoviridae*: re-annotating and defining the core set of iridovirus genes. Virol J 4:11.

Edholm ES, Albertorio Saez LM, Gill AL et al. (2013). Nonclassical MHC class I-dependent invariant T cells are evolutionarily conserved and prominent from early development in amphibians. Proc Natl Acad Sci USA 110:14342-14347.

Elharrar M, Hirth C, Blanc J et al. (1973). Pathogenesis of the toxic hepatitis of mice provoked by FV 3 (frog virus 3): inhibition of liver macromolecular synthesis (author's transl). Biochim Biophys Acta 319:91-102.

Fernandez-Trujillo MA, Garcia-Rosado E, Alonso MC et al. (2013). Mx1, Mx2 and Mx3 proteins from the gilthead seabream (*Sparus aurata*) show in vitro antiviral activity against RNA and DNA viruses. Mol Immunol 56:630-636.

Finlay BB, McFadden G (2006). Anti-immunology: evasion of the host immune system by bacterial and viral pathogens. Cell 124:767-782.

Flajnik MF, Kasahara M (2001). Comparative genomics of the MHC: glimpses into the evolution of the adaptive immune system. Immunity 15:351-362.

Fullerton JN, O' Brien AJ, Gilroy DW (2013). Pathways mediating resolution of inflammation: when enough is too much. J Pathol 231:8-20.

Gantress J, Maniero GD, Cohen N et al. (2003). Development and characterization of a model system to study amphibian immune responses to iridoviruses. Virology 311:254-262.

Gendrault JL, Steffan AM, Bingen A et al. (1981). Penetration and uncoating of frog virus 3 (FV3) in cultured rat Kupffer cells. Virology 112:375-384.

Glaser R, Litsky ML, Padgett DA et al. (2006). EBV-encoded dUTPase induces immune dysregulation: implications for the pathophysiology of EBV-associated disease. Virology 346:205-218.

Goodenow MM, Rose SL, Tuttle DL et al. (2003). HIV-1 fitness and macrophages. J Leukoc Biol 74:657-666.

Gousset K, Ablan SD, Coren LV et al. (2008). Real-time visualization of HIV-1 GAG trafficking in infected macrophages. PLoS Pathog 4:e1000015.

Goyos A, Ohta Y, Guselnikov S et al. (2009). Novel nonclassical MHC class Ib genes associated with CD8 T cell development and thymic tumors. Mol Immunol 46: 1775-1786.

Goyos A, Sowa J, Ohta Y et al. (2011). Remarkable conservation of distinct nonclassical MHC class I lineages in divergent amphibian species. J Immunol 186: 372-381.

Gray MJ, Miller DL (2013). Rise of ranavirus: an emerging pathogen threatens ectothermic vertebrates. Wildl Professional 7:51-55.

Grayfer L, Garcia EG, Belosevic M (2010). Comparison of macrophage antimicrobial responses induced by type II interferons of the goldfish (*Carassius auratus* L.). J Biol Chem 285:23537-23547.

Grayfer L, De Jesús Andino F, Robert J (2014). The amphibian (*Xenopus laevis*) type I interferon response to Frog Virus 3: new insight into ranavirus pathogenicity. J Virol 88(10):5766-5777.

Groot F, Welsch S, Sattentau QJ (2008). Efficient HIV-1 transmission from macrophages to T cells across transient virological synapses. Blood 111:4660-4663.

Gut JP, Anton M, Bingen A et al. (1981). Frog virus 3 induces a fatal hepatitis in rats. Lab Invest 45:218-228.

Hagmann W, Steffan AM, Kirn A et al. (1987). Leukotrienes as mediators in frog virus 3-induced hepatitis in rats. Hepatology 7:732-736.

Haislip NA, Gray MJ, Hoverman JT et al. (2011). Development and disease: how susceptibility to anemerging pathogen changes through anuran development. PLoS One 6:e22307.

Hervas-Stubbs S, Perez-Gracia JL, Rouzaut A et al. (2011). Direct effects of type I interferons on cells of the immune system. Clin Cancer Res 17:2619-2627.

Holopainen R, Tapiovaara H, Honkanen J (2012). Expression analysis of immune response genes in fish epithelial cells following ranavirus infection. Fish Shellfish Immunol 32:1095-1105.

Horton JD, Horton TL, Dzialo R et al. (1998). T-cell and natural killer cell development in thymectomized *Xenopus*. Immunol Rev 166:245-258.

Hoverman JT, Gray MJ, Miller DL (2010). Anuran susceptibilities to ranaviruses: role of species identity, exposure route, and a novel virus isolate. Dis Aquat Organ 89: 97-107.

Hsu E (1998). Mutation, selection, and memory in B lymphocytes of exothermic vertebrates. Immunol Rev 162:25-36.

Hyatt AD, Williamson M, Coupar BE et al.(2002). First identification of a ranavirus from green pythons (*Chondropython viridis*). J Wildl Dis 38:239-252.

Jancovich JK, Jacobs BL (2011). Innate immune evasion mediated by the *Ambystoma tigrinum virus* eukaryotic translation initiation factor 2alpha homologue. J Virol 85:5061-5069.

Joerink M, Forlenza M, Ribeiro CM et al. (2006a). Differential macrophage polarisation during parasitic infections in common carp (*Cyprinus carpio* L.). Fish Shellfish Immunol 21:561-571.

Joerink M, Ribeiro CM, Stet RJ et al.(2006b). Head kidney-derived macrophages of common carp (*Cyprinus carpio* L.) show plasticity and functional polarization upon differential stimulation. J Immunol 177:61-69.

Joerink M, Savelkoul HF, Wiegertjes GF (2006c). Evolutionary conservation of alternative activation of macrophages: structural and functional characterization of arginase 1 and 2 in carp (*Cyprinus carpio* L.). Mol Immunol 43:1116-1128.

Johnston JB, McFadden G (2003). Poxvirus immunomodulatory strategies: current perspectives. J Virol 77:6093-6100.

Jurgens JB, Gartland LA, Du Pasquier L et al.(1995). Identification of a candidate CD5 homologue in the amphibian *Xenopus laevis*. J Immunol 155:4218-4223.

Katze MG (1992). The war against the interferon-induced dsRNA-activated protein kinase: can viruses win? J Interferon Res 12:241-248.

Kawai T, Akira S (2009). The roles of TLRs, RLRs and NLRs in pathogen recognition. Int Immunol 21:317-337.

Kawai T, AkiraS (2010). The role of pattern-recognition receptors in innate immunity: update on Toll-like receptors. Nat Immunol 11:373-384.

Kirn A, Gut JP, Elharrar M (1972). [FV3 (Frog Virus 3) toxicity for the mouse]. Nouv Presse Med 1:1943.

Kirn A, Steffan AM, Bingen A(1980). Inhibition of erythrophagocytosis by cultured rat Kupffer cells infected with frog virus 3. J Reticuloendothel Soc 28:381-388.

Kirn A, Bingen A, Steffan AM et al.(1982). Endocytic capacities of Kupffer cells isolated from the human adult liver. Hepatology 2:216-222.

Kochs G, Reichelt M, Danino D et al. (2005). Assay and functional analysis of dynamin-like Mx proteins. Methods Enzymol 404:632-643.

Kotenko SV (2011). IFN-lambdas.Curr Opin Immunol 23:583-590.

Krieg AM (2002). CpG motifs in bacterial DNA and their immune effects.Annu Rev Immunol 20:709-760.

Krug LT, Pozharskaya VP, Yu Y et al.(2004). Inhibition of infection and replication of human herpesvirus 8 in microvascular endothelial cells by alpha interferon and phosphonoformic acid.J Virol 78:8359-8371.

Landsberg JH, Kiryu Y, Tabuchi M et al.(2013). Co-infection by alveolate parasites and frog virus 3-like ranavirus during an amphibian larval mortality event in Florida, USA.Dis Aquat Organ 105:89-99.

Langland JO, Jacobs BL (2002). The role of the PKR-inhibitory genes, E3L and K3L, in determining vaccinia virus host range.Virology 299:133-141.

Langland JO, Kash JC, Carter V et al.(2006). Suppression of proinflammatory signal transduction and gene expression by the dual nucleic acid binding domains of the vaccinia virus E3L proteins.J Virol 80:10083-10095.

Le Bourhis L, Martin E, Peguillet I et al.(2010). Antimicrobial activity of mucosal-associated invariant T cells.Nat Immunol 11:701-708.

Leibovitz L (1980). Lymphocystis disease.J Am Vet Med Assoc 176:202.

Lester K, Hall M, Urquhart K et al.(2012). Development of an in vitro system to measure the sensitivity to the antiviral Mx protein of fish viruses.J Virol Methods 182:1-8.

Levraud JP, Boudinot P, Colin I et al.(2007). Identification of the zebrafish IFN receptor: implications for the origin of the vertebrate IFN system.J Immunol 178:4385-4394.

Li Z, Strunk JJ, Lamken P et al.(2008). The EM structure of a type I interferon-receptor complex reveals a novel mechanism for cytokine signaling. J Mol Biol 377:715-724.

Li Z, Xu X, Huang L et al.(2010). Administration of recombinant IFN1 protects zebrafish (*Danio rerio*) from ISKNV infection. Fish Shellfish Immunol 29:399-406.

Long S, Wilson M, Bengten E et al.(2004). Identification of a cDNA encoding channel catfish interferon.Dev Comp Immunol 28:97-111.

Lopez C, Aubertin AM, Tondre L et al.(1986). Thermosensitivity of frog virus 3 genome expression: defect in early transcription.Virology 152:365-374.

Lopez-Munoz A, Roca FJ, Meseguer J et al.(2009). New insights into the evolution of

IFNs: zebrafish group II IFNs induce a rapid and transient expression of IFN-dependent genes and display powerful antiviral activities. J Immunol 182: 3440-3449.

Lutfalla G, Roest Crollius H, Stange-Thomann N et al. (2003). Comparative genomic analysis reveals independent expansion of a lineage-specific gene family in vertebrates: the class II cytokine receptors and their ligands in mammals and fish. BMC Genomics 4: 29.

Majji S, LaPatra S, Long SM et al. (2006). *Rana catesbeiana virus* Z (RCV-Z): a novel pathogenic ranavirus. Dis Aquat Organ 73: 1-11.

Maniero GD, Morales H, Gantress J et al. (2006). Generation of a long-lasting, protective, and neutralizing antibody response to the ranavirus FV3 by the frog Xenopus. Dev Comp Immunol 30: 649-657.

Matsuda JL, Gapin L (2005). Developmental program of mouse Valpha14i NKT cells. Curr Opin Immunol 17: 122-130.

Meylan E, Curran J, Hofmann K et al. (2005). Cardif is an adaptor protein in the RIG-I antiviral pathway and is targeted by hepatitis C virus. Nature 437: 1167-1172.

Miller D, Gray M, Storfer A (2011). Ecopathology of ranaviruses infecting amphibians. Viruses 3: 2351-2373.

Morales HD, Robert J (2007). Characterization of primary and memory CD8 T-cell responses against ranavirus (FV3) in *Xenopus laevis*. J Virol 81: 2240-2248.

Morales HD, Abramowitz L, Gertz J et al. (2010). Innate immune responses and permissiveness to ranavirus infection of peritoneal leukocytes in the frog *Xenopus laevis*. J Virol 84: 4912-4922.

Nahrendorf M, Swirski FK, Aikawa E et al. (2007). The healing myocardium sequentially mobilizes two monocyte subsets with divergent and complementary functions. J Exp Med 204: 3037-3047.

Nakajima K, Kunita J (2005). [Red sea bream iridoviral disease]. Uirusu 55: 115-125.

Nakajima K, Maeno Y, Honda A et al. (1999). Effectiveness of a vaccine against red sea bream iridoviral disease in a field trial test. Dis Aquat Organ 36: 73-75.

Nishikawa A, Murata E, Akita M et al. (1998). Roles of macrophages in programmed cell death and remodeling of tail and body muscle of *Xenopus laevis* during metamorphosis. Histochem Cell Biol 109: 11-17.

Oliveros M, Garcia-Escudero R, Alejo A et al. (1999). African swine fever virus dUTPase is a highly specific enzyme required for efficient replication in swine macrophages. J Virol 73:8934-8943.

Panniers R, Rowlands AG, Henshaw EC (1988). The effect of $Mg2^{+}$ and guanine nucleotide exchange factor on the binding of guanine nucleotides to eukaryotic initiation factor 2. J Biol Chem 263:5519-5525.

Pham PH, Lai YS, Lee FF et al. (2012). Differential viral propagation and induction of apoptosis by grouper iridovirus (GIV) in cell lines from three non-host species. Virus Res 167:16-25.

Proud CG (1995). PKR: a new name and new roles. Trends Biochem Sci 20:241-246.

Qi Z, Nie P, Secombes CJ et al. (2010). Intron-containing type I and type III IFN coexist in amphibians: refuting the concept that a retroposition event gave rise to type I IFNs. J Immunol 184:5038-5046.

Raghow R, Granoff A (1979). Macromolecular synthesis in cells infected by frog virus 3. X. Inhibition of cellular protein synthesis by heat-inactivated virus. Virology 98:319-327.

Reeve BC, Crespi EJ, Whipps CM et al. (2013). Natural stressors and ranavirus susceptibility in larval wood frogs (*Rana sylvatica*). Ecohealth 10:190-200.

Robert J, Ohta Y (2009). Comparative and developmental study of the immune system in *Xenopus*. Dev Dyn 238:1249-1270.

Robert J, Morales H, Buck W et al. (2005). Adaptive immunity and histopathology in frog virus 3-infected *Xenopus*. Virology 332:667-675.

Robert J, Abramowitz L, Gantress J et al. (2007). *Xenopus laevis*: a possible vector of Ranavirus infection? J Wildl Dis 43:645-652.

Robert J, George E, De Jesús Andino F et al. (2011). Waterborne infectivity of the Ranavirus frog virus 3 in *Xenopus laevis*. Virology 417:410-417.

Robertsen B (2006). The interferon system of teleost fish. Fish Shellfish Immunol 20:172-191.

Robertsen B, Bergan V, Rokenes T et al. (2003). Atlantic salmon interferon genes: cloning, sequence analysis, expression, and biological activity. J Interferon Cytokine Res 23:601-612.

Rollins-Smith LA (2009). The role of amphibian antimicrobial peptides in protection of amphibians from pathogens linked to global amphibian declines. Biochim

Biophys Acta 1788:1593-1599.

Rollins-Smith LA, Reinert LK, O' Leary CJ et al. (2005). Antimicrobial peptide defenses in amphibian skin.Integr Comp Biol 45:137-142.

Rothenburg S,Chinchar VG,Dever TE (2011). Characterization of a ranavirus inhibitor of the antiviral protein kinase PKR.BMC Microbiol 11:56.

Rowlands AG,Panniers R,Henshaw EC (1988). The catalytic mechanism of guanine nucleotide exchange factor action and competitive inhibition by phosphorylated eukaryotic initiation factor 2.J Biol Chem 263:5526-5533.

Sadler AJ, Williams BR (2008). Interferon-inducible antiviral effectors. Nat Rev Immunol 8:559-568.

Samuel CE (2001). Antiviral actions of interferons.Clin Microbiol Rev 14:778-809.

Seet BT,Johnston JB,Brunetti CR et al.(2003). Poxviruses and immune evasion. Annu Rev Immunol 21:377-423.

Siwicki AK,Pozet F,Morand M et al.(1999). Effects of iridovirus-like agent on the cell-mediated immunity in sheatfish (*Silurus glanis*)—an in vitro study. Virus Res 63:115-119.

Sroller V, Kutinova L, Nemeckova S et al. (1998). Effect of 3-beta-hydroxysteroid dehydrogenase gene deletion on virulence and immunogenicity of different vaccinia viruses and their recombinants.Arch Virol 143:1311-1320.

Sun B,Robertsen B,Wang Z et al.(2009). Identification of an Atlantic salmon IFN multigene cluster encoding three IFN subtypes with very different expression properties.Dev Comp Immunol 33:547-558.

Sutton WB, Gray MJ, Hardman RH et al. (2014). High susceptibility of the endangered dusky gopher frog to ranavirus.Dis Aquat Organ 112(1):9-16.

Svingerud T,Solstad T,Sun B et al.(2012). Atlantic salmon type I IFN subtypes show differences in antiviral activity and cell-dependent expression:evidence for high IFNb/IFNc-producing cells in fish lymphoid tissues. J Immunol 189: 5912-5923.

Toth AM,Zhang P,Das S et al.(2006). Interferon action and the double-stranded RNA-dependent enzymes ADAR1 adenosine deaminase and PKR protein kinase. Prog Nucleic Acid Res Mol Biol 81:369-434.

Trobridge GD, Chiou PP, Leong JA (1997). Cloning of the rainbow trout (*Oncorhynchus mykiss*) Mx2 and Mx3 cDNAs and characterization of trout Mx protein expression in salmon cells.J Virol 71:5304-5311.

Turner RJ, Manning MJ (1974). Thymic dependence of amphibian antibody responses. Eur J Immunol 4:343-346.

Wiegertjes GF, Forlenza M (2010). Nitrosative stress during infection-induced inflammation in fish: lessons from a host-parasite infection model. Curr Pharm Des 16:4194-4202.

Wu YC, Chi SC (2007). Cloning and analysis of antiviral activity of a barramundi (*Lates calcarifer*) Mx gene. Fish Shellfish Immunol 23:97-108.

Wu MS, Chen CW, Liu YC et al. (2012). Transcriptional analysis of orange-spotted grouper reacting to experimental grouper iridovirus infection. Dev Comp Immunol 37:233-242.

Yang SJ, Xiao XH, Xu YG et al. (2012). Induction of antimicrobial peptides from *Rana dybowskii* under Rana grylio virus stress, and bioactivity analysis. Can J Microbiol 58:848-855.

Zhang F, Romano PR, Nagamura-Inoue T et al. (2001). Binding of double-stranded RNA to protein kinase PKR is required for dimerization and promotes critical autophosphorylation events in the activation loop. J Biol Chem 276:24946-24958.

Zhao C, Zhang H, Wong WC et al. (2009). Identification of novel functional differences in monocyte subsets using proteomic and transcriptomic methods. J Proteome Res 8:4028-4038.

Zhu R, Wang J, Lei XY et al. (2013). Evidence for *Paralichthys olivaceus* IFITM1 antiviral effect by impeding viral entry into target cells. Fish Shellfish Immunol 35:918-926.

Ziegler-Heitbrock L (2007). The CD14+ CD16+ blood monocytes: their role in infection and inflammation. J Leukoc Biol 81:584-592.

Zilberg D, Grizzle JM, Plumb JA (2000). Preliminary description of lesions in juvenile largemouth bass injected with largemouth bass virus. Dis Aquat Organ 39:143-146.

Zou J, Secombes CJ (2011). Teleost fish interferons and their role in immunity. Dev Comp Immunol 35:1376-1387.

Zou J, Tafalla C, Truckle J et al. (2007). Identification of a second group of type I IFNs in fish sheds light on IFN evolution in vertebrates. J Immunol 179:3859-3871.

第六章 蛙病毒的比较病理学及诊断技术

黛布拉·L. 米勒[①]，艾伦·P. 伯西尔[②]，保罗·希克[③]和查理德·J. 惠廷顿[④]

1 蛙病毒疾病

1.1 引言

171 早在20世纪60年代，蛙病毒就被检测出来，并诊断为两栖类动物的病原。20世纪80年代，蛙病毒相继被确诊为爬行类动物和鱼类的病原（Duffus等人，2015）。蛙病毒与这三类低等脊椎动物中的许多动物流行性死亡事件存在着关联（Duffus等人，2015）。尽管缺乏长期的种群数据，但是有证据显示，在两栖类动物当中，蛙病毒能造成种群数量的下降（Teacher等人，2010；Beebee，2012；Earl和Gray，2014；Price等人，2014）。而且有报道记载，在鱼类和两栖类动物当中，由于蛙病毒的暴发，几乎造成整个年龄类别的全部死亡（Petranka等人，2003；Todd-Thompson，2010；Waltzek等人，2014；Wheelwright等人，2014）。受全球保护的一些物种也具有很高的易感性（Geng等人，2010；Sutton等人，2014a）。最近，蛙病毒对养殖鱼类和两栖类动物（Mazzoni等人，2009；Waltzek等人，2014）及对休闲渔业（Grizzle和

①D. L. Miller/Center for Wildlife Health and Department of Biomedical and Diagnostic Sciences，University of Tennessee，Knoxville，TN 37996，USA/e-mail：dmille42@ utk. edu.

②A. P. Pessier/Institution for Conservation Research，San Diego Zoo，San Diego，CA 92112，USA/e-mail：APessier@ sandiegozoo. org.

③④P. Hick · R. J. Whittington/Faculty of Veterinary Science，University of Sydney，Sydney，NSW，Australia/e-mail：paul. hick@ sydney. edu. au；richard. whittington@ sydney. edu. au.

Brunner，2003）产生的潜在经济影响也得到了公认。鉴于这些问题，感染两栖类动物的蛙病毒和动物流行性造血器官坏死病毒被世界动物卫生组织 172 （World Organization for Animal Health）列为必须申报的病原（OIE，Schloegel 等人，2010）。

尽管全球对蛙病毒已经有了初步认识，但对蛙病毒的分布及其对宿主种群和国际贸易的影响仍然知之甚少。而且关于影响宿主-蛙病毒相互作用的机制以及导致死亡事件的因素，目前资料还有限（Gray 等人，2009）。各种各样的实地研究和控制实验正在拓展我们对蛙病毒的认识（Duffus 等人，2015；Brunner 等人，2015；Jancovich 等人，2015b）。尽管如此，还是需要进行更多的研究。正确地设计研究（Gray 等人，2015）是研究有关蛙病毒病出现假设的第一步。采用适当的诊断技术是鉴定病毒感染和确定感染对宿主物种影响的关键。了解病毒性疾病的鉴别诊断是很重要的，感染数据能否与病理和环境信息相结合是确认蛙病毒是否是病原性疾病病因的关键因素。

这一章，我们首先概括了与蛙病毒病相关的大体病变和显微镜所观察的病变；然后讨论了目前在世界范围内的研究室和兽医诊断实验室所使用的诊断试验，指出了某些诊断技术的局限性，确定了需要改进的领域；最后，我们简单地讨论了对蛙病毒的治疗和疫苗开发的研究。

1.2　蛙病毒病的病理学

1.2.1　实地和临床发现

在两栖类动物中，蛙病毒病的暴发常见于幼体，最近还见于变态后的动物（Green 等人，2002；Docherty 等人，2003；Balseiro 等人，2009；2010）。两栖类成年动物蛙病毒病的暴发得到越来越多学者的关注（Cunningham 等人，2007；Cheng 等人，2014）。此外，厄尔和格雷（2014）的研究表明，在高度易感的物种当中，幼体或变态体与蛙病毒相关的死亡足以造成种群数量的下降。研究者们已经证明，在实验室用 FV3 样蛙病毒所进行的试验当中，感染流行率和死亡率之间存在着很强的相关性（Haislip 等人，2011；Hoverman 等人，2011；Brenes 等人，2014a）。因此，在野外实地观测到的高感染流行率，可能是即将发生大规模死亡的一个前兆（Gray 等人，2015）。

死亡事件通常是跨多个物种的突然死亡和大规模死亡（Todd-Thompson，2010；Wheelwright 等人，2014）。动物的死亡会持续几周，较晚期的死亡是由于个体死于继发细菌或真菌感染（Jancovich 等人，1997；Cunningham 等人，2007；Miller 等人，2008；Cheng 等人，2014）。在所报道的所有动物类别的感
173 染中，现场的临床症状变化不定，其中，昏睡（lethargy）是最常见的（表1）。龟类动物会有呼吸窘迫（respiratory distress）的症状（Ruder 等人，2010；Farnsworth 和 Seigel，2013）。死亡之前，发病的两栖类动物和鱼类表现出不稳定的游泳（erratic swimming）、浮力丧失并失去翻正反射（righting reflex）（Mao 等人，1999；Bollinger 等人，1999；Zilberg 等人，2000；Geng 等人，2010；Miller 等人，2011）。最近，在巴西养殖的成年牛蛙中，见到了一例前庭综合征（R. Mazzoni，CRMV-GO，Brazil，个人通讯资料），这可以用来解释在运动协调上的这些变化。对这些功能缺乏协调的动物，未来有必要研究其脑内的组织学变化。

1.2.2 大体病理学

单个患病动物表现出病毒呈全身性分布，并具有与之关联的宿主反应。临床疾病通常是急性的，种群中感染的比例相当高。在野生种群中，疾病的急性过程和快速死亡，可能会阻止对发病事件的检测（Gray 等人，2015），疾病的暴发可能导致大量的个体死亡。研究者一般喜欢检测临床患病的个体，这是因为对个体最有可能作出诊断。感染的个体存在着出血（hemorrhages）、水肿（edema）以及坏死（necrosis）等最常见的大体病变。然而，这些代表性的变
174 化会依据感染物种的不同而有所变化，并且不管它们是否接触到环境应激因子或者其他病原，这些变化也照样产生（见表1）。

表1 所能观察到的患有蛙病毒病的个体的野外（临床）体症和大体变化

类别	病变
两栖类幼体	浮力丧失；不稳定的游泳；厌食；躯干、头部、腿和内部软组织肿胀（水肿）；外部出血（特别是排泄孔周围、眼周围、咽部区域以及腿部）；偶尔内部出血（特别是前肾、肝和脾）
无尾类成体	昏睡；厌食；浮力丧失和不稳定的游泳（水生物种）；腿、脚、躯干和内部软组织肿胀（水肿）；皮肤溃疡；皮肤、口腔和内出血（瘀血斑，瘀血点）；器官易脆（坏死）
有尾类成体	昏睡；厌食；浮力丧失和不稳定的游泳（水生物种）；出血（特别是在尾部和足底面）；肿胀（水肿）；皮肤溃疡；内出血（瘀血斑，瘀血点）；器官易脆（坏死）；四肢坏死（中国大鲵）

续表

类别	病变
鱼类	浮力丧失；不稳定的游泳；厌食；鳃红肿；出血（特别是眼周围、脂肪体、鱼鳔）；过度膨胀的鱼鳔；器官易脆（坏死）；肝脏多发性苍白病灶
龟类	呼吸困难；厌食；口腔坏死斑；头、脖子、腿、内部软组织、眼周围肿胀（水肿或极少见到的坏死）；皮肤溃疡；器官易脆（坏死）；出血（特别是内出血）
蜥蜴	昏睡；厌食；口腔坏死斑；皮肤溃疡；器官易脆（坏死）；偶尔内出血和水肿
蛇	昏睡；厌食；口腔和鼻溃疡

两栖类动物

出血包括瘀血斑、瘀血点和红疹（erythema），常见于无尾两栖类和有尾两栖类动物的皮肤［见图 1（a），（b）］。出血常见于排泄孔附近、腿部附近的腹部表面以及咽部区域，但同样也见于眼周围、耳膜、尾部和脚（Balseiro 等人，2009；Cheng 等人，2014；Cunningham 等人，2007；Docherty 等人，2003；Geng 等人，2010；Kik 等人，2011；Meng 等人，2014；Sutton 等人，2014a）。有学者在虎纹钝口螈和中国大鲵中观察到突出的皮肤疹块或者息肉（Jancovich et al，1997；Bollinger 等人，1999；Geng 等人，2010）。也有学者观察到了无尾两栖类和有尾两栖类动物皮肤的其他病变，如溃疡（ulceration）或 175
出现粗糙的灰色区域（Bollinger 等人，1999；Cunningham 等人，2007；Kik 等人，2011）。两栖类动物常常由于组织、淋巴囊（lymph sacs）及体腔积液即水肿引起腿部、身体和头部的肿胀，特别是在幼体中更为明显（Wolf 等人，1968；Miller 等人，2011；Meng 等人，2014）。

在身体内部，常常发现出血和坏死，特别是在脾脏、前肾和中肾（肾脏）以及肝脏［见图 1（c），（d）］。细胞死亡的原因可能与细胞凋亡或者病毒的复制有关（Grayfer 等人，2015）。坏死普遍存在于易脆的器官或作为弥散的苍白病灶散布在整个器官中。也有报道出现脾肿大（splenomegaly）和肝肿大（hepatomegaly）（Kik 等人，2011），并且这些肝脾肿大可能与淤血（congestion）和出血有关。肠道出血见于死亡病例以及实验感染接种的无尾两栖类和有尾两栖类（Bollinger 等人，1999；Geng 等人，2010；Cheng 等人，2014；Meng 等人，2014）。在巴西，出血和坏死见于蛙病毒阳性牛蛙的前庭区（vestibular region），并表现出前庭综合征（R. Mazzoni，CRMV-GO，个人通讯

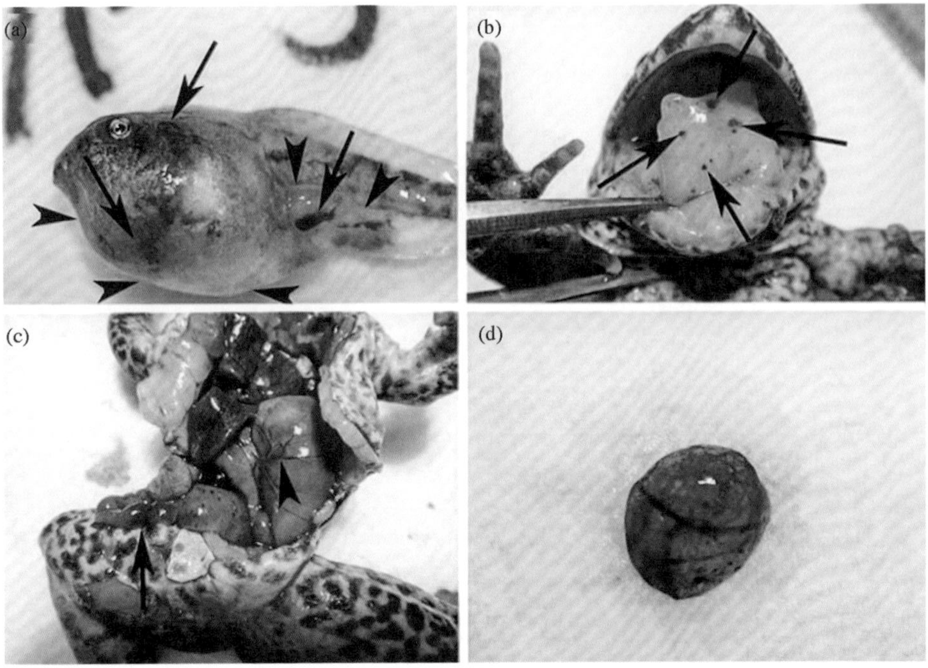

图1 所见到的两栖类蛙病毒病的大体病变。(a) 实验接种感染 FV3 样病毒后林蛙蝌蚪的出血（箭头）和水肿（箭头的头）；(b) 实验接种感染 FV3 样病毒后隙穴蛙多斑亚种（adult dusky gopher frogs，*Lithobates sevosus*）的舌头出血；(c) 实验接种感染 FV3 样病毒后隙穴蛙多斑亚种的肠道出血及血管充血；(d) 实验接种感染 FV3 样病毒后隙穴蛙多斑亚种的棕褐色易脆及出血的脾脏。

资料)。

在欧洲，成年林蛙曾出现过两种综合征，一种综合征是全身性出血，另一种综合征是出现广泛的皮肤溃疡。坎宁安（Cunningham 等人，1996）首次报道了这些死亡病例的综合征，这些死亡病例发生在整个英国，普通大众成员也观察到了这些综合征。后来，这两种综合征通过实验在林蛙中得到了重复（Cunningham 等人，2007）。依据这一结果，作者得出结论，感染的途径和特异性的蛙病毒分离株影响着发病机理（pathogenesis），然而这两种综合征可以在同一蛙中产生。成年林蛙的出血性综合征类似于萨顿（Sutton）等人（2014a）在成年隙穴蛙多斑亚种（*Lithobates sevosus*）所报道的病理变化。

鱼类

在鱼类当中，可见到多发性、随机性皮肤出血［见图 2（a）］（Waltzek 等人，2014），还可见到鳃红肿（Mao 等人，1999）。在身体内部，出血可以发生在任何器官，包括脂肪体和鱼鳔以及易脆的器官［见图2（b）］（Zilberg 等人，2000；Waltzek 等人，2014）。还有报道，鱼鳔出现过度膨胀（Grizzle 和 Brunner，2003）。瑞达克利夫（Reddacliff）和惠廷顿（1996）对 EHNV 在红鲈鲈（*Perca fluviatilis*）和虹鳟（*Oncorhynchus mykiss*）中所引起的病变进行了详细的描述。病鱼发黑，停止进食，有时出现共济失调。大体病变包括腹部肿胀，并伴有脾脏和肾脏的肿大，在肝脏有时可见多发性苍白病灶。西尔贝格（Zilberg）等人（2000）报道，大嘴鲈（*Micropterus salmoides*）实验接种感染 SCRV（Santee-Cooper ranavirus）以后，胃肠道黏膜上皮、鳃以及心脏发生了坏死。

爬行类动物

爬行动物大多数病变已在龟类中作过描述，这些病变包括眼周围肿胀、口腔溃疡和坏死、头和四肢肿大、眼和鼻有分泌物以及偶尔见到皮肤溃疡（见图3）（Johnson 等人，2008；Ruder 等人，2010）。类似地，海特（Hyatt）等人（2002）报道了绿树蟒（*Chondropython viridis*）口腔黏膜溃疡的形成。在蜥蜴中，皮肤病变最为常见，这些皮肤病变包括皮肤颜色变灰，溃疡性和坏死性皮炎以及角化过度（Stöhr 等人，2013）。多灶性到融合性棕褐色易脆坏死区域见于内部器官，特别是在胃肠道和呼吸道。有时，在水龟当中，胃肠道出血是唯一能观察到的变化［见图3（d）］。然而，这些病灶是否是由继发感染造成的，目前还不清楚。 177

1.2.3　组织病理学

造血组织、血管内皮和上皮细胞的坏死，出血以及出现胞浆内嗜碱性包涵体（basophilic inclusion bodies）是所有宿主常见的显微病变（见表2和图4）（Reddacliff 和 Whittington，1996；Cunningham 等人，2007；Allender 等人，2013b；Bayley 等人，2013；Cheng 等人，2014；Waltzek 等人，2014）。在死亡病例中，肝脏、脾脏和肾脏，包括前肾和中肾，是最常见的受到影响的器官，并且既涉及这些器官的造血组织，也涉及这些器官的非造血组织［见图4（a）~（c）］。在两栖类动物当中，有学者观察到广泛的组织趋向性，其他报道的病理变化包括上皮的变性和溃疡的形成（Cunningham 等人，2007；Geng 等人，2010；Cheng 等人，2014；Meng 等人，2014）、胃肠黏膜的坏死（Bollinger 等人，1999）及淋巴组织的坏死（Bollinger 等人，1999；Balseiro 等人，2009；Meng 等人，2014）、神经上皮组织的坏死（Docherty 等人，2003）、骨骼肌变性（Miller 等人，2008）、胰腺坏死（Balseiro 等人，2010；Kik 等人，2011）、多中心出血、血管出血和坏死（R Mazzoni， 178

CRMV-GO，个人通讯资料）、眼部畸形（Burton 等人，2008）。在龟类动物中，常见的病变包括纤维素样血管炎、肌炎、坏死性咽炎、食道炎和胃炎［见图 4（d）］（Johnson 等人，2007，2008；Ruder 等人，2010；Allender 等人，2013b）。类似地，海特等人（2002）曾报道蛇类中的绿树蟒咽黏膜下
179 层的坏死、鼻腔黏膜的溃疡以及肝脏的变性与坏死。有报道记载，在蜥蜴中（见图 5），产生溃疡性坏死舌炎、肝坏死、溃疡性皮炎及继发性感染（Marschang 等人，2005；Behncke 等人，2013；Stöhr 等人，2013）。在龟类动物中，还观察到了坏死性口炎和咽炎的病变与疱疹病毒和腺病毒感染的严重重叠（Johnson 等人，2005；Rivera 等人，2009）。

1.2.4 亚临床感染

亚临床感染在蛙病毒流行病学中起着很重要的作用（Brunner 等人，2015）。值得注意的是，原始两栖动物蛙病毒分离株（FV1、FV2 和 FV3）来自的动物中，除一例具有肾肿瘤以外，其余都外表正常（Granoff 等人，1966）。亚临床感染已在野生两栖类（Gray 等人，2007，2009；Rothermel 等人，2013）、龟类（Allender 等人，2013b；Goodman 等人，2013）和鱼类中检测出来（Goldberg，2002；Whittington 等人，2010）。此外，在两栖类（Brunner 等人，2004；Harp 和 Petranka，2006；Robert 等人，2007，2011）、
180 龟类（Johnson 等人，2007；Brenes 等人，2014a，b）和鱼类（Bang Jensen 等人，2011；Becker 等人，2013；Brenes 等人，2014a，b）实验中，也产生了亚临床感染。亚临床感染的个体有时会具有非特异性的组织学变化，如在肾小管上皮和肝细胞中形成空泡（Miller 等人，2011；Allender 等人，2013b）。安那·巴尔塞罗（Ana Balseiro）（SERIDA，个人通讯资料）曾观察到肾小球免疫组织化学染色阳性的成年普通产婆蟾并没有疾病的临床症状。在大多数上述的研究中，还不清楚亚临床感染能否发展成临床疾病，因为不同的个体都是安乐致死或直接释放到环境当中。在某些情况下，亚临床感染很可能是蛙病毒病的早期阶段，同时，在其他情况下，亚临床感染是感染结束后的结果或是持续的一种感染状态。

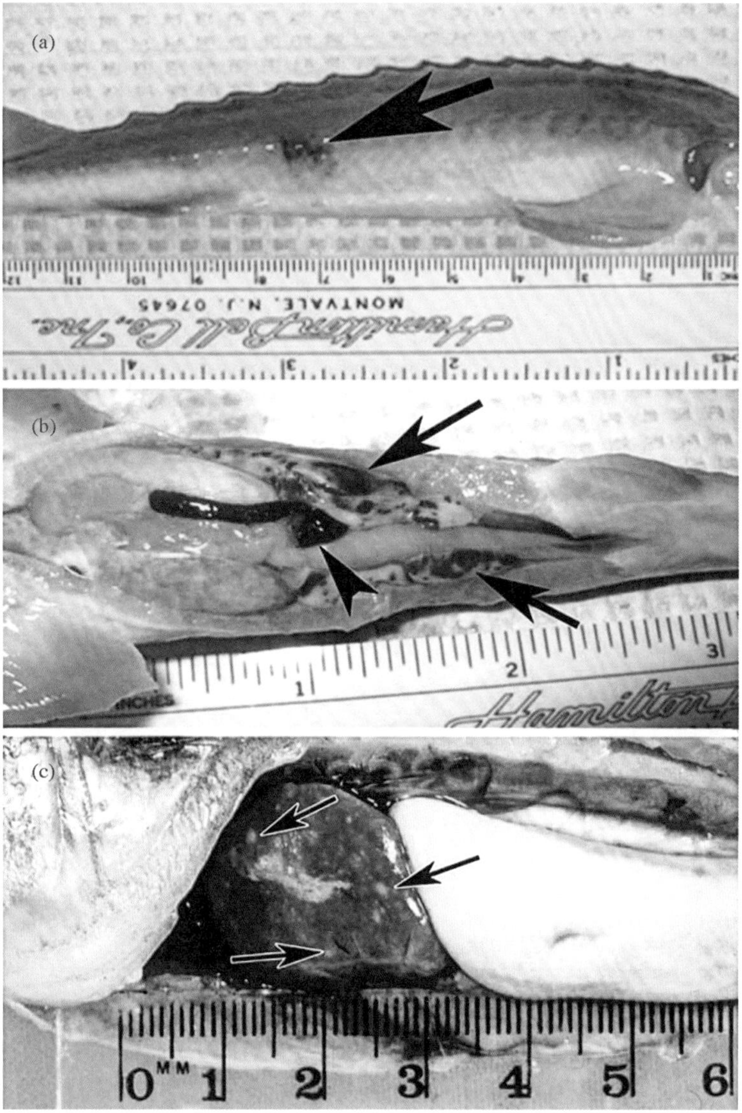

图 2　所见到的鱼类蛙病毒病的大体病变。(a) 由 FV3 样病毒引起的白鲟皮肤出血性瘀斑（箭头），照片由美国佛罗里达大学 Thomas B Waltzek 提供；(b) 由 FV3 样病毒引起的白鲟脂肪体（箭头）和脾脏（箭头的头）出血，照片由美国佛罗里达大学 Thomas B Waltzek 提供；(c) 感染 EHNV 的成年红鳍鲈肝脏中颜色变苍白并且出现区域明显的多病灶肝坏死（箭头）以及腹膜后腔的瘀斑性出血。

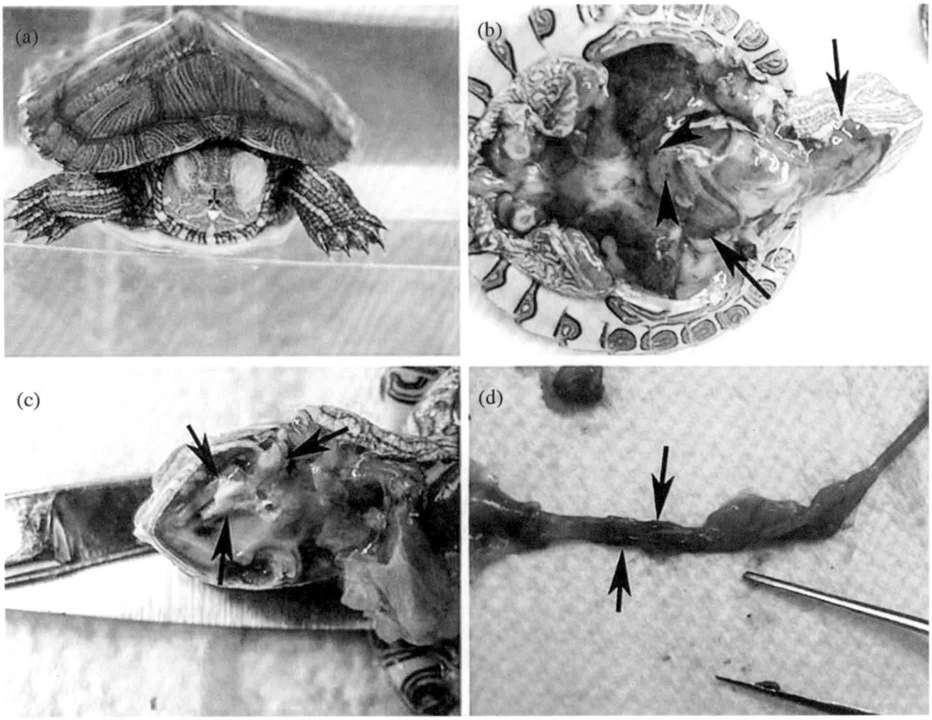

图 3 红耳龟（*Trachemys scripta elegans*）实验感染 FV3 样病毒后产生的大体病变。（a）眼周围肿胀，肿胀是双侧性的，但在病龟左眼周围的肿胀更加突出；（b）内部软组织水肿（箭头）和出血（箭头的头）；（c）口腔黏膜的坏死斑（箭头）；（d）肠道黏膜出血。

表 2 所见到的蛙病毒个体的组织病理学变化的例子

器官	组织病理学变化
肾脏（包括两栖类和鱼类幼体的前肾和成体的中肾）	变性或坏死（管状上皮细胞和肾小球），胞浆内包涵体，造血组织坏死
肝	变性或坏死（血窦，黑色素巨噬细胞中心，肝细胞），胞浆内包涵体，造血组织坏死
脾脏	坏死，胞浆内包涵体
胰	坏死，胞浆内包涵体
肌肉	肌纤维变性，出血
皮肤（特别是蜥蜴、鱼类和成年两栖类）	糜烂，溃疡，出血，胞浆内包涵体

续表

器官	组织病理学变化
胸腺、淋巴组织	衰竭，细胞凋亡，坏死
胃肠组织	细胞凋亡，上皮细胞坏死，胞浆内包涵体
血管	坏死（内皮细胞）
上呼吸道（特别是龟类和蛇类）	上皮细胞坏死，胞浆内包涵体

2　诊断试验

2.1　世界动物卫生组织标准

世界动物卫生组织为各种动物病原体的诊断试验提供了建议和方案（http://www.oie.int）。OIE 的目标之一就是提供各种规程来申报动物的国际运输或者申报一个地点、一个区域或一个国家无特定的病原。OIE 关注的病原称为须申报的病原，这些病原被认为是国际商务和人类健康的一种风险。感染两栖类动物的蛙病毒以及 EHNV 已被列为须申报的病原。世界动物卫生组织所出版的《水生动物疾病诊断手册》中提供了采样和诊断程序（OIE，2012c， 181
d）。OIE 关于蛙病毒的实验室参考指南为蛙病毒的诊断试验以及进行诊断试验提供了试剂和操作规程（见表 3）。格雷等人（2015）讨论了如何遵从 OIE 操作规程，进行蛙病毒引入的风险分析。

OIE 的职责范围就是出于保障国际贸易的目的，进行蛙病毒的检测。但是，它所总结的采样程序和诊断技术可能不适合所有的研究。总的来说，我们推荐遵从格雷等人（2015）的操作规程，来确定所需要样本的大小和设计蛙病毒的相关研究。由 OIE 所推荐的诊断程序因成本高昂同样也不适用于全世界的所有区域。以下内容，我们综述了用于检测蛙病毒感染及确定个体是否处于疾病状态的大多数技术。表 4 所列内容为一般研究方向上应使用什么样的技术提供了指导。总的来说，我们建议研究人员应咨询常规进行蛙病毒诊断的专家，从而选择适当的技术进行研究来达到特定目标。重要的是，适当的样品采集程序可能取决于所采用的技术。全球蛙病毒联盟在其网站上保存了进行常规蛙病毒诊断的实验室列表（http：//www.ranavirus.org/）。

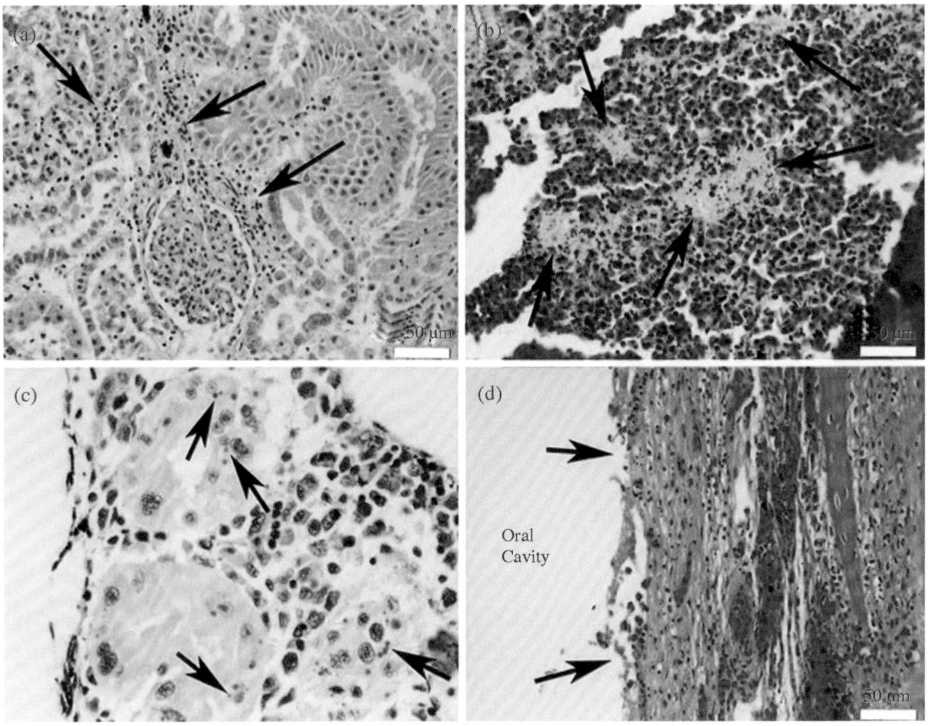

图4 实验接种感染 FV3 样病毒的蛙病毒病个体所见到的病理组织学变化。(a) 白鲟（*Scaphirhynchus albus*）的肾脏造血组织坏死（箭头所指）；（b）隙穴蛙多斑亚种（Dusky gopher frog, *Lithobates sevosus*）的脾坏死（箭头所指）；（c）南方豹蛙（Southern leopard frog, *Lithobates sphenocephalus*）的肾小管上皮细胞的胞浆内包涵体（箭头所指）；（d）红耳龟（Red-eared slider, *Trachemys scripta elegans*）的口腔黏膜坏死（箭头所指）。

2.2 与研究目标匹配的诊断试验

183 对蛙病毒的研究具有各种目的，依据研究目标，有些诊断技术可能比另外一些技术更适合。此外，现场研究与对照研究所需要的诊断可以有所不同。总的来说，大多数研究可以归类为：（1）依照 OIE 指南为促进贸易所进行的检测（Gray 等人，2015）；（2）按病毒的地理分布作图并估计流行率或发病率所进行的检测（Gray 等人，2015）；（3）对病毒分离株进行系统发生的分类（Jancovich 等人，2015a）；（4）病毒的形态学和宿主免疫反应（Jancovich 等人，2015b；Grayfer 等人，2015）；（5）与蛙病毒同时出现的相关生态因子（Brunner 等人，2015）；（6）死亡事件的诊断（Duffus 等人，

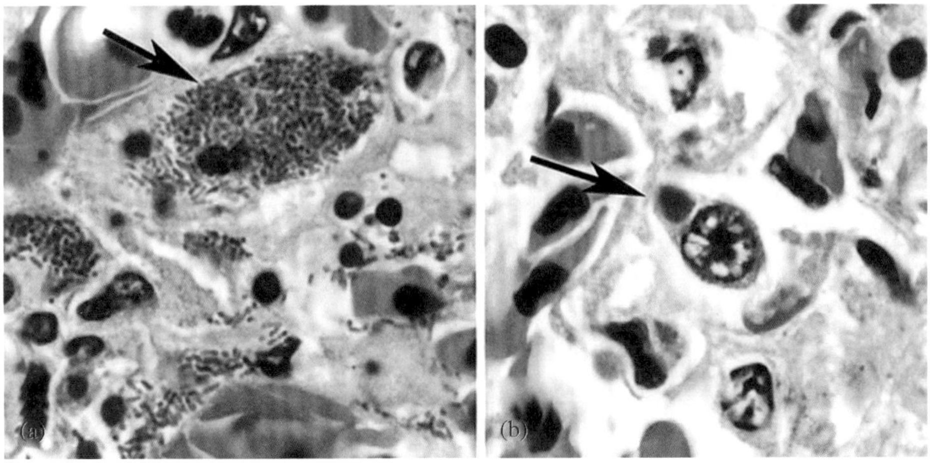

图 5　美国牛蛙（*Lithobates catesbeianus*）蛙病毒感染与继发细菌感染叠加的组织病理学变化。(a) 脾脏中的细菌菌落（箭头所指）可以使蛙病毒感染的微妙证据变得模糊。(b) 仔细检查该个体的组织学切片，结果发现稀少的胞浆内包涵体，从而强烈提示蛙病毒感染。

2015)。尽管诊断技术在不断地发展和改进，但是用于病毒调查的常用诊断工具只包括能够检测到蛙病毒的诊断工具以及能够检测到宿主对感染反应的诊断工具。例如，聚合酶链反应（PCR）只能简单地检测病毒核酸一个特异性序列的存在，而不能提供病毒是否有活性即能够复制并造成发病的证据。然而，病毒的分离显示，存在着传染性的病毒。类似地，其他试验，例如，组织学、细胞学、基因表达和抗体产生等技术，常用于检测宿主感染病毒的细胞反应。诸如电镜（EM）、免疫组化（IHC）和原位杂交这些在一个病灶内显示病毒存在的技术，对于展示病毒的存在与疾病表现之间的联系是特别有用的。总的来说，以检测蛙病毒为主要目标的研究，如监测研究，使用 PCR 或抗原检测技术为宜；进行系统发生研究，则采用病毒分离和基因组测序；进行宿主免疫研究及死亡调查，可以采用所有技术，具体见表 4。另外，检测蛙病毒是否是大规模死亡的病因，其确定需要采用多种技术，同时还需要宿主细胞内病理变化的资料。重要的是，不能只通过感染数据或对大体症状的观察来作出蛙病毒病的诊断。

表 3 OIE 提供的参考实验室蛙病毒试剂、方案和检测联系人员

联系人	联系方式
尼克·穆迪博士 (Dr. Nick Moody)	联邦科学与工业研究组织家禽产业 (CSIRO Livestock Industries) 澳大利亚动物卫生实验室 (Australian Animal Health Laboratory) 赖里街私人邮箱 24 (Private Bag 24，Ryrie Street) 吉朗 (Geelong) 维多利亚 3220 (Victoria 3220) 澳大利亚 (Australia) 电话：+61-352270000 传真：+61-352275555 电子邮箱：nick. moody@ csiro. au 网址：www. csiro. au
理查德·惠廷顿博士 (Dr. Richard Whittington)	悉尼大学 (University of Sydney) 兽医学院 (Faculty of Veterinary Science) 魏诺毕路 425 (425 Werombi Road) 私人邮箱 3 (Private Bag 3) 卡姆登 NSW 2570 (Camden NSW 2570) 澳大利亚 (Australia) 电话：+61-293511619 传真：+61-293511618 电子邮箱：richardw@ camden. usyd. edu. au

表 4　依赖研究目的所使用的诊断试验例子

试验[a]	研究目的					
	检测/监测	研究死亡事件	系统发生研究	生态学研究	病毒形态学研究	宿主免疫研究
常规 PCR	X	X		X	X	X
qPCR	X	X		X	X	X
DNA 测序	X	X	X	X	X	X
抗原捕获 ELISA	X	X		X	X	X
病毒分离		X	X		X	X
血清学	X	X				X
生物测定		X			X	X
组织学；细胞学		X		X	X	X
IHC；ISH；EM		X			X	X

[a]PCR 为聚合酶链反应，qPCR 为定量实时 PCR，IHC 为免疫组化，抗原捕获 ELISA 为抗原捕获酶联免疫吸附试验，ISH 为原位杂交，EM 为电镜。

除了研究目的，样品的类型决定了所进行的试验类型（见表 5）。非死亡样品一般包括棉签样本、尾或趾剪切物和血液，这些样品一般来说容易采集（Greer 和 Collins，2007；Gray 等人，2012）。棉签采样器材应使用塑料或金属丝体（不是木质的）材料，以避免 PCR 抑制剂（Pessier 和 Mendelson，2010）。棉签采样时，应该稳稳地进行擦拭，而不是强有力地沿着被检测表面猛擦（一次到多次）。用棉签样本进行蛙病毒检测的表面通常包括口腔、泄殖腔或皮肤病灶（Pessier 和 Mendelson，2010）。用棉签进行排泄孔采样能提供肠道排毒的证据。尾或趾剪切物同样在检测蛙病毒感染中非常有效，但与棉签样本相比，有可能导致少数假阳性结果（Gray 等人，2012）。对于蝾螈来说，在尾部顶端附近，一般会出现自然断点，这样就可以采取轻压的办法来自动断尾，并在没有剪切的情况下收集组织样品（Sutton 等人，2014b）。棉签样本或所采取组织样本可用乙醇保存或冷冻保存。

采样时，应戴上一次性手套，对不同动物取样时，应及时更换手套，以减 184
少样本的交叉污染，同时阻止病毒在动物之间造成人为的传染。此外，单个动物不要共同饲养，野外处理应在无菌的表面进行，蛙病毒检测的各种样本和所进行的试验类型以及试验结果的局限性见表5。对于尾或趾剪切物来说，必须使用消毒器材，以避免样本的污染。我们建议，每一个动物使用一套不同的消毒器材，如高压蒸汽消毒器材。虽然蛙病毒能通过各种消毒药进行灭活（Byran 等人，2009），但是，若消毒时蛙病毒 DNA 没有降解，则可能发生交叉污染，这样会导致在分子学试验时出现假阳性结果。然而，还没有对蛙病毒 DNA 进行降解的研究，而高压蒸汽消毒、火焰消毒及长时间（>12 h）浸泡在强力漂白剂（6% 次氯酸钠）当中，已广泛用于降解其他病原体的 DNA（Cashin 等人，2008）。有些研究者正在研究使用肝吸取物（liver aspirates）样本（Forzan 和 Wood，2013），然而现场采样的可行性以及采样中需要相应的专业知识限制了它们的应用。可以从多个位置采取血液样本，这些位置包括鱼类、有尾两栖类和蛇类尾静脉，龟类的壳下静脉窦（subcarapacial sinus）或枕静脉窦（occipital venus sinus）（Martínez-Silvestre 等人，2002；Allender 等人，2011），无尾两栖类的腹部及面部静脉（Forzan 和 Wood，2013），以及无尾两栖类和蛇类的心脏。

表5 用于蛙病毒检测的各种样本和所进行的试验类型以及试验结果的局限性

样本	试验	试验结果的局限性
棉签样本	PCR，病毒分离	假阳性（环境污染）；总 DNA 可能最低；不能进行组织学试验
尾或趾剪切物	PCR，病毒分离	假阳性（环境污染）；不能进行组织学试验
整个身体或内脏	PCR，病毒分离，组织学，免疫组化	需要死亡动物
固定后的组织	PCR，组织学，免疫组化	不能进行病毒分离，可以进行电镜观察
血液	PCR，病毒分离	最好从活动物获得，血液样本比较难得，常常不能从小动物个体中获得足量的样本
	假如分离了血清，做 ELISA	
	假如进行了血液涂片，进行细胞分类计数	

关于两栖类的尸体剖检和死亡事件的采样方案已有详细的综述（Green 等人，2009；Pessier 和 Mendelson，2010）。简而言之，最好的做法是将整个垂死的动物连夜送到诊断实验室。假如这样操作不可行，则应采集和提交组织样

本。最重要的是要认识到，一个样本提交后，如果只用于一个检测，那么该检测的结果只会是一个病原体的阳性或阴性。假如是阳性，则说明该病原体在发病和死亡中起作用，或者不起作用。假如是阴性，那么发病和死亡原因仍不能确定。因此，比较理想的是，应提交多个样本，进行多个诊断试验。组织样本的采集至少应包括肝、肾、脾、肺/鳃、心脏、皮肤和消化道等主要器官，以
及能观察到的病灶。采集这样一套组织样本后，应提交新鲜或冷冻的样本，以 186
便进行病原体的检测，还有一套组织样本应进行固定，例如，用10%缓冲福尔马林进行固定，以便进行组织学观察。组织样本应三层装在防漏容器中，按照承运人指南进行运输（http：//www. fedex. com/downloads/hk_ english/packagingtips/pointers. pdf）。自溶的样本几乎没有诊断价值，但是如果它们都是那样，那么就有可能收集到一些信息。最好在提交组织样本之前，先联系诊断实验室，特别是在提交自溶性样本之前。前面已经提及，在 GRC 网站可以找到诊断实验室的一个列表。重要的是，研究人员应对接触过感染动物或含有蛙病毒水体的鞋类和采样器材进行消毒。1%双氯苯双胍己烷（Nolvasan®美国爱荷华州，道奇堡，道奇堡动物健康公司）及4%漂白水是有效的蛙病毒消毒剂，同时对其他病原体如两栖类壶菌也很有效（Bryan 等人，2009，Gold 等人，2014）。

2.3 诊断试验

可以使用各种方法来确定蛙病毒的存在，其中包括病毒分离、电镜、抗原捕获酶联免疫吸附试验（Ag-capture ELISA）、免疫组化和 PCR。在这些试验当中，从死后变化不大的生物所采集的样品，可进行所有试验，从活体采集的样本可进行其中的大多数试验，具体见表 5。有些试验如 PCR，即使样本处于溶解晚期，也能得出结果。对于所有的诊断试验来说，为了使蛙病毒从宿主细胞中释放出来，有必要制备组织样本。结合自动、半自动和手工进行组织匀浆制备的几种验证方法已经作过描述，并对从组织中提取蛙病毒的效率进行了比较（Whittington 和 Steiner，1993；Rimmer 等人，2012）。假如病毒载量在组织中比较低，如亚临床感染，提取效率就显得特别重要。此外，所有的诊断试验应该包括阳性和阴性对照。例如，PCR 阳性对照，一般来说，包括从一个病毒分离株和已知感染动物提取的 DNA，而 PCR 阴性对照，一般来说，包括从已知阴性动物和 DNA 级水中提取的 DNA。在这种情形之下，试验所验证的对照才是有效的，并且具有低污染的可能性。

2.3.1 聚合酶链反应

无论是常规 PCR，还是定量实时 PCR（qPCR），都被用来检测蛙病毒 DNA。PCR 检测可使用新鲜组织、甲醛固定石蜡包埋组织以及棉签样本等多

图 6 展示消毒采样技术的野外采样站。(a) 动物应该放在单独的容器里，例如，含有无菌水的玻璃罐用于水生发育阶段的动物。(b) 可密封的塑料袋用于陆生发育阶段的动物。(c) 一个便携式桌子可能作为一个处理站，在每个野外地点采样后很容易进行消毒。(d) 一套消毒过的器材应专门用于每个动物。对于有些种类（如美国隐鳃鲵，*Cryptobranchus alleganiensis*）可以用无菌一次性手术刀和镊子从背尾上收集一小片样本。(e) 对于许多蝾螈种类，在尾部存在自然断裂点，很容易在袋内而没有移出袋外的情况下，轻压断裂点附近而获得尾部样本。为了检测蛙壶菌是否并发感染，也可在容器内用棉签完成采样。(f) 所有的设备、防水连靴裤以及靴子，在离开野外采样点之前都应进行消毒（如用 1% 双氯苯双胍己烷即 Nolvasan®）。

种样本。在最常用的试验当中，主衣壳蛋白基因是靶基因（Mao 等人，1997；Hyatt 等人，2000；Kattenbelt 等人，2000；Marsh 等人，2002；Pallister 等人，2007）。主衣壳蛋白基因高度保守，该基因具有一个理想的区域作为靶向目标 187
来鉴定蛙病毒的存在。然而在一个特定的蛙病毒种内，假如研究的目标是探索不同基因组或系统发生的微小差别，MCP 的这种保守性就显示出局限性（Jancovich等人，2015a）。PCR 的其他靶向目标包括神经丝三联体 H1 蛋白（Holopainen 等人，2009）、DNA 聚合酶（Holopainen 等人，2009）及基因间的一个可变区（Jancovich 等人，2005）。

PCR 的敏感性依据所检测的样品类型而有所不同。一般认为，内脏器官的组织样本所代表的感染状态要比非死亡采样技术所获得的尾部剪样、趾部剪样和棉签采样好得多（Greer 和 Collins，2007；Gray 等人，2012）。考虑到蛙病毒的趋向性在不同的组织类型之间及宿主感染后的时间长度不同而存在差别（Robert 等人，2011；Ma 等人，2014），通常，PCR 的试验结果取决于所使用的组织样本。在感染 FV3 的蛙类中，一般来说，蛙病毒在肾脏和肠道最先被检出（Robert 等人，2011），然后出现在肝脏、脾和其他主要器官中。因此，对不同组织所进行的检测可以提供感染严重程度的证据。通常来讲，研究者对蛙病毒感染的检测主要使用不同组织的匀浆（Hoverman 等人，2011），从而提高检出的概率。

与整个动物或肝脏样本比较时，非死亡样本蛙病毒的检测，可能导致假阴性和假阳性结果（Greer 和 Collins，2007；Gray 等人，2012）。假阴性结果可能是由棉签样本中病毒量不够造成的，假阳性结果也有可能是由被检动物以外的病毒造成的（Gray 等人，2012）。不管怎样，假如动物采样的条件不可行或动物种群的丰度比较低时，非死亡采样技术对于蛙病毒的检测是非常有用的。

值得注意的是，PCR 检测并不是完美的。这意味着即使没有错误地进行极具分析特色的实验室规程操作，也会出现一些假阳性和假阴性结果。对 PCR 结果的准确解释，需要对试验诊断的敏感性和特异性进行评估，以便计算出结果的阳性和阴性预测值（Greiner 和 Gardner，2000）。当一个试验方案被证实适用于特定的样本类型和特定的宿主物种时，要对这些特征进行评估。有时，当病毒的存在量很低时，同一样本会产生不同结果。我们建议，所有在同一台 qPCR 机器上进行 qPCR 检测的样本都要进行重复检测，最好是做三个平行重复检测。并且，只有具有连续阳性结果的样本，才能宣布是阳性。假如一个样本出现阳性，而另外一个样本是阴性，那么就要进行第三个样本的检测，感染结果的宣布就要依据多数结果。目前，正在进行的研究就是（E Grant，美国地质调查局以及 DLM）用 Picco 等人（2007）的实验方案和双重采样规程来评估 qPCR 检出的概率。

常规 PCR

188 常规 PCR（conventional PCR）已经被用于检测来自监测研究和死亡事件的新鲜或固定组织标本中的病毒感染，同时还用于培养病毒的鉴定（Miller 人，2007；Gray 等人，2009；Meng 等人，2014）。代表着世界各地的病毒分离株，并贮存在 GenBank 中的 DNA 序列系统发生图谱显示出显著的序列一致性（sequence identity，又称同源性），这种一致性已被用来设计 PCR 引物（primers），用于蛙病毒的检测。此外，MCP 序列多态性（polymorphisms）可以用来在病毒属级水平，如 FV3 和 EHNV 进行限制性消化后（restriction digestion），进行常规 PCR 扩增，从而进行一些病毒分离株的鉴定（Marsh 等人，2002；Holopainen 等人，2009）。常规 PCR 对蛙病毒进行的检测及对亚型进行的限制性内切酶分析（restriction endonuclease analysis，REA）已在 OIE《水生动物疾病诊断手册》中进行了描述（OIE 2012c，d）。

用常规 PCR 产物进行 DNA 测序来确认阳性结果是没有必要的，尤其是在流行地区。然而，常用的 MCP 引物组会产生非蛙病毒 DNA 的扩增（A. Pessier，未发表的资料）。因此，当新宿主被感染或在以前认为没有这些病毒的地区检测出了蛙病毒，并出现一系列阳性结果的时候，建议进行测序。PCR 产物的测序也可在病毒属级水平为病毒的初步鉴定提供信息，如 FV3 样病毒或虎纹钝口螈病毒。

定量实时 PCR

定量实时 PCR（quantitative real-time PCR，qPCR）显著推动了各种病原体及其毒力的研究。研究发现，当进行低病毒载量检测时，qPCR 要比常规 PCR 更加敏感，并且在病毒分离中，也更为敏感（Pallister 等人，2007；Jaramillo等人，2012）。帕里斯特（Pallister）等人（2007）发现，qPCR 能够在许多感染鱼类、两栖类和爬行类动物的蛙病毒之间进行鉴别，特别是对欧洲与澳大利亚蛙病毒的鉴定，效果特别好。皮科（Picco）等人（2007）报道了一个对两栖类病毒检测十分有用的实验方案，在该实验方案中，扩增了蛙病毒 MCP 基因的一个 70 bps 片段。阿蓝德等人（2013a）针对龟类开展了一个类似的试验。病毒载量可以在样本中 DNA 定量时进行估计，每个检测样本要使用等量的 DNA。病毒载量可能通过将一个样本的循环阈值（cycle thresholdvalue，又称 Ct 值）代入一个回归方程（又称标准曲线）进行预测，该回归方程与 Ct 值和已知用于 PCR 检测系统的病毒载量呈相关关系（Yuan 等人，2006）。重要的是，不同 PCR 系统的 Ct 值是不相等的。因此，使用标准曲线将 Ct 值转换成标准的病毒数量单位是研究中需要进行解释的必要条件，通常，标准的测量单位是每单位基因组 DNA 或组织中病毒浓度的 $\log_{10}$ 转换值。例如，布雷内斯等人（2014a）报道，病毒的含量为每 0.25 μg 基因组 DNA 噬斑形成单位（plaque-forming units，PFU）。PCR 系统可以在不同的操作间产生差异，每次 189
操作也就是每个板都生成一个标准曲线是比较理想的（Bustin 等人，2010）。标准曲线的斜率和截距参数可以在多个独立操作中得到平均，类似于模型平均，以便得到更可靠的病毒载量估计。标准曲线应该与 qPCR 结果一起发表。

考虑到病原体的负荷通常与动物的发病率呈正相关，qPCR 可以提供对疾病状态（临床与亚临床）的深入认识。但这一点对蛙病毒来说仍需要验证。在解释高 Ct 值（如，Ct 值>35）时必须小心谨慎，这是因为高 Ct 值可能表示扩增的结果，或者是由荧光制品造成的，或者是因交叉污染造成的（Caraguel 等人，2011）。如果接下来还要进行 DNA 测序或病毒分离，那么常规 PCR 可以用来检验具有高 Ct 值样本内的病毒 DNA 是否存在。此外，PCR 系统的标准曲线可以确定一个 Ct 值，这个值被用来确定样本是否为 PCR 阳性。例如布雷内斯等人（2014a，b）认为，如果 Ct 值小于预期病毒量为 0 的 95% 置信区间下限，则应保守地确定为阳性结果。一般来说，关于蛙病毒 qPCR 试验诊断的敏感性和特异性的资料，目前还缺乏。然而，贾拉米洛（Jaramillo）等人（2012）所描述的试验，已在 EHNV 中得到了证实，该试验将在未来新版的 OIE 手册中予以描述。类似地，阿蓝德等人（2013a）在东方箱龟中验证了 qPCR 检测 FV3 样病毒的有效性。

蛙病毒种和病毒株的区别

在大多数常规诊断和实验研究中，基于 MCP 基因所建立的 PCR 实验，对于确定蛙病毒是否存在，或者将一个病毒归类为一个主要的物种类群，例如一个“FV3 样病毒”（Jancovich 等人，2015a），其实验效果良好。然而，在其他例子当中，如流行病学调查，则需要确定不同蛙病毒株的分布，或者确定野生动物的转运程序，此时，对于确定是否在原产地和目的地种群当中存在同一蛙病毒是很重要的，在这种情况下，这些实验不是很有用，因为 MCP 基因是高度保守的（Jancovich 等人，2015a）。作为一个例子，现在有充分的证据表明，由基因组限制性内切酶分析（REA）所确定的不同的 FV3 病毒株，与经常用于诊断性常规 PCR 的 MCP 基因片段，有着完全相同的 DNA 序列（Schock 等人，2008；Duffus 和 Andrews，2013）。

用于区分特异性病毒株的实验室技术，如基因组 REA 之后的病毒分离和纯化，在所有的实验室都是没有的，而且在一些研究中也是不实用的。因此，有必要寻找和验证快速进行病毒株鉴定的方法。一个途径就是运用常规 PCR

进行基因分型及对基因诸如基因间的可变区进行 DNA 测序（Jancovich 等人，2005；Weir 等人，2012）或对包含可变重复区神经丝三联体 H1 样蛋白质（neurofilament triplet H1-like protein）进行测序（Holopainen 等人，2009；
190 Cheng 等人，2014）。通过使用一致的方法来选择靶向目标的测序，这样才能确保研究者对蛙病毒系统发生和流行病学的认识获得更为快速的进展。

2.3.2 抗原捕获 ELISA

应用鱼类组织匀浆检测 EHNV 的抗原捕获 ELISA 的分析敏感性为 $10^3 \sim 10^4$ $TCID_{50}$/mL。抗原捕获 ELISA 对于诊断和监测是非常有用的，因为该方法的应用，与病毒分离和分子试验相比，可以快速而廉价地检测大量的样品。相对于病毒分离来说，抗原捕获 ELISA 的诊断特异性和敏感性分别为 100% 和 60%（Whittington 和 Steiner，1993）。

2.3.3 病毒分离

使用特征良好的细胞系进行病毒分离是确定蛙病毒存在的另一个途径，这些细胞系在细胞库中可商业性购得。此外，细胞培养技术的优点包括展示存活的病毒和对病毒扩增后做进一步鉴定。培养后的样本相对于组织样本来说，能产生更好的 PCR 结果，最终还能产生更好的扩增产物用于测序。EHNV 能在 15 ℃～22 ℃，于许多鱼类细胞系中进行复制，其中包括黑头呆鱼（fathead minnow，FHM）、虹鳟性腺（rainbow trout gonad，RTG）、蓝鳃太阳鱼苗（bluegill fry，BF-2）以及鲑鱼胚胎（chinook salmon embryo，CHSE-214）细胞系（Langdon 等人，1986；Crane 等人，2005；Ariel 等人，2009；OIE，2012c）。欧洲鲶鱼病毒可用鲤鱼上皮乳头瘤（epithilioma papulosum cyprinid，EPC）细胞、FHM 和斑点叉尾鮰卵巢（channel catfish ovary，CCO）细胞，在 15 ℃～25 ℃进行分离。但 BF-2 细胞的敏感性是 EPC 和 CCO 细胞的 10 倍（Ahne 等人，1989；Pozet 等人，1992）。感染大嘴鲈的桑堤-库珀蛙病毒（SCRV）最早是在 FHM 细胞中分离到的（Plumb 等人，1996）。后来于 25 ℃～32 ℃，分别在 BF-2、EPC 和 CCO 细胞中得到了分离（McClenahan 等人，2005）。石斑鱼虹彩病毒能在一系列不同细胞中进行增殖，其中包括 FHM 和 BF-2 细胞，但在石斑鱼胚胎细胞中，能产生更快速的细胞病变效应（rapid cytopathic effect，CPE）和更高的病毒滴度（Qin 等人，2003）。

通常，在两栖动物中发现的蛙病毒，可以在一些鱼类细胞及两栖类动物细胞中进行分离。孵育温度可以有所变化，关键要考虑的是达到最佳效果。例如，ATV 在 FHM、RTG 和蛙舌细胞（bullfrog tongue cells）中，于 25 ℃条件下进行繁殖（Docherty 等人，2003）；FV3 在 FHM 细胞或蛙胚胎成纤维细胞中，于 27 ℃条件下进行繁殖（Cunningham 等人，1996，2007）以及在 EPC 细胞中，于 24 ℃条件下进行繁殖（Ariel 等人，2009）。CMTV 能在 EPC 细胞中，于 15 ℃ 条件下进行繁殖。同时，耿毅等人（2010）发现，中国大鲵病毒（Chinese giant salamander virus，CGSV）又名中国大鲵虹彩病毒（Chinese giant salamander iridovirus，GSIV）分别于 25 ℃ 和 20 ℃ 条件下，能在 EPC 细胞中进行繁殖。

蛙病毒在接种感染培养细胞后，都能产生类似 CPE 的灶状细胞溶解。随 191
后，在几天之内会造成整个单层细胞的破坏（见图 7）。在 FHM 细胞单层中，蛙病毒 CPE 呈现出网状样外观（G Chinchar，个人通讯资料）。然而，所产生的 CPE 外观不足以表示蛙病毒就是其致病原因。在细胞传代后，培养的单层细胞中也可见到 CPE。有好几种技术可以用来确定蛙病毒的存在，其中比较合适的技术包括间接荧光抗体染色（indirect fluorescent antibody stain）、抗原捕获 ELISA 和 PCR（OIE，2012c）。另外，病毒核酸测序或限制性内切酶消化也能展示蛙病毒的存在，并允许在病毒种和病毒株之间进行鉴别（Hyatt 等人，2000；Marsh 等人，2002）。病毒分离技术可以用来确定感染病毒的定量分析，即 50% 组织培养感染剂量（50% tissue culture infective dose，TCID50）（Rojas 等人，2005）。

病毒分离需要从野外到实验室维持病毒的存活。因此，这一技术在整个运输和转运过程中需要一个适当的冷藏链，否则就会产生假阴性结果（OIE，2012d）。同时，还应避免组织的自溶。用细胞培养进行病毒分离具有很高的分析敏感性，但不如定量实时 PCR 敏感性高。例如，用 qPCR 检测 EHNV 的敏感性要比用 BF-2 细胞高 100 倍（Jaramillo 等人，2012）。出于监测目的和检测亚临床感染时，病毒分离的敏感性需要病毒制备技术，以保证组织中与细胞相连的病毒最大程度地释放（Whittington 和 Steiner，1993）。最近，对用于高通量实验室技术的这些方法进行了改进，产生了如珠打破碎技术。该技术用于病毒分离，并增加了 qPCR 对 EHNV 检测的敏感性（Rimmer 等人，2012）。有临 192
床症状的感染鱼类、两栖类及爬行类动物的肝、肾和脾的蛙病毒丰度比较高。对于桑堤-库珀蛙病毒检测来说，建议将鳃、鱼鳔和后肾作为最小样本（Beck 等人，2006），而肾、肝和脾大样本则更适合于 EHNV 的检测。

病毒分离为蛙病毒复制和形态发生的基础研究提供了一个很重要的工具。例如，依据诊断试验包括在 EPC 细胞上进行的病毒分离，在对 GCSV 进行初步描述之后（Geng 等人，2010），有学者用 EM 和蛋白组学技术以及在 EPC 细胞上进行 RNA 敲低技术进行研究，获得了关于病毒的其他重要资料（Li 等人，2014b；Ma 等人，2014）。这些类型的研究提供了关于病毒复制有价值的信息，对开发治疗和控制病毒感染的方法有所帮助。

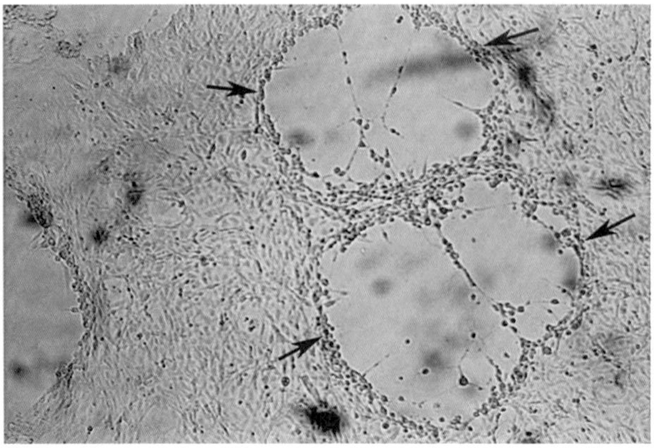

图 7 EHNV 引起的蓝鳃太阳鱼苗（BF-2）细胞典型细胞病理效应（CPE），显示出细胞单层多灶状溶解。注意在溶解区域边缘细胞变圆（箭头所指）。数天之后，可见到整个单细胞层全部破坏。

2.3.4 抗体的检测：血清学

将抗蛙病毒抗体和种特异性抗免疫球蛋白试剂相结合，开发出了检测抗体的一种 ELISA，并用于检测红鲈鲈（*Perca fluviatilis*）、虹鳟（*Oncorhynchus mykiss*）以及海蟾蜍（*Bufo marinus*）（Whittington 等人，1994，1997；Whittington 和 Reddacliff，1995；Whittington 和 Speare，1996）血清中蛙病毒感染的特异性适应性免疫反应。红鲈鲈和虹鳟注射灭活 EHNV 时，结果发生了血清转化（Whittington等人，1994；Whittington 和 Reddacliff，1995）。针对蛙病毒感染的长期适应性免疫反应所产生的血清学方法提供了一种灵敏的方法来区分蛙病毒流行地区的宿主种群和来自无蛙病毒地区的宿主种群。例如，惠廷顿等人（1999）报道，在 EHNV 感染的养殖鱼类种群中，存在着小比例血清阳性的成年虹鳟。类似地，在自由生活的无尾两栖类种群中，同样也检测出了抗蛙病毒抗体（Whittington等人，1997；Zupanovic 等人，1998）。最近，在全面调查非洲爪蟾蛙病毒感染的发病机制时，对体液免疫反应进行了评估（Gantress 等人，2003）。

鉴于在鱼类中 ELISA 检测的成功，ELISA 对抗体的检测已被应用于爬行类动物蛙病毒感染。阿里尔（1997）利用亨斯特伯杰（Hengstberger）等人（1993）所建立的方法，在澳大利亚野生爬行类动物中，检测出了抗博乐虹彩病毒的抗体。最近，在美国，约翰逊等人（2007，2010）和阿蓝德（2012a）将 ELISA 用于监测和实验室调查，检测各种龟类动物血清中抗蛙病毒抗体。

尽管血清学实验具有一定的局限性，但还是一个简单易行且成本效益好的技术。血清学实验的应用受到对所有抗蛙病毒抗体的交叉反应性的限制，并缺

少二次检测方法来评估其敏感性和特异性。由于免疫兔子后不能刺激产生中和 193
EHNV 活性的抗体（Hedrick 等人，1992）或来自小鼠的单克隆抗体，也不能中和 EHNV 活性（Monini 和 Ruggeri，2002），因而病毒的中和实验不能得到应用。目前，来自两栖类、爬行类和鱼类免疫系统的知识表明，血清学技术的局限性是由于缺乏亲和成熟、免疫记忆差以及免疫反应的温度依赖性（McLoughlin 和 Graham，2007）。基于这一原因，目前，OIE 还不支持将血清学作为一个可行的检测手段。

2. 3. 5　动物实验：生物测定

生物测定或实验传播试验是诊断过程的重要组成部分，并用于一种新型蛙病毒的检测或在一个新宿主或一个新的生态环境中检测一种新疾病的情况。这种例子见于华尔兹克（Waltzek）等人（2014）的报道，这个例子报道了由 FV3 样病毒引起白鲟的死亡事件。在这个例子当中，用大规模死亡事件中分离到的病毒进行实验性传染，在实验室感染白鲟，结果导致了白鲟大规模死亡事件的发生，模仿了病毒最先分离的死亡事件，后来进行的病毒测序显示其病原与 FV3 样病毒的亲缘关系最近 。

2. 3. 6　组织检查

组织病理学和细胞学

组织病理学对于调查不明原因的疾病是一个重要的工具。特征性的病变可以提供某些病原的初步证据，同时为选择随后的诊断实验提供指导。组织病理学在监测研究或疾病事件调查中，对确定所检测到的蛙病毒感染的相关性也很重要。一旦生物死亡，很快就会产生组织崩解（tissues breakdown）即自溶，特别是在温暖的环境下，这样就会使显微镜下的详细观察显得模糊不清。因此，所进行的组织学最好是使用在动物还活着时或人为安乐死时所采集的病态样本（AVMA，2013），并且在死亡后立即保存在固定剂中。比率为 9∶1 的 10% 中性缓冲福尔马林溶液一般用于组织的固定（Pessier 和 Mendelson，2010）。乙醇（ethanol）或其他固定剂（fixatives），例如戴维森液（Davidson's）或波恩氏液（ Bouin's），比较适合于鱼类组织的固定。假如要将小动物运送到诊断实验室，可以将整个小动物置入固定剂中，并在体腔作一个 5 mm 的小切口，有助于内脏组织的保存。小鱼可去掉尾和鳃盖进行保存。较大的动物可以打开，进行尸体剖检，并代表性地采集所有组织的样本，参见 2. 2 节。能观察到大体病变的动物组织采样也是如此，并考虑进行鉴别诊断。在尸体剖检时，采集两套组织样品，一套进行新鲜贮存（立即进行检测），另一套进行冷冻保存（用于将来检测），这是很重要的。假如组织病理学结果表
明，存在着与蛙病毒感染一致的病变，则需要进行验证性检测排除并发病原的 194

存在。为了防止交叉污染，应当使用一套不同的器材对每个个体进行采样。

通常情况下，蛙病毒病的病例并不是直接的，需要用其他诊断方法如IHC、EM或分子方法来进行确诊。典型的胞浆内包涵体的观察，其结果会不一致，特别是龟类动物（DeVoe等人，2004；Johnson等人，2007，2008）。在其他动物中，蛙病毒感染的细小病变会因继发感染细菌或真菌，而变得模糊不清（见图5）。

虽然不是对疾病的决定性实验，但细胞学可以通过血细胞计数记录先天性免疫系统的变化，并用于检测病毒包涵体。血液可以按上述方法进行采集（见2.2节），对风干的血液涂片进行染色，可用光学显微镜的油镜进行细胞学的检查，在白细胞的胞浆内可以观察到病毒包涵体（Allender等人，2006）。虽然不是肯定的，但当同时存在对应的临床症状、大体病变以及组织病理学变化时，病毒包涵体可以提示存在蛙病毒感染。应该注意的是，在红细胞中的病毒包涵体可能表示红细胞虹彩病毒，该病毒隶属虹彩病毒科的一个不同属，尤其是在爬行类动物红细胞中，见到的病毒包涵体更是如此（Wellehan等人，2008；Grosset等人，2014）。

病毒可视化实验

关于蛙病毒的致病性有几个长期存在的问题：是蛙病毒造成的病变吗？亚临床感染中病毒在哪里？病毒感染什么细胞，又是如何因宿主物种和发育阶段的不同而发生变化的呢？发病机理中一个关键性因素就是病毒的靶细胞类型，靶细胞类型的鉴定只能通过进行组织学研究来完成。然而，细胞的变化通常是看不见的，例如亚临床感染。另外，由蛙病毒引起的细胞变化是组织学的一个挑战，特别是与其他病原存在并发感染的情况下。因此，对细胞内的病毒进行可视化，对于观察引起细胞变化的蛙病毒是非常有用的（Miller和Gray，2010），并且可视化可以通过免疫组化（IHC）、原位杂交和EM来达到目的。

免疫组化

在免疫组化中，利用酶结合物（enzyme-conjugated）和病毒特异性抗血清，通过一种酶如辣根过氧化物酶或碱性磷酸酶（horseradish peroxidase or alkaline phosphatase）的催化反应，显示病毒抗原在细胞内的位置（见图8）。已开发出运用单克隆抗血清建立的抗原捕获ELISA和免疫电子显微镜技术来检测EHNV和欧洲鲶鱼病毒（ECV）抗原（Steiner等人，1991；Hengstberger等人，1993；Hyatt等人，1991；Whittington和Steiner，1993；Ahne等人，
195 1998）。适用于免疫荧光或酶法检测的第二抗体，能够检测出组织切片和细胞培养中的蛙病毒抗原（Reddacliff和Whittington，1996）。一些抗血清是直接抗MCP的，并且具有交叉反应性，因而所有的蛙病毒都能被检测出来，但是要进行蛙病毒之间的区分是不可能的（Hedrick等人，1992；Ahne等人，1998）。

目前，抗纯病毒子的其他抗血清已开发出来（Balseiro 等人，2009），但是再次强调，用这种方法进行蛙病毒之间的区分是不可能的。有学者对两栖动物进行免疫组化染色的有效应用作了报道（Hyatt 等人，2002；Cunningham 等人，2008；Balseiro 等人，2009，2010；Bayley 等人，2013）。巴尔塞罗等人（2009，2010）使用免疫组化检测各种细胞中存在的蛙病毒，结果发现，在幼蟾肾小球中阳性染色最深，同时，在稚蝾螈的神经节，特别是在骨骼肌阳性染色最深。

尽管发表了这些报道，但 IHC 并没有被广泛应用。在获得抗体过程中所发现的困难，通常被认为是不能将其作为基础实验室检测的部分理由。主要原因是，免疫组化只用于研究，而研究者只制备他们自身的抗体或从其他研究者那里获得抗体（Hyatt 等人，2002；Cunningham 等人，2008；Balseiro 等人，2009，2010；Bayley 等人，2013；Whitley 等人，2010；Chinchar 等人，1984）。这一技术在鉴定坏死区域的包涵体时特别有用，但是当存在大量细胞碎片时，这一技术面临着挑战（见图 8）。免疫组化在坏死和包涵体产生之前同样可能会出现阳性，因而在检测具有亚临床症状动物病毒的存在可能是有用的。由 OIE 参考实验室所推荐的抗体，是抗高度纯化、已验证的蛙病毒株的抗体，并进行过预试验，有已知的抗体滴度和保存期限（Whittington 和 Deece，2004）。目前尚不清楚这是否适用于为 IHC 开发的其他抗体。巴尔塞罗（Balseiro）和米勒（Miller）（未发表的数据）的初步研究表明，蛙病毒分离株之间及宿主之间的 IHC 结果可能存在差异。

在组织固定的情况下，如在大规模死亡事件的诊断调查中，在不可能运输新鲜和冷冻组织的情况下，可以应用 IHC。尽管 PCR 能够使用保存的样本并提供蛙病毒存在的信息，但这一技术不能验证病毒是否参与了特定的病变，是否有可能是死亡的原因。通过显微镜观察，免疫组化的病毒原与病灶的联系可视化，从而为病变的原因提供支持性的证据。

原位杂交

病毒可视化的另外一种方法就是原位杂交（in situ hybridization，ISH），原位杂交就是用分子探针定位固定组织切片内的特异性核酸序列。例如，在感染新加坡石斑鱼虹彩病毒（SGIV）的点带石斑鱼（*Epinephelus malabaricus*）的肾和脾中观察到了特异性染色。SGIV 是一种感染海洋鱼类的虹彩病毒，并与
FV3 的 MCP 基因序列有接近 83% 的同源性（Huang 等人，2004）。虽然用原位 197
杂交检测蛙病毒的报道甚少，但是未来的研究可考虑使用这一技术来记录蛙病毒病的发病机理。

电镜

电镜（electron microscopy，EM）用来目视确认培养病毒的身份，同时用于组织切片内病毒的可视化。培养产物中病毒的确诊涉及负染（negative staining），并需要扫描电子显微镜（scanning electron microscopy，SEM）进行评估。然而，保存的组织则需要透射电子显微镜（transmission electron microscopy，TEM）使组织内的病毒可视化。例如，伯顿（Burton）等人（2008）用透射电子显微镜观察了保存标本中畸形眼的超微结构。电镜观察显示，样本所含的病毒粒子与虹彩病毒科的病毒相一致，这样就允许使用PCR和测序进行进一步鉴定。类似地，组织学检查能揭示细胞内病毒包涵体的结构，SEM可以用来验证这些颗粒是否是病毒的来源（Cunningham等人，1996，2007；Balseiro等人，2010；Cheng等人，2014；Meng等人，2014）。总的来说，只有在病毒科（如虹彩病毒科）水平作出的鉴定才是可靠的。然而，海特等人（2000，2002）报道，虹彩病毒科的病毒成员在大小上存在着差异，这意味着更具体的鉴定是有可能的。

早在20世纪80年代，冷冻电子显微镜就被用来检查病毒表面的超微结构（Adrian等人，1984）。这种技术采用了对纯化病毒进行快速冷冻的办法，避免化学防腐剂引起的改变。在此过程中，需拍摄多个照片，通过重构来生成病毒的3D图像。研究人员一直在使用冷冻电子显微镜来评估病毒组装和复制中各种表面蛋白的作用（Yan等人，2009；Whitley等人，2010；Tran等人，2011）。这些发现有可能为治疗方案或疫苗的开发提供一些认识。

2.4 验证试验和效率

2.4.1 诊断试验的黄金标准和局限性

目前，还没有针对蛙病毒建立的单一黄金标准的诊断试验。依据所提出的问题的不同，蛙病毒的诊断试验会有所变化。在疾病诊断时，实验室结果应与动物的临床状态、病理学发现和观察到的种群影响相结合。例如，从组织中分离到的蛙病毒与相应的尸检发现（包括大体的和组织病理学的发现）以及免疫组化的结果一起，提供强有力的证据，来支持蛙病毒是死亡的病因。类似地，用于检测或认定无感染个体或区域的诊断试验需要一个统计学上有效的采
198 样方案（Gray等人，2015）。评估免于感染所要求的样本大小的计算，需要知道最小预期患病率和一个具有已知敏感性和特异性的诊断试验。例如，EHNV感染虹鳟的患病率可以低至4%（Whittington等人，1994，1999）。格雷等人（2015）建议检测蛙病毒所需要的样本大小，但要给出假定感染患病率的范围后，才能确定。

鉴于诊断试验，例如 qPCR，诊断还不完善。诊断试验的敏感性和特异性的估计对于准确地解释结果是很重要的。实验室用于论证试验结果有效性的方法，在《水生动物疾病诊断手册》的介绍性章节中作过概述（OIE，2012a）。OIE 指南概述了使用实验室质量管理系统的原则，通过记录最小化以及鉴定假阳性和假阴性结果所需要的一整套措施，来确保诊断试验结果的可靠性。总的来说，几个实验室正协作，通过检测相同的样本共同评估假阳性和假阴性结果，并进一步用于评估诊断试验的敏感性和特异性。这是在盲试设计中通过实验室之间分配对照样本（例如已知阳性和阴性）和标准试剂，并在实验室之间估计误差率来实现的。

3　治疗和疫苗的开发

蛙病毒感染的治疗与疫苗接种可能适用于人工养殖的种群，同时，可能对动物学上群集的动物或珍稀物种的保护项目也是很有用的。目前，对蛙病毒感染的治疗受到限制。阿蓝德等人（2012b）报道，使用鸟嘌呤类似物抗病毒药即阿昔洛韦（Acyclovir）和伐昔洛韦（Valacyclovir）治疗龟类虹彩病毒和疱疹病毒感染是有可能的。最近，李鹏飞等人（2014a）报道，DNA 核酸适配体（DNA aptamers）在治疗 SGIV 时，显示出抗病毒活性。热处理灭活许多变温脊椎动物病原如蛙壶菌（*Batrachochytrium dendrobatidis*）是很有效的（Woodhams 等人，2003）。然而，热处理蛙病毒的有效性在宿主物种和病毒株之间存在着差异。罗贾斯（Rojas）等人（2005）报道，ATV 感染的蝾螈，饲养在高温（26 ℃）比那些饲养在低温的蝾螈更有可能生存下来。类似地，阿兰德等人（2013b）发现，在红耳龟（*Trachemys scripta elegans*）中，FV3 样病毒在 22 ℃时的致病性要比在 28 ℃时更高。然而还有几个其他研究报道，在较暖的温度条件下，蛙病毒的复制更快，致病性更高（Whittington 和 Reddacliff，1995；Grant 等人，2003；Ariel 和 Jensen，2009；Ariel 等人，2009；Bayley 等人，2013）。鉴于大多数蛙病毒在高于 32 ℃以后不能复制（Chinchar，2002；Ariel 等人，2009），因而体温升高超过这一阈值对一些宿主物种来说可能是有益的。如果宿主物种和病毒株之间存在差异，那么所做的研究就需要确定热处理的效果以及灭活病毒所需要的时间。

到目前为止，针对虹彩病毒疫苗的开发主要集中在水产养殖业中的鱼种当 199
中。有些疫苗商业有售，但这些疫苗不适用于所有物种或与其他虹彩病毒具有交叉反应性（Oh 等人，2014）。虽然常常使用活疫苗（live vaccines），但 DNA

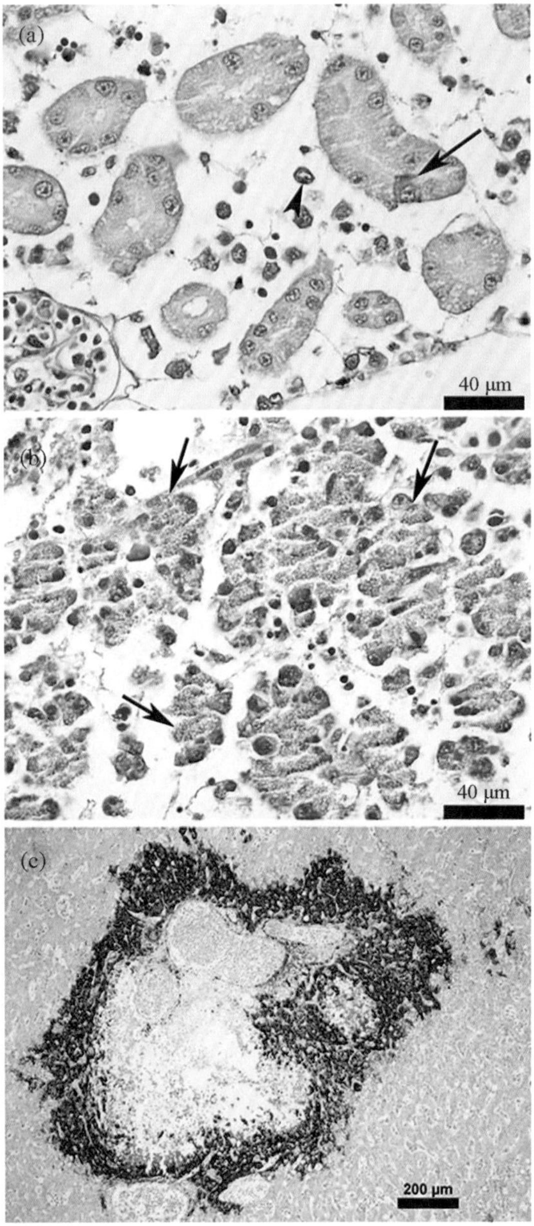

图 8 （a）南方豹蛙（*Lithobates sphenocephala*）肾小管上皮细胞（箭头所指）中蛙病毒的免疫组化染色；（b）在造血细胞（箭头所指）胞浆内包涵体也染成阳性；（c）红鲈鲈（*Perca fluviatilis*）肝脏的阳性染色。

疫苗显示出广阔的前景。例如，张民（Zhang Min）等人（2012）研究发现，在同时感染石鲷虹彩病毒时，接种 DNA 疫苗的大菱鲆的存活率要比未接种 DNA 疫苗的对照鱼高。类似地，蔡庞（Caipang）等人（2006）报道了真鲷（*Pagrus major*）在接种 DNA 疫苗之后，针对真鲷虹彩病毒感染产生免疫力的证据。目前，限制疫苗使用的一个方面就是通过肌肉注射进行疫苗接种，这样劳动强度大。其他研究者正在探索更可行的接种方法，如口服剂型的疫苗接种（Tamaru 等人，2006）。

4　小结与最后的建议

蛙病毒病对易感宿主是毁灭性的，可造成出血、溃疡、水肿以及器官的坏死。虽然在变温脊椎动物不同类别之间，病变存在差别，但是内皮细胞的坏死伴随后来的出血，是发生在所有动物类别当中的共同变化。新的技术已经使一些诊断技术得到了应用，如免疫组化（IHC），IHC 允许我们对组织内的病毒抗原进行可视化观察。技术进步包括激光对病变的解剖，然后进行 IHC 和 PCR 以及 3D 连续断层扫描（3D sequential tomography）。

由于所有的诊断试验都存在局限性，因而调查人员和研究人员使用多种诊断试验来进行准确的诊断，并在亚临床感染和疾病之间作出鉴别显得特别重要。何种方法最适合诊断试验，取决于需要回答的问题。例如，假如想要确定一个动物是否携带蛙病毒，那么 PCR、qPCR、IHC 及病毒分离的所有试验也许能给出答案。重要的是，病原检测与确定一个病原是否与所观察到的临床疾病之间存在联系是有区别的。目前，一些推荐的不同试验正用于确定目前或过去感染的流行，如分别进行 qPCR 和抗原捕获 ELISA。确定感染状态的一种典型方法，是将已知敏感性和特异性的试验应用于统计学上有效采样的种群中，来展示种群是否免于感染或疾病的最小预期患病率（Cameron 和 Baldock，1998）。高通量的 qPCR 实验室方法非常适合处理这些调查所需要的大样本。然而，我们仍然没有试验来验证单个动物是否能免受蛙病毒感染，特别是在非死亡采样的情况下。例如，如果一个动物通过 PCR 检测呈阳性，但在两周后再次检测时呈阴性，目前还不能判断两周后，受检动物是否已经清除了病毒或者病毒是否仍然隐藏在动物体内，或者是否存在蛙病毒的潜伏感染（见第 6 章）。

我们还需要一种成本核算的、经过反复验证的方法来检测和鉴定不同的蛙
病毒株。目前能这样做的方法还不是很普遍，并受到成本的限制。然而，未来 200
动物再引进的成功或动物转运项目的成功取决于我们区分原产地和目的地动物

种群中不同病毒株的能力。同样，病毒株的鉴定包括未来蛙病毒的报告和蛙病毒地理分布作图将是很重要的，因为这些资料可以用于流行病学研究。

目前，有许多不同的实验方案用于各种实验，在不同的实验室之间，OIE制定了标准化的实验指南。由于蛙病毒存在比较宽的宿主范围，因而在蛙病毒研究期间会遇到不同的样品类型。因此，诊断技术开发和共享验证数据是很重要的。我们建议，从事蛙病毒研究的实验室应携手合作，使检测方案标准化。这一任务可能通过环试（ring trial）来完成。在环试中，盲样（blinded samples）同时送到不同的国际性实验室进行检测。GRC是一个能够引导技术如何标准化大讨论的实体，可以帮助和促进环试组织及数据验证的协调。

关于蛙病毒病的病理以及病毒感染和疾病的诊断检测，还有很多需要探索的地方。鉴于我们目前对发病机理的了解及最近在蛙病毒遗传学和宿主免疫反应方面的进展，我们应该制定有效的管理和治疗方法，应用于保护计划、商业和动物园的设施以及水族馆。此外，我们对蛙病毒病过程的理解，再加上我们对蛙病毒生态学和流行病学越来越多的认识，为制定水生生态系统的管理规划奠定了一定的基础。

致谢

我们要感谢安娜·巴尔塞罗（Ana Balseiro）和亚历克斯·海特（Alex Hyatt）对这一章进行的仔细评审。开放存取出版物的发行得到了田纳西大学（林业、野生动物与渔业系，研究与交流办公室，农业研究所）、华盛顿州立大学图书馆、戈登州立大学（学术事务办公室）、两栖爬行动物兽医协会以及两栖爬行动物保护协会的资金赞助。

开放存取

参考文献

Adrian M, Dubochet J, Lepault J, McDowall AW (1984). Cryo-electron microscopy of viruses. Nature 308: 32-36.

Ahne W, Schlotfeldt HJ, Thomsen I (1989). Fish viruses: isolation of an icosahedral cytoplasmic deoxyribovirus from sheatfish *Silurus glanis*. J Vet Med B 36: 333-336.

AhneW, Bearzotti M, Bremont M, Essbauer S (1998). Comparison of European systemic piscine and amphibian iridoviruses with epizootic haematopoietic necrosis virus and frog virus 3. J Vet Med 45:373-383.

Allender MC (2012a). Characterizing the epidemiology of ranavirus in North American chelonians: diagnosis, surveillance, pathogenesis, and treatment. Dissertation, University of Illinois at Urbana-Champaign, Urbana-Champaign, p 219.

Allender MC, Fry MM, Irizarry AR, Craig L, Johnson AJ, Jones M (2006). Intracytoplasmic inclusions in circulating leukocytes from an eastern box turtle (*Terrepene carolina carolina*). J Wildl Dis 42:677-684.

Allender MC, Abd-Eldaim M, Schumacher J, McRuer D, Christian LS, Kennedy M (2011). PCR prevalence of Ranavirus in free-ranging eastern box turtles (*Terrapene carolina carolina*) at rehabilitation centers in three southeastern US states. J Wildl Dis 47:759-764.

Allender MC, Mitchell MA, Yarborough J, Cox S (2012). Pharmacokinetics of a single oral dose of acyclovir and valacyclovir in North American box turtles (*Terrapene* sp). J Vet Pharmacol Ther 36:205-208.

Allender MC, Bunick D, Mitchell MA (2013a). Development and validation of Taqman quantitative PCR for detection of frog virus 3-like virus in eastern box turtles (*Terrepene carolina caro-lina*). J Virol Methods 188:121-125.

Allender MC, Mitchell MA, Torres T, Sekowska J, Riskell EA (2013b). Pathogenicity of frog virus 3-like virus in red-eared slider turtles at two environmental temperatures. J Comp Pathol 149:356-367.

Ariel E (1997). Pathology and serological aspects of *Bohle iridovirus* infections in six selected water-associated reptiles in North Queensland. Ph. D. Dissertation, James Cook University, Townsville City, p 214.

Ariel E, Jensen BB (2009). Challenge studies of European stocks of redfin perch, *Perca fluviatilis* L., and rainbow trout, *Oncorhynchus mykiss* (Walbaum), with epizootic haematopoietic necrosis virus. J Fish Dis 32(12):1017-1025.

Ariel E, Nicolajsen N, Christophersen MB, Holopainen R, Tapiovaara H, Jensen BB (2009). Propagation and isolation of ranaviruses in cell culture. Aquaculture 294:159-164.

AVMA (2013). AVMA guidelines for the euthanasia of animals:2013 edition. American Veterinary Medical Association, Schaumburg, p102.

Balseiro A, Dalton KP, del Cerro A, Marquez I, Cunningham AA, Parra F, Prieto JM, Casais R (2009). Pathology, isolation and molecular characterization of a ranavirus from the common midwife toad *Alytes obstetricans* on the Iberian Peninsula. Dis Aquat Organ 84:95-104.

Balseiro A, Dalton KP, del Cerro A, Marquez I, Parra F, Prieto JM, Casais R (2010). Outbreak of common midwife toad virus in alpine newts (*Mesotriton alpestris cyreni*) and common midwife toads (*Alytes obstetricans*) in Northern Spain: a comparative pathological study of an emerging ranavirus. Vet J 186:256-258.

Bang Jensen B, Holopainen R, Tapiovaara H, Ariel E (2011). Susceptibility of pike-perch *Sander lucioperca* to a panel of ranavirus isolates. Aquaculture 313: 24-30.

Bayley AE, Hill BJ, Feist SW (2013). Susceptibility of the European common frog *Rana temporaria* to a panel of ranavirus isolates from fish and amphibian hosts. Dis Aquat Organ 103:171- 183.doi:10.3354/dao02574.

Beck BH, Bakal RS, Brunner CJ, Grizzle JM (2006). Virus distribution and signs of disease after immersion exposure to largemouth bass virus. J Aquat Anim Health 18:176-183.

Becker JA, Tweedie A, Gilligan D, Asmus M, Whittington RJ (2013). Experimental infection of Australian freshwater fish with epizootic haematopoietic necrosis virus (EHNV). J Aquat Anim Health 25: 66-76. doi: 10. 1080/08997659. 2012.747451.

Beebee T (2012). Impact of Ranavirus on garden amphibian populations. Herpetol Bull 120:1-3.

Behncke H, Stöhr AC, Heckers K, Ball I, Marschang RE (2013). Mass-mortality in green striped tree dragons (*Japalura splendida*) associated with multiple viral infections. Vet Rec 173:248.doi:10.1136/vr.101545.

Bollinger TK, Mao J, Schock D, Brigham RM, Chinchar VG (1999). Pathology, isolation, and preliminary molecular characterization of a novel iridovirus from tiger salamanders in Saskatchewan. J Wildl Dis 35:413-429.

Brenes R, Gray MJ, Waltzek TB, Wilkes RP, Miller DL (2014a). Transmission of ranavirus between ectothermic vertebrate hosts. PLoS One 9(3):e92476.doi:10. 1371/journal pone.0092476.

Brenes R, Miller DL, Waltzek TB, Wilkes RP, Tucker JL, Chaney JC, Hardman RH, Brand MD, Huether RR, Gray MJ (2014b). Susceptibility of fish and turtles to three ranaviruses isolated from different ectothermic vertebrate classes. J Aquat Anim Health 26(2):118-126.doi:10.10 80/08997659.2014.886637.

Brunner JL, Schock DM, Davidson EW, Collins JP (2004). Intraspecific reservoirs: complex life cycles and the persistence of a lethal ranavirus. Ecology 85: 560-566.

Brunner JL, Storfer A, Gray MJ, Hoverman JT (2015). Ranavirus ecology and evolution: from epidemiology to extinction. In: Gray MJ, Chinchar VG (eds) Ranaviruses: lethal pathogens of ectothermic vertebrates. Springer, New York.

Bryan LK, Baldwin CA, Gray MJ, Miller DL (2009). Efficacy of select disinfectants at inactivating Ranavirus. Diseases of Aquatic Organisms 84(2):89-94.

Burton EC, Miller DL, Styer EL, Gray MJ (2008). Amphibian ocular malformation associated with frog virus 3. Vet J 177:442-444.

Bustin SA, Beaulieu JF, Huggett J, Jaggi R, Kibenge FSB, Olsvik PA, Penning LC, Toegel S (2010). MIQE precis: practical implementation of minimum standard guidelines for fluorescence-based quantitative real-time PCR experiments. BMC Mol Biol 11:74.doi:10.1186/1471-2199-11-74.

Caipang CMA, Takano T, Hirono I, Aoki T (2006). Genetic vaccines protect red seabream, *Pagrus major*, upon challenge with red seabream iridovirus (RSIV). Fish Shellfish Immunol 21:130-138.

Cameron AR, Baldock FC (1998). A new probability formula for surveys to substantiate freedom from disease. Prev Vet Med 34:1-17.

Caraguel CGB, Stryhn H, Gagne N, Dohoo IR, Hammell KL (2011). Selection of a cutoff value for real-time polymerase chain reaction results to fit a diagnostic purpose: analytical and epidemiologic approaches. J Vet Diagn Invest 23:2-15.

Cashins S, Skerratt FL, Alford RA (2008). Sodium hypochlorite denatures the DNA of the amphibian chytrid fungus *Batrachochytrium dendrobatidis*. Dis Aquat Organ 80:63-67.

Cheng K, Jones MEB, Jancovich JK, Burchell J, Schrenzel MD, Reavill DR, Imai DI, Urban A, Kirkendall M, Woods LW, Chinchar VG, Pessier AP (2014). Isolation of a Bohle-like iridovirus from boreal toads housed within a cosmopolitan aquarium collection. Dis Aquat Organ 111(2):139-152.

Chinchar VG, Metzger DW, GranoffA, Goorha R (1984). Localization of frog virus 3 proteins using monoclonal antibodies. Virology 137:211-216.

Chinchar VG (2002). Ranaviruses (family *Iridoviridae*): emerging cold-blooded killers. Arch Virol 147:447-470.

Crane MSJ, Young J, Williams L (2005). Epizootic haematopoietic necrosis virus (EHNV): growth in fish cell lines at different temperatures. Bull Eur Assoc Fish Pathol 25:228-231.

Cunningham AA, Langton TES, Bennett PM, Lewin JF, Drury SEN, Gough RE, Macgregor SK (1996). Pathological and microbiological findings from incidents of unusual mortality of the common frog (*Rana temporaria*). Philos Trans R Soc Lond B Biol Sci 351:1539-1557.

Cunningham AA, Hyatt AD, RussellP, Bennett PM (2007). Emerging epidemic diseases of frogs in Britain are dependent on the source of ranavirus agent and the route of exposure. Epidemiol Infect 135:1200-1212. doi:10.1017/S0950268806007679.

Cunningham AA, Tems CA, Russell PH (2008). Immunohistochemical demonstration of Ranavirus antigen in the tissues of infected frogs (*Rana temporaria*) with systemic haemorrhagic or cutaneous ulcerative disease. J Comp Pathol 138:3-11.

DeVoeR, Geissler K, Elmore S, Rotstein D, Lewbart G, Guy J (2004). Ranavirus-associated morbidity and mortality in a group of captive eastern box turtles (*Terrepene carolina carolina*). J Zoo Wildl Med 35:534-543.

Docherty DE, Meteyer CU, Wang J, Mao J, Case ST, Chinchar VG (2003). Diagnostic and molecular evaluation of three iridovirus-associated salamander mortality events. J Wildl Dis 39(3):556-566.

Duffus AL, Andrews AM (2013). Phylogenetic analysis of a frog virus 3-like ranavirus found at a site with recurrent mortality and morbidity events in south-eastern Ontario, Canada: partial major capsid protein sequence alone is not sufficient for fine-scale differentiation. J Wildl Dis 49:464-467.

Duffus ALJ, Waltzek TB, Stöhr AC, Allender MC, Gotesman M, Whittington RJ, Hick P, Hines MK, Marschang RE (2015). Distribution and host range of ranaviruses. In: Gray MJ, Chinchar VG (eds) Ranaviruses: lethal pathogens of ectothermic vertebrates. Springer, New York.

Earl JE, Gray MJ (2014). Introduction of ranavirus to isolated wood frog populations could cause local extinction. Ecohealth. doi:10.1007/s10393-014-0950-y.

Farnsworth SD, Seigel FA (2013). Responses, movements, and survival of relocate box turtles during construction of the Intercounty Connector Highway in Maryland. J Transp Res 2362:1-8.

Forzan MJ, Wood J (2013). Low detection of ranavirus DNA in wild postmetamorphic green frogs, *Rana* (*Lithobates*) *clamitans*, despite previous or concurrent tadpole mortality.J Wildl Dis 49:879-886.

Gantress J, Maniero GD, Cohen N, Robert J (2003). Development and characterization of a model system to study amphibian immune responses to iridoviruses. Virology 311:254-262.

Geng Y, Wang KY, Zhou ZY, Li CW, Wang J, He M, Yin ZQ, Lai WM (2010). First report of a ranavirus associated with morbidity and mortality in farmed Chinese giant salamanders (*Andrias davidianus*). J Comp Pathol. doi: 10.1016/j.jcpa.2010.11.012.

Gold KK, Reed PD, Bemis DA, Miller DL, Gray MJ, Souza MJ (2013). Efficacy of common disinfectants and terbinafine HCl at inactivating the growth of *Batrachochytrium dendrobatiditis* in culture. Dis Aquat Organ 107:77-81.

Goldberg TL (2002). Largemouth bass virus: an emerging problem for warmwater fisheries? Am Fish Soc Symp 31:411-416.

Goodman RM, Miller DL, Ararso YT (2013). Prevalence of ranavirus in Virginia turtles as detected by tail-clip sampling versus oral-cloacal swabbing. Northeastern Nat 20:325-332.

Granoff A, Came PE, Breeze DC (1966). Viruses and Renal Carcinoma of *Rana Pipiens*.I.Isolation and properties of virus from normal and tumor tissue. Virology 29:133-148.

Grant EC, Philipp DP, Inendino KR (2003). Effects of temperature on the susceptibility of largemouth bass to largemouth bass virus.J Aquat Anim Health 15:215-220.

Gray MJ, Miller DL, Schmutzer AC, Baldwin CA (2007). Frog virus 3 prevalence in tadpole populations inhabiting cattle-access and non-access wetlands in Tennessee, U.S.A.Dis Aquat Organ 77:97-103.

Gray MJ, Miller DL, Hoverman JT (2009). First report of ranavirus infecting lungless salamanders.Herpetol Rev 40(3):316-319.

Gray MJ, Miller DL, Hoverman JT (2012). Reliability of non-lethal surveillance methods for detecting ranavirus infection. Dis Aquat Organ 99: 1-6. doi: 10.3354/dao02436.

Gray MJ, Brunner JL, Earl JE, Ariel E (2015). Design and analysis of ranavirus studies: surveillance and assessing risk. In: Gray MJ, Chinchar VG (eds) Ranaviruses: lethal pathogens of ectothermic vertebrates.Springer, New York.

Grayfer L, Edholm E-S, De Jesús Andino F, Chinchar VG, Robert J (2015). Ranavirus host immunity and immune evasion. In: Gray MJ, Chinchar VG (eds) Ranaviruses: lethal pathogens of ectothermic vertebrates. Springer, New York.

Green DE, Converse KA, Schrader AK (2002). Epizootiology of sixty-four amphibian morbidity and mortality events in the USA, 1996—2001. Ann NY Acad Sci 969: 323-339.

Green DE, Gray MJ, Miller DL (2009). Disease monitoring and biosecurity. In: Dodd CK (ed) Amphibian ecology and conservation: a handbook of techniques. Oxford University Press, Oxford.

Greer AL, Collins JP (2007). Sensitivity of a diagnostic test for amphibian ranavirus varies with sampling protocol. J Wildl Dis 43: 525-532.

Greiner M, Gardner IA (2000). Epidemiologic issues in the validation of veterinary diagnostic tests. Prev Vet Med 45: 3-22.

Grizzle JM, Brunner CJ (2003). Review of largemouth bass virus. Fisheries 28(11): 10-14. doi: 10.1577/1548-8446(2003)28[10:ROLBV]2.0.CO;2.

Grosset C, Wellehan JF, Owens SD, McGraw S, Gaffney PM, Foley J, Childress AL, Yun S, Malm K, Groff JM, Paul-Murphy J, Weber ES (2014). Intraerythrocytic iridovirus in central bearded dragons (*Pogona vitticeps*). J Vet Diagn Invest 26: 354-364.

Haislip NA, Gray MJ, Hoverman JT, Miller DL (2011). Development and disease: how susceptibility to an emerging pathogen changes through anuran development. PLoS ONE 6(7): e22307. Doi: 10.1371/journalpone.

Harp EM, Petranka JW (2006). Ranavirus in wood frogs (*Rana sylvatica*): potential sources of transmission within and between ponds. J Wildl Dis 42: 307-318.

Hedrick RP, McDowell TS, Ahne W, Torhy C, De Kinkelin P (1992). Properties of three iridovirus-like agents associated with systemic infections of fish. Dis Aquat Organ 13: 203-209.

Hengstberger SG, Hyatt AD, Speare R, Coupar BEH (1993). Comparison of epizootic haematopoietic necrosis and Bohle iridoviruses, recently isolated Australian iridoviruses. Dis Aquat Organ 15: 93-107.

Holopainen R, Ohlemeyer S, Schutze H, Bergmann SM, Tapiovaara H (2009). Ranavirus phylogeny and differentiation based on major capsid protein, DNA polymerase and neurofilament triplet H1-like protein genes. Dis Aquat Organ 85: 81-91.

Hoverman JT, Gray MJ, Haislip NA, et al. (2011). Phylogeny, life history, and ecology contribute to differences in amphibian susceptibility to ranaviruses. Ecohealth 8: 301-319.

Huang C, Zhang X, Gin KYH, Qin QW (2004). In situ hybridization of a marine fish virus, Singapore grouper iridovirus with a nucleic acid probe of major capsid protein. J Virol Methods 117: 123-128.

Hyatt AD, Eaton BT, Hengstberger S, Russel G (1991). Epizootic haematopoietic necrosis virus: detection by ELISA, immunohistochemistry and immunoelectronmicroscopy. J Fish Dis 14: 605-617.

Hyatt AD, Gould AR, Zupanovic Z, Cunningham AA, Hengstberger S, Whittington RJ, Kattenbelt J, Coupar BEH (2000). Comparative studies of piscine and amphibian iridoviruses. Arch Virol 145: 301-331.

Hyatt AD, Williamson M, Coupar BEH, Middleton D, Hengstberger SG, Gould AR, Selleck P, Wise TG, Kattenbelt J, Cunningham AA, Lee J (2002). First identification of a ranavirus from green pythons (*Chondropython viridis*). J Wildl Dis 38(2): 239-252.

Jancovich JK, Davidson EW, Morado JF, Jacobs BL, Collins JP (1997). Isolation of a lethal virus from the endangered tiger salamander *Ambystoma tigrinum stebbinsi*. Dis Aquat Organ 31: 161-167.

Jancovich JK, Davidson EW, Parameswaran N, Mao J, Chinchar VG, Collins JP, Jacobs BC, Storfer A (2005). Evidence for emergence of an amphibian iridoviral disease because of human-enhanced spread. Mol Ecol 14: 213-224.

Jancovich JK, Steckler N, Waltzek TB (2015a). Ranavirus taxonomy and phylogeny. In: Gray MJ, Chinchar VG (eds) Ranaviruses: lethal pathogens of ectothermic vertebrates. Springer, New York.

Jancovich JK, Qin Q, Zhang Q-Y, Chinchar VG (2015b). Ranavirus replication: molecular, cellular, and immunological events. In: Gray MJ, Chinchar VG (eds) Ranaviruses: lethal pathogens of ectothermic vertebrates. Springer, New York.

Jaramillo D, Tweedie A, Becker JA, Hyatt A, Crameri S, Whittington RJ (2012). A validated quantitative polymerase chain reaction assay for the detection of ranaviruses (Family *Iridoviridae*) in fish tissue and cell cultures, using EHNV as a model. Aquaculture 356: 186-192.

Johnson AJ, PessierAP, Wellehan JF, Brown R, Jacobson ER (2005). Identification of a novel herpesvirus from a California desert tortoise (*Gopherus agassizii*). Vet

Microbiol 111:107-116.

Johnson AJ, Pessier AP, Jacobson ER (2007). Experimental transmission and induction of ranaviral disease in western ornate box turtles (*Terrapene ornata ornata*) and red-eared sliders (*Trachemys scripta elegans*). Vet Pathol 44:285-297.

Johnson AJ, Pessier AP, Wellehan JFX, Childress A, Norton TM, Stedman NL, Bloom DC, Belzer W, Titus VR, Wagner R, Brooks JW, Spratt J, Jacobson ER (2008). Ranavirus infection of free-ranging and captive box turtles and tortoises in the United States. J Wildl Dis 44:851-863.

Johnson AJ, Wendland L, Norton TM, Belzer B, Jacobson ER (2010). Development and use of an indirect enzyme-linked immunosorbent assay for detection of iridovirus exposure in gopher tortoises (*Gopherus polyphemus*) and eastern box turtles (*Terrapene carolina carolina*). Vet Microbiol 142:160-167.

Kattenbelt JA, Hyatt AD, Gould AR (2000). Recovery of ranavirus dsDNA from formalin-fixed archival material Dis Aquat Organ 39:151-154

Kik M, Martel A, Spitzen-vander Sluijs A, PasmansF, Wohlsein P, Grone A, Rijks JM (2011). Ranavirus-associated mass mortality in wild amphibians, The Netherlands, 2010: a first report. Vet J 190:284-286.

Langdon JS, Humphrey JD, Williams LM, Hyatt AD, Westbury HA (1986). First virus isolation from Australian fish: an iridovirus-like pathogen from redfin perch, *Perca fluviatilis* L. J Fish Dis 9:263-268.

Li P, Yan Y, Wei S, Weiv J, Gao R, Huang X, Huang Y, Jiang G, Qin Q (2014a). Isolation and characterization of a new class of DNA aptamers specific binding to Singapore grouper iridovirus (SGIV) with antiviral activities. Virus Res 188: 146-154.

Li W, Zhang X, Weng S, Zhao G, He J, Dong C (2014b). Virion-associated viral proteins of a Chinese giant salamander (*Andreas davidianus*) iridovirus (genus *Ranavirus*) and functional study of the major capsid protein (MCP). Vet Microbiol 172:129-139.

Ma J, Zeng L, Zhou Y, Jiang N, Zhang H, Fan Y, Meng Y, Xu J (2014). Ultrastructural morphogenesis of an amphibian iridovirus isolated from Chinese giant salamander (*Andrias davidianus*). J Comp Pathol 150:325-331.

Mao J, Hedrick RP, Chinchar VG (1997). Molecular characterization, sequence analysis, and taxonomic position of newly isolated fish iridoviruses. Virology 229: 212-220.

Mao JD, Green E, Fellers G, Chinchar VG (1999). Molecular characterization of iridoviruses isolated from sympatric amphibians and fish. Virus Res 63:45-52.

Marschang RE, Braun S, Becher P (2005). Isolation of a ranavirus from a gecko (*Uroplatus fimbriatus*). J Zoo Wildl Med 36:295-300.

Marsh IB, Whittington RJ, O'Rourke B, Hyatt AD, Chisholm O (2002). Rapid differentiation of Australian, European and American ranaviruses based on variation in major capsid protein gene sequence. Mol Cell Probes 16:137-151.

Martínez-Silvestre A, Perpiñán D, Marco I, Lavín S (2002). Venipuncture technique of the occipital venous sinus in freshwater aquatic turtles. J Herpetol Med Surg 12:31-32.

Mazzoni R, de Mesquita AJ, Fleury LFF, deBrito WMED, Nunes IA, Robert J, Morales H, Coelho ASG, Barthasson DL, Galli L, Catroxo MHB (2009). Mass mortality associated with a frog virus 3-like Ranavirus infection in farmed tadpoles *Rana catesbeiana* from Brazil. Dis Aquat Organ 86:181-191.

McClenahan SD, Beck BH, Grizzle JM (2005). Evaluation of cell culture methods for detection of largemouth bass virus. J Aquat Anim Health 17:365-372.

McLoughlin MF, Graham DA (2007). Alphavirus infections in salmonids — a review. J Fish Dis 30:511-531.

Meng Y, Ma J, Jiang N, Zeng LB, Xiao HB (2014). Pathological and microbiological findings from mortality of the Chinese giant salamander (*Andrias davidianus*). Arch Virol 159:1403-1412. doi:10.1007/s00705-013-1962-6.

Miller DL, Rajeev S, Gray MJ, Baldwin C (2007). Frog virus 3 infection, cultured American bull-frogs. Emerg Infect Dis 13:342-343.

Miller DL, Rajeev S, Brookins M, Cook J, Whittington L, Baldwin CA (2008). Concurrent infection with *Ranavirus*, *Batrachochytrium dendrobatidis*, and *Aeromonas* in a captive anuran colony. J Zoo Wildl Med 39:445-449.

Miller DL, Gray MJ (2010). Amphibian decline and mass mortality: The value of visualizing ranavirus in tissue sections. The Veterinary Journal 186:133-134.

Miller DL, Gray MJ, Strofer A (2011). Ecopathology of ranaviruses infecting amhibians. Viruses 3:2351-2373. doi:10.3390/v3112351.

Monini M, Ruggeri FM (2002). Antigenic peptides of the epizootic haematopoietic necrosis virus. Virology 297:8-18.

Oh SY, Oh MJ, Nishizawa T (2014). Potential for a live red seabream iridovirus (RSIV) vaccine in rock bream *Oplegnathus fasciatus* at a low rearing temperature. Vaccine 32: 363-368.

OIE (2012a). Chapter 1.1.2 Principles and methods of validation of diagnostic assays for infectious diseases. In: Manual of diagnostic tests for aquatic animals (World Organisation for Animal Health). http://www. oie. int/international-standard-setting/aquatic-manual/access-online/.

OIE (2012b). Chapter 2.1.2 Quality management in veterinary testing laboratories (World Organisation for Animal Health). In: Manual of diagnostic tests for aquatic animals. http:// www. oie. int/international-standard-setting/aquatic-manual/access-online/.

OIE (2012c). Chapter 2.3.1 Epizootic Haematopoietic necrosis. In: Manual of diagnostic tests for aquatic animals (World Organisation for Animal Health). http://www.oie.int/international-standard-setting/aquatic-manual/access-online/.

OIE (2012d). Chapter 2.1.2 Infection with ranavirus. In: Manual of diagnostic tests for aquatic animals (World Organisation for Animal Health). http://www.oie.int/international-standard-setting/aquatic-manual/access-online/.

Pallister J, Gould A, Harrison D, Hyatt A, Jancovich J, Heine H (2007). Development of real-time PCR assays for the detection and differentiation of Australian and European ranaviruses. J Fish Dis 30: 427-438.

Pessier AP, Mendelson JR (2010). A manual for control of infectious diseases in amphibian survival assurance colonies and reintroduction programs. IUCN/SSC Captive Breeding Specialist Group, Apple Valley, MN.

Petranka JW, Murray SM, Apple Valley MN, Kennedy CA (2003). Response of amphibians to restoration of a southern Appalachian wetland: perturbations confound post-restoration assessment. Wetlands 23: 278-290.

Picco AM, Brunner JL, Collins JP (2007). Susceptibility of the endangered California tiger salamander, *Ambystoma californiense*, to ranavirus infection. J Wildl Dis 43: 286-290.

Plumb JA, Grizzle JM, Young HE, Noyes AD, Lamprecht S (1996). An iridovirus isolated from wild largemouth bass. J Aquat Anim Health 8: 265-270.

Pozet F, Morand M, Moussa A, Torhy C, de Kinkelin P (1992). Isolation and preliminary characterization of a pathogenic icosahedral deoxyribovirus from the catfish *Ictalurus melas*. Dis Aquat Organ 14: 35-42.

Price, SJ, Garner TWJ, Nichols RA, et al. (2014). Collapse of amphibian communities due to an introduced Ranavirus. Curr Biol 24: 2586-2591. http://www.cell.com/

currentbiology/pdfEx-tended/S0960-9822(14)01149-X.

Qin QW, Chang SF, Ngoh-Lim GH, Gibson-Kueh S, She C, Lam TJ (2003). Characterization of a novel ranavirus isolated from grouper *Epinephelus tauvina*. Dis Aquat Organ 53:1-9.

Reddacliff LA, Whittington RJ (1996). Pathology of epizootic haematopoietic necrosis virus (EHNV) infection in Rainbow Trout (*Oncorhynchus mykiss* Walbaum) and Redfin Perch (*Perca fluviatilis* L.).J Comp Pathol 115:103-115.

Rimmer AE, Becker JA, Tweedie A, Whittington RJ (2012). Validation of high throughput methods for tissue disruption and nucleic acid extraction for ranaviruses (family *Iridoviridae*).Aquaculture 338:23-28.

Rivera S, Wellehan JF, McManamon R, Innis CJ, Garner MM, Raphael BL, Gregory CR, Latimer KS, Rodriguez CE, Diaz-Figueroa O, Marlar AB, Nyaoke A, Gates AE, Gilbert K, Childress AL, Risatti GR, Frasca S (2009). Systemic adenovirus infection in Sulawesi tortoises (*Indotestudo forsteni*) caused by a novel siadenovirus.J Vet Diagn Invest 21:415-426.

Robert J, Abramowitz L, Gantress J, Morales HD (2007). *Xenopus laevis*: a possible vector of ranavirus infection? J Wildl Dis 43:645-652.

Robert J, George E, De Jesús AF, Chen G (2011). Waterborne infectivity of the ranavirus frog virus 3 in *Xenopus laevis*.Virology 417(2):410-417.

Rojas S, Richards K, Jancovich JK, Davidson E (2005). Influence of temperature on Ranavirus infection in larval salamanders *Ambystoma tigrinum*.Dis Aquat Organ 63:95-100.

Rothermel BB, Travis ER, Miller DL, Hill RL, McGuire JL, Yabsley MJ (2013). High occupancy of stream salamanders despite high Ranavirus prevalence in a Southern Appalachians Watershed.Ecohealth 10:184-189.

Ruder MG, Allison AB, Miller DL, Keel MK (2010). Pathology in practice: ranaviral disease in a box turtle.J Am Vet Med Assoc 237:783-785.

Schloegel LM, Daszak P, Cunningham AA, Speare R, Hill B (2010). Two amphibian diseases, chytridiomycosis and ranaviral disease, are now globally notifiable to the World Organization for Animal Health (OIE): an assessment. Dis Aquat Organ 92:101-108.

Schock DM, Bollinger TK, Chinchar VG, Jancovich JK, Collins JP (2008). Experimental evidence that amphibian ranaviruses are multi-host pathogens. Copeia 2008:133-143.

Steiner KA, Whittington RJ, Petersen RK, Hornitzky C, Garnet H (1991). Purification of epizootic haematopoietic necrosis virus and its detection using ELISA.J Virol Methods 33:199-210.

Stöhr AC, Blahak S, Heckers KO, Wiechert J, Behncke H, Mathes K, Gunther P, Zwart P, Ball I, Marschang RE RB (2013). Ranavirus infections associated with skin lesions in lizards. Vet Res 44: 84, http://www. veterinaryresearch. org/content/44/1/84.

Sutton WB, Gray MJ, Hardman RH, Wilkes RP, Kouba A, Miller DL (2014a). High susceptibility of the endangered dusky gopher frog to ranavirus. Dis Aquat Organ 112(1):9-16.

Sutton WB, Gray MJ, Hoverman JT, Secrist RG, Super P, Hardman RH, Tucker JL, Miller DL (2014b). Trends in ranavirus prevalence among plethodontid salamanders in the Great Smoky Mountains National Park. Ecohealth, doi: 10.1007/s10393-014-0994-z.

TamaruY, Ohtsuka M, Kato K, Manabe S, Kuroda K, Sanada M, Ueda M (2006). Application of the arming system for the expression of the 380R antigen from red sea bream iridovirus (RSIV) on the surface of yeast cells: a first step for the development of an oral vaccine. Biotechnol Prog 22:949-953.

Teacher AGF, Cunningham AA, Garner WJ (2010). Assessing the long-term impact of ranavirus infection in wild common frog populations. Anim Conserv 13: 514-522.

Todd-Thompson M (2010). Seasonality, variation in species prevalence, and localized disease for ranavirus in Cades Cove (Great Smoky Mountains National Park) amphibians. Master Thesis, University of Tennessee, Knoxville. http://trace.tennessee.edu/utk_gradthes/665. Accessed 2 Sep 2014.

Tran BN, Chen L, Liu Y, Wu J, Velazquez-Campoy A, Sivaraman J, Hew CL (2011). Novel histone H3 binding protein ORF158L from the Singapore grouper iridovirus. J Virol 85:9159-9166. doi:10.1128/JVI.02219-10.

Waltzek TB, Miller DL, Gray MJ, Drecktrah B, Briggler JT, MacConnell B, Hudson C, Hopper L, Friary J, Yun SC, Malm KV, Weber ES, Hedrick RP (2014). Expansion of the host range of frog virus 3 into hatchery-reared pallid sturgeon *Scaphirhynchus albus*. Dis Aquat Organ 111:219-227. doi:10.3354/dao02761.

Weir RP, Moody NJ, Hyatt AD, Crameri S, Voysey R, Pallister J, Jerrett IV (2012). Isolation and characterization of a novel Bohle-like virus from two frog species in the Darwin rural area, Australia. Dis Aquat Organ 99:169-177.

Wellehan JFX, Strik NI, Stacy BA, Childress AL, Jacobson ER, Telford SR (2008). Characterization of an erythrocytic virus in the family *Iridoviridae* from a peninsula ribbon snake (*Thamnophis sauritus sackenii*). Vet Microbiol 131: 115-122.

WheelwrightNT, Gray MJ, Hill RD, Miller DL (2014). Sudden mass die-off of a large population of wood frog (*Lithobates sylvaticus*) tadpoles in Maine, USA, likely due to ranavirus. Herpetol Rev 45:240-242.

Whitley DS, Yu K, Sample RC, Sinning A, Henegar J, Norcross E, Chinchar VG (2010). Frog virus 3 ORF 53R, a putative myristoylated membrane protein, is essential for virus replication *in vitro*. Virology 405:448-456.

Whittington R, Deece K (2004). Aquatic animal health subprogram: development of diagnostic and reference reagents for epizootic haematopoietic necrosis virus of finfish. The University of Sydney and Fisheries Research and Development Corporation, Sydney and Canberra.

Whittington RJ, Reddacliff GL (1995). Influence of environmental temperature on experimental infection of redfin perch (*Perca fluviatilis*) and rainbow trout (*Oncorhynchus mykiss*) with epizootic haematopoietic necrosis virus, an Australian iridovirus. Aust Vet J 72:421-424.

Whittington RJ, Steiner KA (1993). Epizootic haematopoietic necrosis virus (EHNV): improved ELISA for detection in fish tissues and cell cultures and an efficient method for release of antigen from tissues. J Virol Methods 43:205-220.

Whittington RJ, Philbey A, Reddacliff GL, Macgown AR (1994). Epidemiology of epizootic haematopoietic necrosis virus (EHNV) infection in farmed rainbow trout, *Oncorhynchus mykiss* (Walbaum): findings based on virus isolation, antigen capture ELISA and serology. J Fish Dis 17:205-218.

Whittington RJ, Speare R (1996). Sensitive detection of serum antibodies in the cane toad*Bufo marinus*. Dis Aquat Organ 26:59-65.

Whittington RJ, Kearns C, Speare R (1997). Detection of antibodies against iridoviruses in the serum of the amphibian *Bufo marinus*. J Virol Methods 68:105-108.

Whittington RJ, Reddacliff LA, Marsh I, Kearns C, Zupanovic Z, Callinan RB (1999). Further observations on the epidemiology and spread of epizootic haematopoietic necrosis virus (EHNV) in farmed rainbow trout (*Oncorhynchus mykiss*) in southeastern Australia and a recommended sampling strategy for surveillance. Dis

Aquat Organ 35:125-130.

Whittington RJ, Becker JA, Dennis MM (2010). Iridovirus infections in finfish — critical review with emphasis on ranaviruses.J Fish Dis 33:95-122.doi:10.1111/j.1365-2761.2009.01110.x.

Wolf K, Bullock GL, Dunbar CE, Quimby MC (1968). Tadpole edema virus: a viscerotropic pathogen for anuran amphibians.J Infect Dis 118:253-262.

Woodhams DC, Alford RA, Marantelli G (2003). Emerging disease of amphibians cured by elevated body temperature.Dis Aquat Organ 55:65-67.

Yan X, Yu Z, Zhang P, Battisti AJ, Holdaway HA, Chipman PR, Bajaj C, Bergoin M, Rossmann MG, Baker TS (2009). The capsid proteins of a large, icosahedral dsDNA virus.J Mol Biol 385:1287-1299.doi:10.1016/j.jmb.2008.11.002.

Yuan JS, Reed A, Chen F, Stewart CN (2006). Statistical analysis of real-time PCR data.BMC Bioinformatics 7:85.doi:10.1186/1471-2105-7-85.

Zhang M, Hu YH, Ziao ZZ, Sun Y, Sun L (2012). Construction and analysis of experimental DNA vaccines against megalocytivirus.Fish Shellfish Immunol 33:1192-1198.

Zilberg D, Grizzle JM, Plumb JA (2000). Preliminary description of lesions in juvenile largemouth bass injected with largemouth bass virus.Dis Aquat Organ 39:143-146.

Zupanovic Z, Lopez G, Hyatt AD, Green B, Bartran G, Parkes H, Whittington RJ, Speare R (1998). Giant toads *Bufo marinus* in Australia and Venezuela have antibodies against 'ranaviruses'.Dis Aquat Organ 32:1-8.

第七章　蛙病毒研究的设计与分析：监测和风险评估

马修·J. 格雷[①]，杰西·L. 布伦纳[②]，朱莉娅·E. 厄尔[③]和埃伦·阿里尔[④]

1　引言

越来越多的证据显示，蛙病毒能影响变温脊椎动物种群，并导致种群数量 209
的下降。研究者所进行的研究设计可以确定蛙病毒的地理分布和流行率、蛙病毒引入未感染地区的风险以及蛙病毒对种群数量的影响。正确的研究设计，依赖于现场数据、实验室的试验以及定量分析的全面结合，这种结合一般需要具有足够资源的不同专家团队。评估蛙病毒风险的财务成本是巨大的。例如，RANA 项目即欧洲鱼类和水生生态系统新出现的系统性虹彩病毒病的风险评估的成本，按 2012 年欧元计，接近 140 万欧元（Evira，2013）。目前，美国马里兰自然资源部（Maryland Department of Natural Resources，MDNR）正进行一项跨美国 7 个州，按 2014 年美元计，价值达 17.8 万美元的监测研究（Smith 等人，2014）。按照全球蛙病毒联盟（Global Ranavirus Consortium，GRC）网站发布的资料，基因组 DNA 的提取和定量实时 PCR 的平均成本，按 2014 年美元
计，每个样本大约 25 美元（http://www.ranavirus.org/）。为了检测蛙病毒并 210

---

①M. J. Gray/Center for Wildlife Health，Department of Forestry，Wildlife and Fisheries，University of Tennessee，274 Ellington Plant Sciences Building，Knoxville，TN 37996-4563，USA/e-mail：mgray11@utk.edu.

②J. L. Brunner/School of Biological Sciences，Washington State University，Pullman，WA 99164-4236，USA.

③J. E. Earl/National Institute for Mathematical and Biological Synthesis，University of Tennessee，Knoxville，TN37996-3410，USA.

④E. Ariel/College of Public Health，Medical and Veterinary Sciences，James Cook University，Townsville，QLD 4811，Australia.

获得精确感染流行率的估计，就恰当的样本数量而言，可以非常大（$n > 60$），因而与蛙病毒检测有关的实验费用相当高。在大的地理区域，调动现场工作人员的成本同样也相当大，例如，马里兰自然资源部用于上述研究的财政预算，超过95%的费用是专门用于个人、野外供给以及差旅费。因此，对评估野外和养殖种群蛙病毒风险感兴趣的组织应进行适当的资源投资，做好重要的规划，确保采集到足够量的样本，并做到样本的污染最小化（Miller 等人，2015）。同时，确保所收集到的信息成为符合预期的可测量而又能交付的成果。这一章为评估蛙病毒风险的研究设计提供了一些基本的知识。此外，我们鼓励组织与正在进行蛙病毒研究的专家进行合作，GRC 可以提供你所在地区蛙病毒专家的资料。

2 蛙病毒的监测

传染病的出现必须动员不同大学和组织机构来确定野生种群中病原体的风险。为了对风险量化，有必要在分子水平（Jancovich 等人，2015a）和生物个体水平（Brunner 等人，2015）对宿主-病原系统有一个基本的认识。风险的评估常常是从确定一个病原是否是新出现的病原开始的。这意味着病原在地理分布上正在扩大、在种群中的流行率正在增加或宿主范围正在扩大（Wobeser，2006）。在这一章，我们将蛙病毒暴发定义为蛙病毒发病率的增加超出了其背景值，而背景值常常是未知的。由于对传染病的流行率和发病率的估计被用作风险管理决策的依据，因此，为了达到这一目的，常常实施病原监测项目。假如设计得当，监测项目在检测病原、对发病率和流行率进行精确的估计将是十分有效的，从而提供必要的数据以确定病原是否对种群或物种构成威胁。

2.1 对感染数据的解释

PCR 的方法越来越多地被用来检测蛙病毒的感染，但是其他方法对于直接检测病毒如细胞培养中病毒的分离、电子显微镜、抗原捕获 ELISA，或展示
211 感染的证据如组织学、血清学方法也是非常重要的（Miller 等人，2015）。通过定量实时 PCR 或细胞培养的方法对病毒载量进行估计如噬斑试验或 50% 组织培养感染剂量（50% tissue culture infective dose，TCID50）与其他诊断工具如组织学一起，能够提供感染和发病强度或严重程度的信息资料（Miller 等人，2015）。搞清楚这些方法当中每一种方法要检测的是什么（例如，PCR 检测蛙病毒的存在，而细胞培养中病毒的分离则是展示传染性病毒粒子的存在。）和

认识到每一种方法的局限性，这是很重要的（Miller 等人，2015）。一般来说，检测出一种传染病的能力，会随着感染接种后的时间延长、感染的严重程度以及检测方法敏感性的增加而增强（Miller 等人，2015）。假如一个实验的敏感性和特异性已知，那么这些值应该用来调整流行率和发病率的估计值。反之，则应谨慎地解释感染数据。

在监测研究中，所测量的最常见变量是感染流行率和发病率。感染流行率等于感染的个体数量除以样本数量。感染流行率所估计的是在特定的时间内种群感染的比例。另外，还有一个相关的变量就是血清阳性率，该变量所估计的是在感染病原之前就有血清学证据的种群血清阳性比例。另一方面，发病率是指在规定的时间期限内已感染个体在种群中所占的比例（Wobeser，2006）。然而，发病率常常表示为单位时间内新病例的数量。通常，发病率在大多数情况下用来表示个体平均发病率，例如每 1 000 个个体中存在风险的个体数量。

然而，在监测研究过程中，估计流行率要比估计发病率更普遍。流行率是在特定时间内对感染负荷的一个“快照”。因此，在缺少生物学背景资料的情况下，很难对其作出解释。熟悉物种对蛙病毒的相对易感染性，有助于对病毒流行资料的解释。例如，假如实验感染接种显示，一个物种在接种蛙病毒后迅速死亡，那么高流行率与在一个流行病高峰期采样的结果是最一致的。有好几个研究报道了在控制条件下物种水平的易感性（Hoverman 等人，2011；Brenes 等人，2014b；Brunner 等人，2015）。生物学背景资料可收集种群密度以及与生物体物候学（phenology）有关的调查时间。例如，在早春，观察到的低流行率和高密度两栖类幼体种群，更符合病毒是最近引入的，而不是由已经存在的病毒引起的暴发。

几项监测研究表明，在两栖类和龟类种群中，存在蛙病毒的背景流行率值小于 5%（Todd-Thompson，2010；Hoverman 等人，2012；Allender 等人，2006；Forzán 和 Wood，2013；Hamed 等人，2013；Sutton 等人，2014）。与 FV3 样病毒病相关的死亡率和感染流行率之间存在明显的相关性（Haislip 等人，2011；Hoverman 等人，2011；Brenes 等人，2014b）。在两栖动物种群中，流行率超过 40% 可能预示着一种疾病的暴发（Gray 和 Miller，2013）。尽管这些经验法则在解释流行程度上很有用，但是我们强烈要求对超出生物学背景以外的流行数据的解释要特别小心。同样值得注意的是，蛙病毒引起的大规模死亡发生很 212
快，一般少于 2 周（Todd-Thompson，2010；Waltzek 等人，2014）。因此，频繁的采样对检测和认识蛙病毒的流行病学是很有必要的。

最后值得一提的是，要认识到感染个体比未感染个体更容易被发现或捕获，这样就会使流行率的估计发生偏差（Cooch 等人，2012）。例如，濒临死亡的鱼和蝌蚪常常发现于水面附近，因而很容易被发现，从而会放大流行率。而病龟很少活动，因而会降低检出概率，导致流行率的低估。随着时间的推移，检出概率也有所变化，例如在不同的发育阶段，并且在不同的地点，流行率也明显不同，但这并不能反映感染比例的实际差别。然而，我们还不知道有何种蛙病毒监测研究能对检测概率作出解释，我们认为，如果做到了这样，将会大大加深我们对蛙病毒生物学的认识。

当对蛙病毒在不同地区和宿主之间的分布进行描述时，感染流行率是非常有用的，但感染流行率并不传达风险或感染率的信息。感染发病率是指感染病原的个体所占的比例，即在特定的时间期限内，新病例发生的数量（Wobeser，2006）。在人工饲养的小种群当中，在短时间内，有可能确定有多少动物个体已经感染。例如，最初的调查发现 50 个动物当中有 2 个感染，而第二次调查是在月末，结果发现有 10 个动物发生了感染，那么，每个月的发病率为 16.7%（8 个新病例/48 个处于风险中的动物）。需要注意的是，在研究开始时，被感染的个体不会有感染的风险，所以在估计发病率时，它们不包括在内。假如种群不是封闭的，即存在迁入或迁出，则发病率的计算应根据风险时间作出相应调整（详见 Dohoo 等人，2003）。

估计野生种群的发病率很困难，因为我们通常无法追踪个体的命运或感染状况。有两种方法可用于估计野生种群的感染发生率。第一种方法是运用“哨兵”（sentinels），这些“哨兵”实际上是引入环境中的未感染个体，例如，网箱中的蝌蚪放入池塘中，并且定期筛查感染。“哨兵”物种应该是对蛙病毒高度易感的，如北美林蛙（Hoverman 等人，2011）。第二种方法是捕获-标记-再捕获（CMR）研究，在这个方法中，对动物个体进行独特的标记，然后放流，在随后的捕获期间，例如诱捕（trapping）和网捕（netting），研究人员记录新捕获和再捕获的个体，并确定它们的感染状态。鉴于个体要放流，必须用非死亡方法来确定感染状态（St-Amour 和 Lesbarrères，2007；Gray 等人，2012）。最后值得一提的是，CMR 模型所估计的是不断改变感染状态的个体概率，因此，CMR 模型会导致对病原体的检测不充分以及再捕获宿主的概率不准确。

2.2 监测计划

213 在一个时间段内，对多个种群进行采样的横向研究，最适合搞清楚蛙病毒的分布。而随着时间的推移，在同一种群中进行采样的纵向研究，对搞清楚蛙病毒的流行病学及其对种群的影响也是非常有用的。对于刚开始进行监测项目的组织来说，我们建议，从横向研究开始，横向研究涉及跨多个类群进行广泛采样，目的在于鉴定具有高水平蛙病毒感染的地点。如果受到资金的限制，那么对蛙病毒具有高易感性的物种或受保护的物种可以作为采样目标；死亡样本（如器官组织）与非死亡样本（如棉签样本、尾部剪切物）比较，将有可能导致更高的检出率（Gray 等人，2012）。一旦种群或蛙病毒流行率高的地点被鉴定，那么一项更深入的纵向研究可以通过年周期采样来了解季节和年度趋势。当宿主存在时，每两周采样一次，足以发现大多数疫情（Todd-Thompson，2010）。

自然因素或人为因素可以导致蛙病毒的暴发（Gray 等人，2009）。一些已知的自然因素包括宿主的密度、物种的构成、温度以及宿主的发育（Gray 等人，2009；Brunner 等人，2015）。人为因素可能与应激因子有关，例如农药（Kerby 等人，2011），或者与新病毒株的引进有关，如病原体的污染（Storfer 等人，2007）。因此，为了确定疾病暴发的各种因素，在监测项目实施过程中，应该测量理想的宿主密度和发育阶段、水温、室温和水质。假如检出蛙病毒，则可以从新鲜组织或冷冻组织中分离病毒（Miller 等人，2015），并进行基因组比较，看看所分离的病毒是否是一种被引入的新型病毒株（Jancovich 等人，2015b）。

了解蛙病毒对种群的影响是种群保护的一个基本问题（Duffus 等人，2015）。为了了解蛙病毒对种群的影响，有必要连续几年对相同地点进行采样（Price 等人，2014）。除对蛙病毒感染进行个体采样以外，标记-再捕获方法也可用来估计宿主种群的大小（Williams 等人，2002）。对流行率、发病率和宿主种群丰度的估计，对正式确定蛙病毒的影响以及鉴定疫情暴发的原因是很有必要的，只有这样，才便于实施干预策略。

3　研究设计

在设计监测研究时，除一些特定的地点需要作为采样目标以外，比如说，具有重要保护价值的地点，采样地点应随机选定。随机抽样可以根据不同的地理区域或与病毒出现有关的假设如人类的土地利用（如农业与造林）进行分 214
层。随机地点的选择，避免了无意的偏见和潜在的混淆因素。越是容易进入的地方，越容易采样，但也可能会有更大的他人访问率，例如人们钓鱼，这样就会增加蛙病毒的引入率或增加应激因子的强度。采样点的数量取决于研究的目的和可用的资源。很显然，当采样点空间范围增加时，得出的结论就会更有说服力。然而，在几个地点进行密集采样也同样存在好处，特别是在那些大规模死亡复发的地点。

采样时，应该随机采集个体，理想的做法是，所捕获的个体应该分开放到带有编号的容器内，同时，被处理的个体应该使用随机数字表或统计软件来选择。个体不应该被关在一起，因为蛙病毒在它们之间可以迅速发生传播（Brunner 等人，2007；Robert 等人，2011）。另一种方法是，当捕获的个体达到目标样本大小时，再对其进行处理。重要的是，被处理的个体不应在一个类群中随意地选择，因为会引入偏差（Gotelli 和 Ellison，2004）。假如观察到发病的个体，这些个体则应作为诊断的采样目标（Miller 等人，2015）。对具有蛙病毒病可能症状的个体进行采样时，会过高估计流行率和发病率。另外，监测的目的是确定一个地点是否是蛙病毒区（见本章第 7 部分），针对明显患病的个体采样，可以增加检出病原的概率。

最后，在野生种群中进行的监测研究，对获悉蛙病毒的分布和对宿主种群的影响是很重要的。然而，鉴定引起野生种群疫情暴发的因素可能会面临重大挑战。实验室研究和中试研究在鉴定引起疾病发生的自然因素和人为因素是非常有用的。来自对照研究的资料可用于设计监测研究，其目标就是针对蛙病毒出现的特定假设。此外，对照研究可以告诉现场人员，连同感染状态和种群丰度，应该测量哪些因素（例如水质）。

4 必要样本数

确定采样必要样本数是设计一个监测项目的第一步，必要样本数取决于你的研究目的是检测病原还是获得流行率的精确估计，并用于统计学推断。为了估计检测一个病原所需要的样本大小，你需要：（1）一个流行率的先前估计值或流行率的假定水平；（2）宿主种群大小的估计值；（3）在检测病原中规
215 定一个置信水平，一般来说为 95%（Amos 1985；Thoesen，1994）。当病原的流行率下降时，宿主种群会增加，那么检测一个病原体所需要的样本数也随之增加（见表 1）。因此，在检测蛙病毒暴发时，只需要一个小样本（$n \leqslant 10$）。然而，假如流行率很低（≤5%），检测蛙病毒时，则需要一个大样本。总的来说，我们建议，在所有的监测项目中，为了检测蛙病毒，每个地点的最小样本为 30 个（见表 1）。对所关注的地点，应采集较大的样本，这些地点需要精确的估计值来鉴定与新疾病出现相关的各种因素。

表 1　结合种群规模和假定感染流行率的一个宿主种群在 95% 置信区间检测蛙病毒的必要样本数[a]

所估计的种群规模	假定蛙病毒感染流行率			
	20%	10%	5%	2%
50	5	20	35	50
100	8	23	45	75
250	11	25	50	110
500	13	26	55	130
2 000	15	27	60	145
>100 000	15	30	60	150

按照阿摩斯（Amos，1985）和多伊森（Thoesen，1994）的方法所进行的计算。

为了确定必要样本数，从而获得精确的流行率估计，则需要：（1）一个先前流行率的估计值（$\hat{p}$）；（2）一个规定的误差水平（d），这个误差水平是在流行率估计中能够容忍的；（3）在流行率估计中规定的置信区间水平，一般为 95%。因此，样本数可以估计为：

$$n = \hat{p}(1-\hat{p})\left(\frac{1.96}{d}\right)^2 \tag{1}$$

式中 1.96 是标准正态曲线在 95% 置信区间的临界值。假如不能获得先前流行率的估计值 $\hat{p}$，则可用 $\hat{p}=0.5$。因此，假如 $\hat{p}=0.85$ 且 $d=0.05$，则要求 $n=196$。然而，假如愿意接受更大的误差，在估计 $\hat{p}$ 时（例如，10% = 0.10），则要求 $n=49$，此时 $\hat{p}=0.85$。此外，当 p 接近 0.5 时，对于精确估计 $\hat{p}$，必要样本数则要增加。在先前的例子当中，式中 $d=0.10$，要求 $n=96$，对应 $\hat{p}=0.5$。

用统计学检验两个地点发病率的显著性差异，需要一个大样本（见表 2）。例如，当检验条件设定置信区间 α 为 95%，统计功效 β 为 80%，两个地点发病率的显著性差异为 10%，则样本量 n 需为 219~408，同时显著性差异检验取决于置信区间 α 和统计功效 β 这两个比例的值。有好几个网站和软件包可以用于在考虑到两个比例值（http：//epitools. ausvet. com. au/content. php? page =2Proportions）即 α 和 β 最小可检测差异的前提下，设计必要样本数。当最小可检测差异增加时，必要样本数则降低，同时，置信水平和统计学检验功效也 216
下降。

表 2　在 95%置信区间（α=0.05）和 80%统计功效（β=0.80）前提下检验两个比例差异性的必要样本数[a]

比例 1	比例 2										
	0.05	0.1	0.2	0.3	0.4	0.5	0.6	0.7	0.8	0.9	0.95
0.05		474	88	43	27	19	14	11	8	7	6
0.1	474		219	72	38	25	17	13	10	8	7
0.2	88	219		313	91	45	28	19	13	10	8
0.3	43	72	313		376	103	49	29	19	13	11
0.4	27	38	91	376		408	107	49	28	17	14
0.5	19	25	45	103	408		408	103	45	25	19
0.6	14	17	28	49	107	408		376	91	38	27
0.7	11	13	19	29	49	103	376		313	72	43
0.8	8	10	13	19	28	45	91	313		219	88
0.9	7	8	10	13	17	25	38	72	219		474
0.95	6	7	8	11	14	19	27	43	88	474	

[a] 所提供的样本数是根据每个比例并使用网站所提供的软件进行计算的。http://epitools.ausvet.com.au/content.php?page=2Proportions.

5　数据分析

5.1　置信区间

即使有大的样本量，任何关于流行率或发病率的估计也都存在不确定性。置信区间是表达这些估计确定程度的一个常见度量。它们也可以用于统计学比较流行率的估计，不重叠的置信区间意味着统计上的差异。为了构建发病率的置信区间，通常用比率除以时间间隔（time interval）。

估计置信区间的一个常见方法就是计算标准正态近似值（standard normal approximation）。这个过程包括计算一个比率的标准差，乘以标准正态分布 95%置信值的临界值 1.96，再加减样本流行率的估计结果（Brown 等人，2001）：

$$\hat{p} \pm 1.96\sqrt{\frac{\hat{p}(1-\hat{p})}{n}} \tag{2}$$

这一近似值的前提是样本量较大，要求 $n > 20$ 且 $0.10 < \hat{p} < 0.90$。否则，217
置信区间会超出 0 和 1 之外，这样是毫无意义的（Brown 等人，2001）。有几个更好的方法可以用来估计置信区间（参阅 Brown 等人，2001 的综述）。我们推荐威尔逊分数区间（Wilson score interval）（Wilson，1927），因为当 $\hat{p}$ 在 0 和 1 附近时，它在较低的样本容量下表现良好，并且，像一些连续性校正方法一样，它不是过于保守，这一方程是：

$$\frac{1}{1+\frac{1.96^2}{n}}\left[\hat{p}+\frac{1.96^2}{2n} \pm 1.96\sqrt{\frac{1}{n}\hat{p}(1-\hat{p})+\frac{1.96^2}{4n^2}}\right] \tag{3}$$

该公式与公式（2）具有相同的变量。然而，进行手工计算非常耗时，许多统计包对威尔逊分数区间进行估计（例如 R“二项式”统计包）。还有一些网站也可以查找（http：//vassarstats. net/prop1. html）。附录 1 提供了在 R 统计包中用于估计置信区间的示例代码。

5.2　比率的比较

在估计流行率和发病率时，对可信度进行描述是很有用的。我们更感兴趣的是组间和种群间比较的估计。卡方检验（Chi-square tests）经常用于种群间的比率比较，最常见的是皮尔逊的卡方检验：

$$\chi^2 = \sum_{i=1}^{n} \frac{(O_i - E_i)^2}{E_i} \tag{4}$$

式中 O_i 是种群 i 中所观察到的感染数量。E_i 是根据零假设（null hypothesis），种群 i 中预期的感染数量。一般来说，零假设是指在不同的种群中，流行率是相等的。例如，考虑这样一种情况，在一个种群中，35 个动物中有 10 个动物检验呈阳性。而在另一个种群中，45 个动物中有 20 个动物检验呈阳性，那么列联表为：

	种群 A	种群 B	总计
感染数	10	20	30
未感染数	25	25	50
总计	35	45	80

假设种群之间没有差异，预期的感染流行率应是（10+20）/（35 + 45）=

0. 375。因此，第一个种群中所预期的感染数应是 0. 375×35 = 13. 125，而在第二
218 个种群中预期感染数则是 0. 375×45 = 16. 875。χ^2 检验的统计值是种群 i 观察值和预期值差值的平方除以种群 i 预期值的总和（i =1，2，3，…，n）。这个统计值与卡方分布的（第 1 行）×（第 1 列）自由度获得的临界值相比较，作为在一个种群中感染流行率存在差异的证据。列联表有两行，一行是感染的，另一行是未感染的。同时列联表还有两列，即两个种群。因此，自由度=（2−1）×（2−1）=1。要使检验有意义，则种群数量需要大于 2，后续的成对比较可以按照相同的方法进行，并进行实验误差率的适当校正，例如进行邦费罗尼校正（Bonferroni correction）。卡方检验要求，不超过 20% 的预期计数小于 5，这有可能无法实现，特别是在低感染流行率的种群中。假如列联表中的一个边缘值被固定，例如，在我们的实验中，采样点 A 和采样点 B 的样本数，在实验前分别设定为 35 和 45，那么，巴纳德精确检验（Barnard's exact test）是卡方检验强有力的替代方法，从而避免低期望计数的问题（Martín Andrés 等人，2004）。并可用 R 语言巴纳德软件包（"Barnard" package in R）运行计算。附录 1 提供了软件包的示例代码，用于检验比率之间的差异性。

逻辑回归（logistic regression）在比较特定环境或宿主特征明显的个体或种群之间感染概率或死亡概率时，是一个可靠而灵活的检验体系。逻辑模型（logistic model）预测二元结果的对数转换概率（如感染率、死亡率）表示为一个或多个预测变量的线性函数：

$$\operatorname{logit}(p_i) = \ln\left(\frac{p_i}{1 - p_i}\right) = \beta_0 + \beta_1 x_{1,\ i} + \cdots + \beta_m x_{m,\ i} \tag{5}$$

式中：β_0，…，β_m表示截距；x_1，…，x_m表示预测变量的回归系数；对数转换（logit transform）就是概率比（odds ratio）的对数转换；式中概率比通过公式（5）两边取幂进行计算：

$$\exp(\operatorname{logit}(p_i)) = \left(\frac{p_i}{1 - p_i}\right) = e^{(\beta_0+\beta_1 x_{1,\ j}+\cdots+\beta_m x_{m,\ j})} \tag{6}$$

假如 x_1是一个分类预测变量（如雄性与雌性），exp(β_1）的值可以解释为雄性相对于雌性感染概率的增加（或减少）。假如 x_1不是一个连续变量（例如动物的长度），那么，exp(β_1）的值表示，随着预测变量 1 个单位的增加或减少，感染概率的增加或减少。当解释相对于测量单位如毫米与厘米的概率比系数（odds-ratio coefficients）时，以及在测量值范围内的情况下，要特别小心。例如，在每 1 厘米的风险中，一个大的预测风险增加，可能会让人印象深刻，但是如果所有测量的动物都在 0. 1 厘米以内，那么实际影响的大小就没那么明显了。

逻辑回归也可以用来估计种群间，与蛙病毒发生相关的风险因素（Gahl和Calhoun，2008；Greer and 和 Collins，2008）。例如，你感兴趣的是发现蛙病毒感染的预测概率或者在特定种群中发生大规模死亡的预测概率。用逻辑回归 219
模型来预测种群 i 或类似的个体 i 的这一概率也是有可能的：

$$p_i = \frac{e^{\beta_0+\beta_1 x_{1,\ i}+\cdots+\beta_m x_{m,\ i}}}{1+e^{\beta_0+\beta_1 x_{1,\ i}+\cdots+\beta_m x_{m,\ i}}} = \frac{1}{1+e^{-(\beta_0+\beta_1 x_{1,\ i}+\cdots+\beta_m x_{m,\ i})}} \tag{7}$$

假如我们将预测湿地中蛙病毒发生一个逻辑回归模型（logistic regression model）拟合为离道路最近距离的一个函数。在这个函数中，截距（intercept）$\beta_0=-0.5$，距离的斜率参数 $\beta_1=-0.1$，在这种情况下，一个距离公路 10 公里的池塘，可以预测出其蛙病毒发生的概率为：1/（1 + exp［-（-0.5+-0.1×10）］）= 0.182，而一个距离为 5 公里远的种群所具有的一个预测概率为 1/（1+exp［-（-0.5+-0.1×5）］）= 0.269。大多数统计软件包都提供了来自逻辑回归模型的预测概率（predicted probabilities）和置信区间。附录 1 为逻辑回归提供了 R 语言的示例代码。

5.3　病毒滴度

以上统计方法将感染归为二元系，个体要么感染，要么没有感染。然而，感染是从亚临床感染到临床感染的一个连续过程，后者可以引起疾病，并且可以导致死亡（Miller 等人，2015）。定量实时 PCR 和基于细胞培养的方法（如噬斑试验和 TCID50）是估计组织中病毒滴度的常用技术（Miller 等人，2015）。由于组织中的病毒滴度与感染的严重程度相关，因而这些数据提供了蛙病毒对种群影响的其他认识。例如，在一个耐受物种的种群中（例如，美国牛蛙，*Lithobates catesbeianus*），随着时间的推移，测量蛙病毒的流行率和病毒滴度（Hoverman 等人，2011），有人可能会观察到，蛙病毒的流行率很高，但病毒滴度非常低。如果改变条件（如升高温度）使这一物种更易感，那么就会看到病毒滴度随着温度的升高而升高，而感染的流行率仍保持不变。

有报道记载，病毒滴度常常表示为每单位基因组或每单位组织病毒浓度的 $\log_{10}$-转换值。这种转换滴度总体上是呈正态分布的，并适合简单线性模型（即回归分析、方差分析），例如，上面所讨论的关系可以通过病毒滴度在温度上的线性回归进行检验。

因为 0 没有对数，所以在取对数之前，对所有数值包括 0，加 1 或加上一个表示最低检测水平的数值，这是最常见的做法。假如数据集中存在许多 0，即检验的个体是阴性的，由此所产生的结果则不呈正态分布。然而，如果你感兴趣的只是被感染动物的病毒滴度分布，为未受感染的个体排除 0 是合适的。 220

或者你可以使用零膨胀模型（zero-inflated models）。这些模型解释了用等价逻辑斯蒂模型（equivalent of a logistic model）所表示的感染动物的概率，在一个动物感染的条件下，用典型的泊松分布或负二项分布（Poisson or negative binomial distribution）预测病毒粒子的数量。这些模型还可以应用于其他监测数据，如一个种群中感染动物的数量。如果没有感染，或者是被感染动物在取样时被遗漏了，则数值为0。我们要求读者直接阅读 Dohoo 等人（2003）和 Zuur 等人（2012）的文献，以寻求相关方面的指导。

5.4 生存数据分析

虽然我们通常只对感染概率或死亡概率感兴趣，但是了解死亡发生的时间和死亡率也是很有用的。在生存分析（survival analyses）中，特定个体的命运会随着时间的推移而定期且频繁地发生改变。因此，这些设计对人工饲养的种群是最合适的，如动物园、实验室的研究。在人工饲养条件下，每一个个体都要定期接受检查。随着时间的推移，当所有个体的命运都知道的时候，在整个研究期间，生存可以用一个范围为0~100%的一个曲线表示。

在研究期间，当一些个体的命运未知的时候，就要进行删失，同时，还必须在生存分析中予以说明。当个体的命运在某个时间点之后没有被观察到的时候，就要进行右删失。个体在最后一次观察后也要进行删失。在研究期间或在收集诊断信息的研究结束时，对动物个体进行安乐死时，同样要进行右删失。如果一个动物在研究开始前的某个未知时间被感染，它就必须进行左删失。在野外研究中，在第一个采样日之后，将个体添加到一个研究当中，是很常见的，这称为交错输入。

每个时间点处于风险个体的信息，不包括那些被删失的个体，被用来估计时间特异性生存概率 $S(t)$（time-specific survival），并用各种统计包进行分析（例如，MARK 程序，网址为 http：//www. phi-dot. org/software/mark/）。最常用的生存估计函数之一就是 Kaplan-Meier（K-M）函数：

$$S(t) = \prod_{t_i < t} \frac{n_i - d_i}{n_i} \tag{8}$$

式中 $S(t)$ 是时间 $t = t_i$时的生存概率，n_i是生存的个体数量，并在时间 i 之前没有被删失；d_i是时间 t_i时的死亡数量（Jager 等人，2008）。到时间 t 的生存概率等于当前和之前生存概率的乘积。

K-M 生存估计可用 Mantel-Haenszel 检验对两组样本进行比较，而这对于列
221 联表法是最基本的（5.2 节），其期望值和偏差可以通过时间来进行计算，列联表为：

	A 组	B 组	总计
事件数	d_{A_i}	d_{B_i}	d_i
非事件数	$n_{A_i}-d_{A_i}$	$n_{B_i}-d_{B_i}$	n_i-d_i
风险数	n_{A_i}	n_{B_i}	n_i

表中 i 指的是时间 t_i；下标 A 和 B 分别表示 A 组和 B 组。假如两个组按照其生存函数是完全相同的，那么 A 组在时间 t_i的死亡期望值为：

$$\hat{e}_{A_i}(d_i \times n_{A_i})/n_i \tag{9}$$

现在，预期的死亡数可以与 A 组的实际死亡数相比较，重复 $i=1$，2，3，…，$m_i=1$，2，3，…，m 个样本周期。在比较两组的情况下，一个组的期望值可以计算出来，因为 A 组的偏差（deviations）就意味着 B 组的偏差，检验统计量（test statistic）为：

$$Q = \frac{\left(\sum_{i=1}^{m} d_{A_i} - \sum_{i=1}^{m} \hat{e}_{A_i}\right)^2}{\sum_{i=1}^{m} \hat{V}(\hat{e}_{A_i})} \tag{10}$$

式中 $\hat{V}$ 是死亡期望值的方差，检验统计量是具有一个自由度的卡方分布（Dohoo 等人，2003；Hosmer 等人，2008）。

虽然 Mantel-Haenszel 检验在计算上相对容易，但是它不能超过两个组或连续的预测因子或协变量。Cox PH 模型（Cox Proportional Hazard model）分析是一种更通用的方法，用来检验不同组之间或在具有连续协变量（continuous covariates）的个体之间的生存曲线差异，例如身体的大小。Cox PH 估计的是基准风险函数（baseline hazard function）（见方框图 1），并可测出检验组中的个体相对于基值（baseline）来说是否具有更高或更低的风险（Hosmer 等人，2008）。Cox PH 非参数估计一个基准风险函数 h_0（t），具有协变量 x_1，x_2，…，x_n的个体风险是：

$$h_0(t)\,e^{[\beta_1 x_1+\beta_2 x_2+\cdots+\beta_n x_n]} \tag{11}$$

当模型方括号里的线性部分等于 0 时，指数项（exponential term）就等于 1，风险就等于 h_0（t），该值就是基准风险值；假如方括号项的总和>0，那么风险就增加了一定比例；假如方括号项的总和<0，那么风险就降低了一定比例。例如，假设相对雄性而言，雌性动物的系数 $\beta_{Female}=0.693$，那么雌性动物 222
所具有的风险 exp（0.693）= 2×高于雄性动物的风险值。在 Cox PH 分析中，关注的焦点往往是组与组之间生存比例的差异，然而，在大多数统计软件包中都提取了基准风险。

Cox PH 模型也存在着局限性。第一，它不能兼容左删失观察；第二，它假设不同组之间（例如雄性与雌性）风险的比率差异随着时间的推移是不变的。因此，如果绘制生存曲线，它们不应该交叉或发散，而应该是随着时间的推移接近平行的。如果这些假设当中的任何一个都是不合理的，读者应该参考一篇关于生存分析的文章（Dohoo 等人，2003；Hosmer 等人，2008）或咨询统计学家关于其他替代方法。

Cox PH 模型的一个替代方法就是加速失效时间模型（accelerated failure time models，AFT 模型），又称参数生存模型（parametric survival models）（Hosmer 等人，2008）。AFT 模型和 Cox PH 模型有两个重要的区别：第一，在 AFT 模型中，基准风险是函数形式，而不是比率被提前规定，也不是依数据来估计（见方框图 1）。例如，一个恒定的风险会用指数模型（exponential model）来建模（Hosmer 等人，2008）。第二，在 AFT 模型中，由于风险的形式被提前设定，并且只能进行模型参数估计，因而生存评估可以在观察时间期限之外进行预测，并且可能具有更大的统计功效。

5.5 标记-重捕获研究

在野生动物种群中，对流行病学相关参数的估计，如个体的命运、种群的大小，存在许多固有的困难，通常使用 CMR 即捕获-标记-再捕获（capture-mark-recapture，CMR）方法予以解决（Cooch 等人，2012）。这是一个活跃的
223 研究领域，也是文献报道最多的研究领域（Amstrup 等人，2005；Thomson 等人，2009）。因此，我们将简单地概述一下对理解蛙病毒流行病学有帮助的方法，并引导读者了解上述文献内容。

方框图 1

风险函数（hazard function）$h(t)$ ——在时间 t 的瞬时死亡率。累积风险表示为$H(t)$。

生存函数（survival function）$S(t)$ ——超过时间 t 的生存概率。

概率密度函数（probability density function）$f(t)$ ——预期的死亡时间分布。

$$h(t) = \frac{f(t)}{S(t)} = -\frac{\partial \ln S(t)}{\partial t}$$

$$f(t) = S(t)h(t)$$

$$S(t) = \exp\left[-\int_0^t h(t)\right] = \exp[-H(t)]$$

封闭的种群模型对于估计种群规模（或密度）和感染流行率特别有用。这些模型起始时间段和后续重捕获时间段在时间上要足够近，因而我们可以假设没有出生、死亡、迁出或迁入。在简单例子中，有两种情况，种群的大小 $\hat{N}$ 通过个体的实际计数 C 进行估计，并对检测概率 $\hat{p}$ ，即 Lincoln-Peterson 估计函数进行调整：

$$\hat{N} = C/\hat{p} \tag{12}$$

检测概率的计算为最初标记过并被再捕获的个体分数。可以将此模型扩展到解释多个捕获时间段以及组间差异（例如雄性与雌性）或状态（例如感染与未感染）。重要的是，检测概率可以为不同的组或状态分别建模。从而允许对在蛙病毒感染与未感染之间，或有症状与无症状之间检测概率的差异予以解释（参见 2. 1 节）。

开放 CMR 模型（Open CMR models）没有假设种群对于种群统计变化是封闭的，一般来说，该模型更适合做重复监测和估计种群参数，特别是表面生存率（apparent survival）。在开放 CMR 模型中，参数 S 可以在组间分别建模（例如感染和未感染）或作为一个协变量，如年龄和大小的函数，从而为估计自然环境中疾病对个体的影响提供一种手段。例如，我们可以确定蛙病毒感染鱼与未感染鱼之间的表面生存率是否存在差异，以及这些差异在成体和稚体之间是否是恒定的。在一个类似的模型框架中，有可能用一个疾病发生率和流行率的函数来估计种群的增长率（Cooch 等人，2012）。

多态模型（multi-state models）是 CMR 模型的扩展。该模型允许个体在不同状态之间进行转换，例如在未感染和感染状态之间进行转换。这种强大的建模方法为评估从未感染状态到感染状态（如发病率）的比率或概率提供了一种手段，反之亦然。这些模型的前提是生存率和状态之间的转换暂时是分离的，例如，个体首先存活，然后被感染。此外，在对立事件（encounter events）之间的转换只能发生一次，例如，从未感染状态到感染状态之间的转换。因此，精心设计 CMR 研究是很有必要的。在研究中，可以对这些模型及相关的模型进行扩展，以解释错误的状态分类，例如，感染状态没有完全被测定或部分可观测，即对个体进行了观测，但其感染状态没有确定。考虑到使用 CMR 模型的复杂性，我们建议，在研究设计和分析期间咨询统计学家。

6 动态模型的运用

动态模型在研究宿主-病原体相互作用时是非常有用的。宿主内模型 224

(within-host models) 可以阐明导致宿主感染和疾病的生理机制（Mideo 等人，2008，2011；Woodhams 等人，2008）。相比较而言，宿主间模型（between-host models）重点关注的是病原体引入或循环传播（circulating）时个体和种群的命运（Hastings，1997）。在这部分，我们将重点阐述在预测病原体对种群的影响方面发挥作用的宿主间模型。到目前为止，针对蛙病毒的动态模型（dynamic models）不多（Duffus，2009）。因此，我们的几个例子都来自野生动物疾病文献及针对新兴病原体毒蛙真菌（*Batrachochytrium dendrobaditis*，简称 *Bd*）的建模工作。

6.1 SI/SIR 模型：传播

易染-感染-恢复模型（susceptible-infected-recovered models，简称 SIR 模型），用一系列的常微分方程（ordinary differential equations）来检验传播动态学，这些常微分方程对下述三种结果中的一种进行建模和预测：病原体灭绝（pathogen extinction）、宿主灭绝（host extinction）和病原体-宿主共存（pathogen-host persistence）（Allen，2006）。在许多简单情况下，整个宿主种群可以分成三个亚群（subpopulations）：传染病易感个体（individuals susceptible to infection，简称 S）、传染性个体（infected individuals，简称 I）以及传染病恢复个体（recovered individuals，简称 R）和不能再感染的个体或者暂时具有免疫力的个体。传染病恢复个体 R 也可以是从种群中移除的个体。一个更简化的模型就是个体不能获得免疫力，即易染-感染模型（susceptible-infected model，简称 SI 模型）（Allen，2006）。在这个模型中，前提是个体清除了感染，它们再次变得易染。在这里，我只描述基础 SIR 模型（易染-感染-恢复模型）。

在这个最简单的连续时间 SIR 模型中，整个种群的大小（N）假定是恒定的：

$$N = S + I + R \tag{13}$$

式中 S、I 和 R 分别表示每一个亚种群的个体数量（Hastings，1997）。在时间 t 内，每个亚种群的变化率可建模为：

$$dS/dt = -\beta SI \tag{14}$$

$$dI/dt = \beta SI - \gamma I \tag{15}$$

$$dR/dt = \gamma I \tag{16}$$

式中 β 是宿主相互之间接触感染和传播感染的比率，γ 是宿主的恢复比率（recovery rate）或者消除比率（removal rate）。在这里，假定传播是密度依赖
225 的（density-dependent），那么传播就表示为 βSI。有证据表明，蛙病毒的传播可能与密度无关（Harp 和 Petranka，2006），并可建模为 $\beta I/N$。麦卡勒姆

（McCallum）等人（2001）提供了另外几种形式的传播函数（transmission functions），包括不依赖密度的传播函数（density-independent transmission）和非线性密度函数（nonlinear functions of density）。该模型没有包括种群统计（出生、死亡、迁出或迁入）。因此，只有所有个体都属于易染类（当 $I=0$ 时）时，平衡才能产生。对于即将发生的疾病流行来说，感染个体的数量必须增加，即 $dI/dt > 0$。一个疾病的繁殖数量 R_0（reproductive number of a disease）是继发感染病例的数量，是一个感染个体在一个易染种群中所产生的平均继发病例数量，可表示为：

$$R_0 = \beta S / \gamma \qquad (17)$$

如果 $R_0 > 1$，则种群中的感染数量在不断增长，表示疾病在流行。然而，由于这一模型具有密度依赖性，因而对于一种流行病的发生以及所有易染个体全部感染之前流行病的结束，就存在着一个最小种群数量（minimum population size），即种群数量的临界值 $N_T = \gamma/\beta$（Hastings 1997）。针对流行病建模时，假定时间尺度是足够短的，可以忽略宿主种群中的出生和其他形式的死亡。在更为复杂的模型中，通过将出生数加到易染亚种群当中以及将自然死亡数加到一个种群中的每一亚种群中。

对于蛙病毒和大多数自然种群来说，SIR 基本模型有可能过于简化。达弗斯（Duffus，2009）运用离散时间 SI 模型（discrete-time SI model）研究显示，蛙病毒只能在英国林蛙成蛙之间保持传播。她的模型中包括自然死亡和疾病所引起的死亡以及早期生活阶段的补充。疾病的传播率由成体之间的接触率及接触后被感染的可能性决定。达弗斯（2009）的研究同样也显示，在单一种群中，蛙病毒在成体之间的传播只能维持两个症状，即溃疡的形成和出血。这些模型还显示了导致蛙病毒在林蛙种群中持续存在的条件，这些模型参数估计需要其他数据，以更好地认识整个系统和预测特定种群的结果（Duffus，2009）。另一种模式是用于林蛙发育的，该模型研究了特定发育阶段的易感性和水源性传播，以重现自然种群中观察到的大规模死亡模式（JLB，未发表资料）。

其他扩展模型对于预测自然种群中的蛙病毒动态学特别有用。例如，存在于群落中的大多数蛙病毒宿主种群，很可能与其他易染物种相互作用，这些易染物种可能来自不同类别的变温脊椎动物（Gray 等人，2009）。布雷内斯等人（2014a）研究显示，蛙病毒在不同动物类别间有可能是通过水进行传播的。他的研究同样也显示，蛙病毒病的结果取决于两栖类种群中物种的组成，以及哪个物种最先感染蛙病毒（Brenes，2013）。这些研究可以作为一个起点，用来确定多个物种水生群落的传播概率。

在其他疾病系统中，传播模型中加入多物种，对重点宿主种群会产生影 226

响，但取决于宿主作为一个贮藏宿主的能力及该物种在群落中的主导地位（Keesing 等人，2006）。多宿主物种的添加可能使 SIR 模型的分析面临着持续的挑战。到目前为止，大多数模型只包括两个物种和病原体（Keesing 等人，2006）。因此，对于某些蛙病毒-宿主系统来说，这可能是不现实的。多布森（Dobson，2004）通过对多物种模型中大量参数的处理，将参数推衍（scaling）为依赖宿主体大小的异速生长函数（allometric functions）。然而，目前还不清楚这种关系是否存在于蛙病毒的传播中。勒鲁（Lélu 等人，2013）提供了一个模型例子，包括一种寄生虫即刚地弓形虫（*Toxoplasma gondii*）从老鼠到猫的营养转移以及在猫中的垂直传播。类似的复杂相互作用关系肯定存在于蛙病毒宿主物种中，例如，有学者提出假设，这种相互作用有捕食（predation）或食尸（necrophagy）以及通过蚊子进行机械性传播（Allender 等人，2006；Johnson 等人，2007；Kimble 等人，2014）。虽然，在蛙病毒-宿主系统中，可能存在大量的相互作用，但是在蛙病毒的流行病学中，有几个相互作用可能并不重要。一种策略是创建几个相互竞争的模型，并将这些模型与自然种群或中试研究中的动态数据进行拟合，从而确定最重要的传播机制。

对于运用 SIR 模型研究蛙病毒感兴趣的研究者来说，我们推荐奥托（Otto）和戴（Day，2007）所著的专著《生态学和进化论中数学建模的生物学家指南》（*A Biologist's Guide to Mathematical Modeling in Ecology and Evolution*）。该专著综述了数学建模，并描述了构建和分析模型的必要过程以及分析模型的主要常微分方程。懂得计算机编程和软件（例如 Matlab、Maple、Mathematica 以及 R 语言）运用对建模和仿真并用于大多数分析是很有必要的。附录 2 为 SIR 简单模型提供了 Matlab 示例代码。

6.2 基于个体的模型/针对模式的建模

基于个体的模型（individuals-based models，IBM），有时又称基于主体的模型（agent-based models，ABMs），对于研究疾病动态学同样非常有用。IBM 是基于仿真的，在每一个步骤中都有一组规则或涉及每个个体的概率事件。对于生物学家来说，构建 IBM 常常要比构建 SIR 模型容易得多，因为他们不需要解微分方程。然而，IBM 可能很复杂，并需要计算机编程技巧。这些模型常常运用序列方程组（sequential equations）、一系列循环（for-loops）语句以及如果-则语句（If-then statements）对实施的一套项目时间编制目录进行运行，从而确定一个个体的行为或命运。对于疾病 IBM 来说，每个个体的疾病状态都被记录下来，它们的感染风险取决于它们与其他个体或与环境的相互作用。同样，还有其他使用微分方程的 IBM 类型。例如，布里格斯（Briggs）等人

（2010）开发了一个具有微分方程的 IBM，该模型清晰地融入了个体的毒蛙壶
菌载量，并进一步研究了一种病原贮藏宿主（pathogen reservoir）和一种长时
间生活阶段的蝌蚪是如何影响青蛙种群长期与毒蛙壶菌共存或经历局部灭绝。
被开发用于蛙病毒的类似模型包括病毒载量以及排毒量，以更好地认识病毒是
如何与宿主以及引起大规模死亡的因素相互作用的。IBM 一个吸引人的方面是 227
它们明确地包含了动物的行为，对蛙病毒来说，研究人员可能感兴趣的是，不
同的行为如聚群（schooling）或食尸，是如何影响宿主种群以及病原体持续存
在的。

创建 IBM 以及确定合理相互作用的一项有用的技术就是称为针对模式的建模（pattern-oriented modeling，POM）。在 POM 中，数据被用来确定在一个感兴趣的自然系统中所见到的几个突出模式（salient patterns），这些模式构成了模型评估的基础。所创建的 IBM 多种模型，代表了宿主-病原体相互作用的不同假设。对不同 IBM 的评估，是基于它们重新创建突出模式的能力（Grimm 等人，2005；Grimm 和 Railsback，2012）。当一个模型与多个模式匹配的时候，那么这个模型在构建上是符合现实的（Wiegand 等人，2003），并且能够产生经得起检验的预测。在运用 POM 时，研究人员还可以对不同的假设进行对比，确定一个有用的模型结构，并减少参数的不确定性。

对结构开发 IBM 感兴趣的研究者来说，我们推荐两本著作，一本是格林姆（Grimm）和雷尔斯巴克（Railsback）（2005）所著的《基于个体的建模与生态学》（*Individual-based Modeling and Ecology*）以及雷尔斯巴克和格林姆（2011）所著的《基于主体和基于个体的建模：实用建模介绍》（*Agent-based and Individual-based Modeling*：*A Practical Introduction*）。这两本书都描述了一种称为面向对象设计和描述的“最佳模型实践”，这是描述 IBM 各个方面的标准格式。后一本书通过案例和代码展现了构建 IBM 的全过程，这是一个对用户友好和免费的称为 NetLogo 的程序（http://ccl. northwestern. edu/netlogo/index. shtml）。NetLogo 包括一个预构建模型库，这个模型库还包括艾滋病（AIDS）、Solo 病（disease Solo）以及病毒模型，并形成了开发蛙病毒模型的基础。NetLogo 网站还包括了一个建模共享（modeling commons）界面，在这个界面中，用户可以共享他们的模型，从而帮助他们自身模型的开发。其他软件如 Matlab 和 R 语言同样也能用于开发和分析 IBM。

6.3　种群矩阵模型

种群矩阵模型（population matrix model）所考查的是随着时间推移种群大小和年龄结构的变化。这些模型包括每个年龄组之间转换概率的参数。为了把

疾病融入模型之中，在蛙病毒感染之后，每一个年龄组都记录了蛙病毒感染后的存活情况。厄尔和格雷（2014）开发了一个发育阶段结构矩阵模型（stage-structured matrix model）来预测一个林蛙封闭种群的卵、孵化、幼体及变态发育阶段蛙病毒感染的影响。这个研究结合了林蛙种群模型的资料（Harper 等人，2008）与实验接种感染的数据（Haislip 等人，2011）来预测种群的结果。附录 2 为厄尔和格雷（2014）之后的一个矩阵模型提供了 Matlab 中的示例代码。

种群矩阵模型也可以与传播模型结合，更为现实地同时构建两个动态模
228 型，例如，布里格斯等人（2005）融合了黄腿蛙（*Rana muscosa*）的种群模型和一个基于当前传播和死亡率的信息所构建的毒蛙壶菌传播动态学 SIR 模型。这个模型将年间离散时间的种群动态模型和每年的持续时间传播动态模型结合在一起。通过运行具有不同参数值的模型，布里格斯等人（2005）确定了导致蛙种群灭绝、病原体灭亡及蛙种群和病原体存活的条件。

种群模型同样也能进行尺度放大，应用于复合种群过程（metapopulation processes）。一个复合种群是一组空间结构上的局部种群，这些局部种群通过分散而相互作用（Marsh 和 Trenham，2001；Smith 和 Green，2005）。几个蛙病毒宿主在结构上有可能构成复合种群。复合种群模型将局部种群之间的分散概率参数以及每一个局部种群的种群统计参数进行了合并。复合种群模型有助于弄清楚病原体在种群中的空间传播，并用于检验疾病干预策略的有效性（Hess，1996）。在两栖类动物中，蛙病毒的暴发已部分归结为亚临床感染的稚体或返回到繁殖场的成体、排放的病毒以及感染的幼体（Brunner 等人，2004）。对于种群矩阵模型感兴趣的个人，我们推荐阅读卡斯韦尔（Caswell，2000）所著的《矩阵种群模型：构建、分析与解释》（*Matrix Population Models*：*Construction*，*Analysis*，*and Interpretation*）及 Hanski（1999）所著的《复合种群生态学》（*Metapopulation Ecology*），这两本书将有助于那些对研究蛙病毒对复合种群动态学影响感兴趣的人。

6.4 疾病干预策略建模

宿主-病原体动态学建模的一个目标就是确定防止疾病暴发的干预策略。目前，几乎没有针对蛙病毒的建议性控制方案，但未来疫苗的开发使得疾病的控制成为可能（Miller 等人，2011）。如果疾病有可能导致受保护的宿主动物种群发生极高的死亡率，那么就要选择其他的措施，包括对个体或种群进行检疫、捕杀以及对新引入的种群进行隔离圈养。模型也可用于确定宿主-病原体循环（host-pathogen cycle）中的脆弱点（vulnerable points），以便用干预策略

中断宿主-病原体循环。例如，疾病的暴发如果是由密度引起的，那么湿地的挺水植被（emergent vegetation）会减少两栖类幼体间疾病传播的概率（Greer 和 Collins，2008）。假如在水生环境中的应激因子（stressors）如高氮水平可以导致疾病的再次暴发，那么就要运用改善水质的策略。全面了解蛙病毒病暴发的因素和蛙病毒-宿主系统，对制定合理的干预策略是至关重要的。在有些情况下，对自然种群实施干预策略是不可行的，或是成本过高，或是结果不理想。反之，如果策略是可行的，则可以运用模型来确定策略的时间和频率，以达到最佳效果。SIR 模型及其变化模型可用于探索疫苗接种策略（Hethcote，2000）和其他控制技术如捕杀（Lloyd-Smith 等人，2005）。如受资金上的限 229
制，疾病控制成本可以纳入模型之中，以确定最佳策略（Fenichel 等人，2010）。伍德汉姆斯（Woodhams）等人（2011）讨论了毒蛙壶菌的可能干预策略，并提供了它们对具有适应性免疫和没有适应性免疫个体建模的效益结果。这些模型的运行也同样显示正在减少的宿主种群可以防止灭绝，即降低传播的概率。对实施优化控制模型感兴趣的研究人员，我们推荐勒哈特（Lenhart）和沃克曼（Workman）（2007）的著作《应用于生物模型的优化控制》（*Optimal Control Applied to Biological Models*）。该著作重点阐述了连续常微分方程模型的控制，同时还包括计算机程序 Matlab 示例代码。优化控制同样也可应用于 IBM，但是有效的技术仍在开发中（Federico 等人，2013）。

6.5　模型参数化和评估

有许多方法可以参数化建模，并将模型与数据进行集成。通常建模者通过搜索文献来选择参数值，但通常不是所有的参数值都可用。另一种方法是构建一个模型，并将模型的输出结果与现有的数据序列进行拟合。在病毒建模的例子中，预测要与监测数据进行拟合，其中包括感染和未感染的个体丰度，或者大规模死亡的程度和发生的时间。在模型与数据进行拟合以后，参数值要达到最佳拟合或者要与多模式相匹配（与 POM 一样）。如果已知一些参数，并且研究人员对其他参数的可能范围有一个很好的了解，那么就可以探讨这些值的范围以确定这些值是如何改变模型输出结果的。评估参数值改变的影响被称为敏感性分析（Cariboni 等人，2007）。假如模型对某一个参数特别敏感，则表明更好的参数估计将是一个有价值的研究方向（Biek 等人，2002；Cariboni 等人，2007），特别是在参数估计不是基于可靠数据的情况下（例如低的样本数）。卡里博尼（Cariboni）等人（2007）提出了敏感性分析最佳办法，对自然种群疾病建模参数的评估进行了全面的综述，并可在库奇（Cooch）等人（2012）的文献中找到。

模型评估的目的是确定模型是否全面代表了自然系统的特征，从而代表了预期的动态学。这常常涉及确定它们能否用于作出准确的预测。通常，模型评估、模型验证和模型测试这些术语可以互换使用，因为模型的构建是在假设和简化的基础之上的，它们从来都不是真正“有效”或“正确”的。奥古西亚克（Augusiak）等人（2014）曾提出，“评估”这一术语表示评估模型质量和可靠性的过程。一个模型适当的“评估”包括六个要素：

- 评估用于构建模型的数据质量；
- 评价构建模型的简化假设；
- 验证模型是否被正确实施；

230 • 验证模型的输出结果与设计模型所使用的数据是否匹配；

- 探讨模型对参数值改变的敏感性；
- 评估模型是否能与独立数据集即没有用于原始模型公式的数据进行拟合。

建议模型的构建和“评估”遵循一个称为 TRACE 的文件编制程序即透明和综合生态文件（TRAnsparent and Comprehensive Ecological documentation，简称 TRACE），设计该文件是用来确保模型的可靠性以及将科学理论与实际应用联系起来（Grimm 等人，2014）。

7 蛙病毒引入未感染地区的风险分析

进口风险分析（import risk analysis，IRA）用来确定一个病原体进入一个系统并带来威胁的过程。病原体引入的后果可以直接监测（见本节第 2 部分）或使用模型模拟（见本节第 6 部分）。IRA 指导原则主要是从两个国家或地区的贸易角度开发出来的，评估与进口活陆地生产动物相关的疾病风险。然而，同样的原则也适用于评估野生或人工饲养种群中病毒的危险性。一般来说，进口风险分析重点关注同一类群中一个物种或多个物种感染的可能性。正如达弗斯等人（2015）所讨论的那样，蛙病毒是多宿主病原体，具有感染三类不同脊椎动物的能力，这样就使得进口风险分析在蛙病毒中应用更为复杂。进口风险分析可用于建立或修订野生动物的贸易或运输的指导原则，这些野生动物可能感染一种病原并处于亚临床状态（Smith 等人，2009）。世界动物卫生组织列出了感染两栖动物的病毒，这些病毒是必须申报的病原体，也就是说，涉及国际贸易两栖动物的子样本，应该在运输之前被证实为蛙病毒阴性（Schloegel 等人，2010）。目前，这些规定在大多数国家都没有被执行（Kolby 等人，

2014）。下面列出的程序是根据 OIE 的原则和建议而制定的，有一些例子说明它们如何被应用到 IRA 的某些部分，以研究一种蛙病毒被引入到未感染的区域。

7.1　风险的界定

IRA 的第一步是定义一个感兴趣的区域。这个区域包含感兴趣的种群，例如一个包含不常见物种并容易受到蛙病毒感染的种群。这个区域通常是一个地理区域或一个国家（Rödder 等人，2009；OIE 2014）。一般来说，区域是根据动物运动或病原体转移人为的或自然的屏障来界定的（OIE，2014）。例如，蛙病毒的病毒粒子可以流向支流的下游，而相关的洪泛区通常是动物运动的廊道。因此，在区域中应当通过水域界定激流系统（lotic systems）。静水系统 231
（lentic systems）有可能包含蛙病毒宿主的凹洼湿地（depressional wetlands）或湖泊，可以被界定为感兴趣的区域，其前提条件是，在水文上是封闭的，并且被陆地景观包围。在动物环境中，感兴趣的典型区域是人工饲养场（OIE，2014）。

IRA 的第二步是确定感兴趣区域内蛙病毒的存在。本章第 2 节讨论了监测研究以及由 OIE（2014）所提供的其他指导原则。检测蛙病毒的最小样本量（minimum sample size）取决于好几个因素（见第 2 节表 1）。另外，不频繁的取样可能导致检测的缺乏。托德·汤普森（Todd-Thompson，2010）的研究显示，大烟山国家公园的古尔利池塘（Gourley Pond of the Great Smoky Mountains National Park）的蛙病毒，除在晚春的 3 周时间内以外，其他时间都不存在。在晚春的这 3 周时间内，发生了一场蛙病毒的大暴发，导致了多个物种的普遍死亡。因此，当宿主存在时，每两周在采样点采样一次，这样就会有一个大的样本容量（$n>30$），也就能获得高的检测率。如果资源有限，那么当宿主存在时，每年至少需在 4 个时期进行采样。使用这个采样频率，霍韦尔曼等人（2012）在 40 个地点中，检出 33 个地点有蛙病毒。考虑到这种病毒在本研究中的所有地点都存在，用他们的采样频率，蛙病毒检测概率为 82.5%～100%。采样应该持续数年，以验证一个地点是否为蛙病毒阴性。对于感兴趣的大区域，应当在间隔不小于宿主的平均分散距离的多个地点进行采样，对于两栖类动物这个距离大约为 1 km（Wells，2007）。因此，对不同的种群进行采样不会造成它们之间的巨大差异。假如检出了蛙病毒，那么就没有理由进行 IRA，除非担心一个外来蛙病毒株被引进。

虽然对蛙病毒被引入到一个区域感兴趣的主要原因，通常是针对某种特定种类的关注，但是重要的是，在进口风险分析中，所有蛙病毒宿主都在考虑之列。如同布伦纳等人（2015）所讨论的一样，有一些宿主的功能是作为病毒的储藏宿主而保持亚临床感染状态，导致低种群流行率。同时，其他物种作为扩增宿主（amplification host）可以启动疾病的暴发。假如资金受到限制，一个可行的策略就是检测扩增宿主，因为这些宿主对蛙病毒的抵抗力较低，所以检出概率相对较大。达弗斯等人（2015）提供了一个已知蛙病毒宿主的列表以及几个感染接种的研究（Hoverman 等人，2011；Brenes 等人，2014b），可以为不同物种之间易感性的相对差别提供更深入的认识。

7.2 风险评估

风险评估（risk assessment）涉及三个主要步骤：（1）确定引入的途径；（2）确定引入的结果；（3）风险的评估。在风险评估中，绘制流程图来说明评估的每一步都是很有用的（见图 1 和图 2）。为了描述这一过程，下面我们提供一个通过进口感染蛙病毒的水产养殖鱼类，来评价野生两栖类动物风险的例子。

7.2.1 引入途径

232 引入途径包括宿主的传播路径和病毒附着于非宿主即鸟类和哺乳类（Gray 等人，2009）的污染物上转移。人类在污染地与非污染地之间移动，对于引进蛙病毒起着很大的作用。蛙病毒在未消毒的水和土壤环境中的存活时间至少有一周（Nazir 等人，2012）。因此，没有消毒鞋或装置而在水域之间移动的休闲者，有可能是引进蛙病毒的主要传染源（Gray 等人，2009）。鱼类孵化场是已知的蛙病毒暴发的场所（Waltzek 等人，2014），因而临床感染的和亚临床感染的鱼或从孵化场排出的污水是蛙病毒的另外一个主要传染源。对于一个感兴趣的特定区域，确定蛙病毒最有可能的引入途径是很重要的。通常把病毒引进的
233 途径分成三个阶段：进口、释放和感染。在进口水产养殖鱼类时，下列步骤界定了鱼类进口的重要阶段：

- 从一个感染地区进口的鱼感染了蛙病毒；
- 感染鱼通过边境控制未被检出；
- 感染鱼被分销到零售商；
- 感染鱼被销售到研究区域中的渔场。

假设感染蛙病毒的鱼饲养在水产养殖池塘中，那么可以通过几种途径将蛙病毒释放到邻近的水生环境中：

- 排放蛙病毒污染的水体；
- 感染鱼的逃跑；
- 禽类或哺乳动物捕食者转运带毒的活鱼或死鱼；

阶段	控制点	事件	前提条件
进口	进口原产地		进口鱼感染了蛙病毒
		进口到研究区域	
	边境检查	↓	在托运货物中没有检出蛙病毒
		鱼被销售给进口商	
	进口商/零售商	↓	在托运货物中没有检出蛙病毒
		鱼被卖给渔场	
	渔场	↓	在托运货物中没有检出以及在鱼场中也没检出蛙病毒
释放		病毒释放到环境中	
	环境	↓	易感两栖类物种栖息在这一环境中
感染	感染	感染	易感动物被感染

图 1　野外蛙病毒传播给未感染两栖类易感种群的路径流程图

- 蛙病毒宿主如两栖类或爬行类可以进入池塘，发生感染并扩散；
- 机械传播媒介如宠物或人类可以转运蛙病毒污染物。

最后，感染蛙病毒可以通过几个直接或间接的途径而发生（Gray 等人，2009）。水体是一个有效的传播媒介，宿主可以在水中感染蛙病毒，易感动物可通过直接接触或吞食被感染的宿主而感染蛙病毒（Miller 等人，2011）。有证据显示，蛙病毒可以不依赖密度而进行传播，这样有可能增加了宿主种群灭绝的概率（Brunner 等人，2015）。

7.2.2　结果评估

一个物种感染蛙病毒的结果，按照直接结果或间接结果，可以进行定性或定量描述。直接结果是蛙病毒对物种的影响，这种影响一般包括对种群数量下降和灭绝可能性的估计（见本章第 6 节）。高易感性的珍稀物种，特别是这些物种与蛙病毒其他宿主并发感染的时候，具有最大的灭绝概率（Earl 和 Gray，2014）。间接结果是与病原体监测相关的成本（即现场调查成本和诊断费用）及在灭绝之后可能出现的种群遣返（repatriation of populations）。

7.2.3 风险估计

如果蛙病毒沿着所确定的感染动物到易感动物这条途径进行传播，在很多情况下，蛙病毒传入带来的结果是不能接受的。因此，应该沿着上述所确定的传入途径建立一系列的关键控制点（critical control points，CCPs），这些关键控制点可以使蛙病毒在中途截断而中止传播。在每一个关键控制点，通过解决几个问题，可以对蛙病毒突破关键控制点而进行传播的概率进行估计。这个过
234 程可以总结为一个情景树，在这个情景树中，每一个关键控制点都具有一个“是”和一个“否”分支，并赋予一个检测的概率值（见图 2）。在图 2 中，对于每一个边防检查站，关键控制点 CCP 1 和 CCP 4 都进行预先设定。而 CCP 3 将取决于检查人员的培训和经验，CCP 2 则可能受到病毒载量、水温和动物健康的影响。CCP 5 是在实验室试验中进行病原检测，该检测是两个过程的一个函数：样本的大小（第 2 部分）和分子试验的性能，包括 PCR 的敏感性和特异性试验（Miller 等人，2015）。蛙病毒 PCR 的敏感性和特异性是正在研究的一个方向（Miller 等人，2015），并受到样本类型的影响，即受到采集死亡与非死亡样本的影响（Gray 等人，2012）。一般来说，有人认为，肝和肾组织能够提供最可靠的检测估计，其次是尾巴、脚趾和血液（Miller 等人，2015）。
235 如果 PCR 的敏感性和特异性最理想，那么运用表 1 中必要样本数所检测到的蛙病毒概率大约是 95%。在进口的货物中没有检测到蛙病毒的风险，可以计算为：1 减去所有 CCP 的检测概率的乘积（见图 2）。

7.3 风险管理与交流

为了管理蛙病毒引入的风险，进行风险-结果评估是很有用的。如果风险很低，但对目标物种的影响很大，那么风险管理的优先级别也就很高。然而，如果引入蛙病毒的风险很高，但对目标物种的影响很小，那么风险管理的优先级别也就很低。如果进口风险分析表明，风险很高，则建议管理应集中在各个关键控制点以及如何增加检测的概率和以一种成本很低的方式消除被感染的货物运输。

利益相关者之间需要进行有效沟通，无论是收集信息提供给 IRA，还是告知终端用户的发现、管理选择以及它们的实施。风险沟通常以政府层面为中心，但是个人组织如渔民或爬行动物学协会也能研究和实施他们自身的检疫和监测指导原则，并具有合格的诊断支持。各层次的合作和认识的提高，将大大

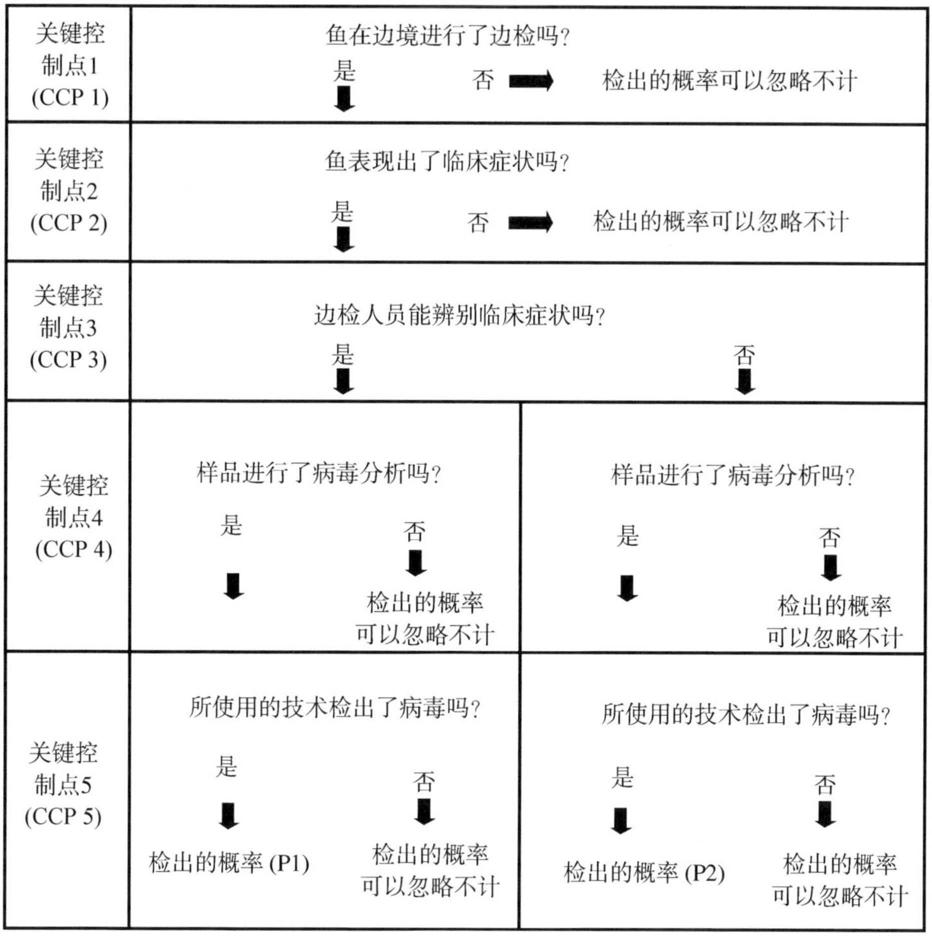
关键控制点1 (CCP 1)
鱼在边境进行了边检吗？
是
否
检出的概率可以忽略不计
关键控制点2 (CCP 2)
鱼表现出了临床症状吗？
是
否
检出的概率可以忽略不计
关键控制点3 (CCP 3)
边检人员能辨别临床症状吗？
是
否
关键控制点4 (CCP 4)
样品进行了病毒分析吗？
是
否
检出的概率可以忽略不计
样品进行了病毒分析吗？
是
否
检出的概率可以忽略不计
关键控制点5 (CCP 5)
所使用的技术检出了病毒吗？
是
否
检出的概率 (P1)
检出的概率可以忽略不计
所使用的技术检出了病毒吗？
是
否
检出的概率 (P2)
检出的概率可以忽略不计

图 2　在边境检查中于托运的感染鱼中检测蛙病毒的情景树。关键控制点（CCP）1~5 是识别点，也是病毒能够检出和未来传播的终止点；P1 是在情景树左分支回答“是”的概率乘积；P2 是在情景树右分支的两个回答“是”的概率乘积；在边境蛙病毒不能检出的概率是 1-（P1+P2）。

降低蛙病毒引入到未感染地区的风险。

许多需要进行全面进口风险分析的事实可见于已发表的许多科学文献之中，这些事实被用来证明风险分析的建议正确与否。在进行风险分析之前，先考虑已发表文献的适用性和质量是很重要的。发表的数据来自不同的物种、时间或大陆。如果你所研究的物种或地区没有发表的数据，那么就需要进行先导性研究（pilot study）来获得数据。另外，根据德尔菲法（Delphi method），获得专家意见是风险分析中确保进行初步估计的一种方法（Helmer，1967；Vose，2000）。因此，我们建议，所有对使用 IRA 感兴趣的组织都要咨询研究蛙病毒的专家。全球蛙病毒联盟（Global Ranavirus Consortium，GRC）是由科学家、兽医和专业人员组成的一个集体，可以为进口风险分析提供指导。每个大洲都有一个区域 GRC 代表，可以提供帮助或与该地区的专家建立必要的联系。

致谢

我们欣然感谢杰森·罗尔（Jason Rohr，南佛罗里达州立大学）和谢丽尔·布里格斯（Cheryl Briggs，加州大学圣塔芭芭拉分校）对我们这一章早期草稿的评阅，并提出了具有帮助性的评审意见。J. E. 厄尔（J. E. Earl）的稿件是其在国家数学和生物合成研究所于博士后时所撰写的。该研究所得到了国家科学基金会（项目号#EF 0832858）、美国国土安全部以及美国农业部的资助以及来自诺克斯维尔田纳西大学的额外资助。

开放存取出版物的发行得到了田纳西大学（林业、野生动物与渔业系，
236 研究与交流办公室，农业研究所）、华盛顿州立大学图书馆、戈登州立大学（学术事务办公室）、两栖爬行动物兽医协会以及两栖爬行动物保护协会的资金赞助。

开放存取

附录 1

下面的链接是使用统计程序 R 来估计置信区间的指南，进行卡方分析和逻辑回归检验，并绘制适当的图形的网站。

http：//fwf. ag. utk. edu/mgray/RanavirusBook/Chap8/SampleCode_ 8. 5. html

http：//fwf. ag. utk. edu/mgray/RanavirusBook/Chap8/SampleCode_ 8. 5. R

http：//fwf. ag. utk. edu/mgray/RanavirusBook/Chap8/Data. csv

附录 2

下面的链接是用于下载 MatLab 程序来做 SIR 和发育阶段-结构模型模拟的网址。

SIR 模型网址：

http：//fwf. ag. utk. edu/mgray/RanavirusBook/Chap8/Example_ SIR_ Model. m

http：//fwf. ag. utk. edu/mgray/RanavirusBook/Chap8/Example_ SIR_ Model. txt

http：//fwf. ag. utk. edu/mgray/RanavirusBook/Chap8/SIR. m

http：//fwf. ag. utk. edu/mgray/RanavirusBook/Chap8/SIR. txt

种群模型网址：

http：//fwf. ag. utk. edu/mgray/RanavirusBook/Chap8/PopulationModelDetails. pdf

http：//fwf. ag. utk. edu/mgray/RanavirusBook/Chap8/PopulationModelCode. txt

http：//fwf. ag. utk. edu/mgray/RanavirusBook/Chap8/PopulationExampleModel. m

参考文献

Allen LJS (2006). An introduction to mathematical biology. Pearson, New York.

Allender MC, Fry MM, Irizarry AR et al. (2006). Intracytoplasmic inclusions in circulating leukocytes from an eastern box turtle (*Terrapene carolina carolina*) with iridoviral infection. J Wildl Dis 42:677-684.

Amos KH (1985). Procedures for the detection and identification of certain fish pathogens, 3rd edn. American Fisheries Society, Corvallis.

Amstrup SC, Mcdonald TL, Manly BFJ (2005). Handbook of capture-recapture analysis. Princeton University Press, Princeton.

Augusiak J, Van Den Brink PJ, Grimm V (2014). Merging validation and evaluation of ecological models to "evaludation": a review of terminology and a practical approach. Ecol Model 280:117-128.

Biek R, Funk WC, Maxwell BA et al. (2002). What is missing in amphibian decline research: insights from ecological sensitivity analysis. Conserv Biol 16:728-734.

Brenes R (2013). Mechanisms contributing to the emergence of ranavirus in ectothermic vertebrate communities. Ph.D. Dissertation, University of Tennessee.

Brenes R, Gray MJ, Waltzek TB et al. (2014a). Transmission of ranavirus between ectothermic vertebrate hosts. PLoS One 9:e92476.

Brenes R, Miller DL, Waltzek TB et al. (2014b). Susceptibility of fish and turtles to three ranaviruses isolated from different ectothermic vertebrate classes. J Aquat Anim Health 26(2):118-126.

Briggs CJ, Vredenburg VT, Knapp RA et al. (2005). Investigating the population-level effects of chytridiomycosis: an emerging infectious disease of amphibians. Ecology 86:3149-3159.

Briggs CJ, Knapp RA, Vredenburg VT (2010). Enzootic and epizootic dynamics of the chytrid fungal pathogen of amphibians. Proc Natl Acad Sci U S A 107: 9695-9700.

Brown LD, Cal TT, Dasgupta A (2001). Interval estimation for a binomial proportion. Stat Sci 16:101-117.

Brunner JL, Schock DM, Davidson EW et al. (2004). Intraspecific reservoirs: complex life history and the persistence of a lethal ranavirus. Ecology 85:560-566.

Brunner JL, Schock DM, Collins JP (2007). Transmission dynamics of the amphibian ranavirus *Ambystoma tigrinum virus*. Dis Aquat Organ 77:87-95.

Brunner JL, Storfer A, Gray MJ, Hoverman JT (2015). Ranavirus ecology and evolution: from epidemiology to extinction. In: Gray MJ, Chinchar VG (eds) Ranaviruses: lethal pathogens of ectothermic vertebrates. Springer, New York.

Cariboni J, Gatelli D, Liska R et al. (2007). The role of sensitivity analysis in ecological modelling. Ecol Model 203:167-182.

Caswell H (2000). Matrix population models: construction, analysis, and interpretation, 2nd edn. Sinauer, Sunderland.

Cooch EG, Conn PB, Ellner SP et al. (2012). Disease dynamics in wild populations: modeling and estimation: a review. J Ornithol 152(2):485-509.

Dobson A (2004). Population dynamics of pathogens with multiple host species. Am Nat 164:S64-S78.

Dohoo I, Martin W, Stryhn H (2003). Veterinary epidemiologic research. AVC, Charlottetown.

Duffus ALJ (2009). Ranavirus ecology in common frogs (*Rana temporaria*) from the United Kingdom: transmission dynamics, alternate hosts, and host-strain interactions. Ph.D., University of London.

Duffus ALJ, Waltzek TB, Stöhr AC, Allender MC, Gotesman M, Whittington RJ, Hick P, Hines MK, Marschang RE (2015). Distribution and host range of ranaviruses. In: Gray MJ, Chinchar VG (eds) Ranaviruses: lethal pathogens of ectothermic vertebrates. Springer, New York.

Earl JE, Gray MJ (2014). Introduction of ranavirus to isolated wood frog population could cause local extinction. EcoHealth 11:581-592.

Evira (2013). Risk assessment of new and emerging systemic iridoviral diseases for European fish and aquatic ecosystems (RANA).http://www.evira.fi/portal/en/about+evira/about+us/ operation+areas/scientific+research/projects/previous/

risk+assessment+of+new+and+emergin g+systemic+iridoviral+diseases+for+european+fish+and+aquatic+ecosystems++rana+/. Accessed 21 May 2014.

Federico P, Gross LJ, Lenhart S et al. (2013). Optimal control in individual-based models: implications from aggregated methods. Am Nat 181:64-77.

Fenichel EP, Horan RD, Hickling GJ (2010). Management of infectious wildlife diseases: bridging conventional and bioeconomic approaches. Ecol Appl 20:903-914.

Forzán MJ, Wood J (2013). Low detection of ranavirus DNA in wild postmetamorphic green frogs, *Rana* (*Lithobates*) *clamitans*, despite previous or concurrent tadpole mortality. Journal of Wildlife Diseases 49:879-886.

Gahl MK, Calhoun AJK (2008). Landscape setting and risk of ranavirus mortality events. Biol Conserv 141:2679-2689.

Gotelli NJ, Ellison AM (2004). A primer of ecological statistics. Sinauer, Sunderland.

Gray MJ, Miller DL (2013). The rise of ranavirus: an emerging pathogen threatens ectothermic vertebrates. Wildl Prof 7:51-55.

Gray MJ, Miller DL, Hoverman JT (2009). Ecology and pathology of amphibian ranaviruses. Dis Aquat Organ 87:243-266.

Gray MJ, Miller DL, Hoverman JT (2012). Reliability of non-lethal surveillance methods for detecting ranavirus infection. Dis Aquat Organ 99:1-6.

Greer AL, Collins JP (2008). Habitat fragmentation as a result of biotic and abiotic factors controls pathogen transmission throughout a host population. J Anim Ecol 77:364-369.

Grimm V, Railsback SF (2005). Individual-based modeling and ecology. Princeton University Press, Princeton.

Grimm V, Railsback SF (2012). Pattern-oriented modelling: a "multi-scope" for predictive systems ecology. Philos Trans R Soc Lond B Biol Sci 367:298-310.

Grimm V, Revilla E, Berger U et al. (2005). Pattern-oriented modeling of agent-based complex systems: lessons from ecology. Science 310:987-991.

Grimm V, Augusiak J, Focks A et al. (2014). Towards better modelling and decision support: documenting model development, testing, and analysis using TRACE. Ecol Model 280:129-139.

Haislip NA, Gray MJ, Hoverman JT et al. (2011). Development and disease: how susceptibility to an emerging pathogen changes through anuran development. PLoS One 6:e22307.

Hamed MK, Gray MJ, Miller DL (2013). First report of ranavirus in plethodontid salamanders from the Mount Roger's National Recreation Area, Virginia. Herpetological Review 44:455-457.

Hanski I (1999). Metapopulation ecology. Oxford University Press, New York.

Harp EM, Petranka JW (2006). Ranavirus in wood frogs (*Rana sylvatica*): potential sources of transmission within and between ponds. J Wildl Dis 42:307-318.

Harper EB, Rittenhouse TG, Semlitsch RD (2008). Demographic consequences of terrestrial habitat loss for pool-breeding amphibians: predicting extinction risks associated with inadequate size of buffer zones. Conserv Biol 22:1205-1215.

Hastings A (1997). Population biology: concepts and models. Springer, New York.

Helmer O (1967). Analysis of the future: the Delphi method. RAND, Santa Monica.

Hess G (1996). Disease in metapopulation models: implications for conservation. Ecology 77:1617-1632.

Hethcote HW (2000). The mathematics of infectious disease. SIAM Rev 42:599-653.

Hosmer DW, Lemeshow S, May S (2008). Applied survival analysis: regression modeling of time to event data. Wiley, Hoboken.

Hoverman JT, Gray MJ, Haislip NA et al. (2011). Phylogeny, life history, and ecology contribute to differences in amphibian susceptibility to ranaviruses. Ecohealth 8: 301-319.

Hoverman JT, Gray MJ, Miller DL et al. (2012). Widespread occurrence of ranavirus in pond-breeding amphibian populations. Ecohealth 9:36-48.

Jager KJ, Van Dijk C, Zoccali C et al. (2008). The analysis of survival data: the Kaplan-Meier method. Kidney Int 74:560-565.

Jancovich JK, Qin Q, Zhang Q-Y, Chinchar VG (2015a). Ranavirus replication: molecular, cellular, and immunological events. In: Gray MJ, Chinchar VG (eds) Ranaviruses: lethal pathogens of ectothermic vertebrates. Springer, New York.

Jancovich JK, Steckler N, Waltzek TB (2015b). Ranavirus taxonomy and phylogeny. In: Gray MJ, Chinchar VG (eds) Ranaviruses: lethal pathogens of ectothermic vertebrates. Springer, New York.

Johnson AJ, Pessier AP, Jacobson ER (2007). Experimental transmission and induction of ranaviral disease in western ornate box turtles (*Terrapene ornata ornata*) and red-eared sliders (*Trachemys scripta elegans*). Vet Pathol 44:285-297.

Keesing F, Holt RD, Ostfeld RS (2006). Effects of species diversity on disease risk. Ecol Lett 9:485-498.

Kerby JL, Hart AJ, Storfer A (2011). Combined effects of virus, pesticide, and predator cue on the larval tiger salamander (*Ambystoma tigrinum*).Ecohealth 8: 46-54.

Kolby JE, Smith KM, Berger L et al. (2014). First evidence of amphibian chytrid fungus (*Batrachochytrium dendrobatidis*) and ranavirus in Hong Kong amphibian trade.PLoS One 9:e90750.

Lélu M, Langlais M, Poulle M et al.(2013). When should a trophically and vertically transmitted parasite manipulate its intermediate host? The case of *Toxoplasma gondii*.Proc Biol Sci 280:20131143.

Lenhart S, Workman JT (2007). Optimal control applied to biological models. Chapman & Hall/ CRC, Boca Raton.

Lloyd-Smith JO, Cross PC, Briggs CJ et al. (2005). Should we expect population thresholds for wildlife disease? Trends Ecol Evol 20:511-519.

Marsh DM, Trenham PC (2001). Metapopulation dynamics and amphibian conservation.Conserv Biol 15:40-49.

Martín Andrés A, Silva Mato A, Tapia García JM et al. (2004). Comparing the asymptotic power of exact tests in 2×2 tables. Comput Stat Data Anal 47: 745-756.

McCallum H, Barlow N, Hone J (2001). How should pathogen transmission be modeled.Trends Ecol Evol 16:295-300.

Mideo N, Barclay VC, Chan BHK et al. (2008). Understanding and predicting strain-specific patterns of pathogenesis in the rodent malaria, *Plasmodium chabaudi*.Am Nat 172:E214-E238.

Mideo N, Savill NJ, Chadwick W et al. (2011). Causes of variation in malaria infection dynamics: insights from theory and data.Am Nat 178:174-188.

Miller DL, Gray MJ, Storfer A (2011). Ecopathology of ranaviruses infecting amphibians. Viruses 3:2351-2373.

Miller DL, Pessier AP, Hick P, Whittington RJ (2015). Comparative pathology of ranaviruses and diagnostic techniques. In: Gray MJ, Chinchar VG (eds) Ranaviruses: lethal pathogens of ectothermic vertebrates.Springer, New York.

Nazir J, Spengler M, Marschang RE (2012). Environmental persistence of amphibian and reptilian ranaviruses.Dis Aquat Organ 98:177-184.

OIE (2014). Aquatic Animal Health Code (online access).Office International des Epizooties, Paris. http://www.oie.int/international-standard-setting/aquaticcode/.

Accessed 21 May 2014.

Otto SP, Day T (2007). A biologist's guide to mathematical modeling in ecology and evolution. Princeton University Press, Princeton.

Price, SJ, Garner TWJ, Nichols RA, et al.(2014). Collapse of amphibian communities due to an introduced Ranavirus. Curr Biol 24:2586-2591. http://www.cell.com/current-biology/pdfEx-tended/S0960-9822(14)01149-X.

Railsback SF, Grimm V (2011). Agent-based and individual-based modeling: a practical introduction. Princeton University Press, Princeton.

Robert J, George E, De Jesús Andino F et al.(2011). Waterborne infectivity of the Ranavirus frog virus 3 in *Xenopus laevis*. Virology 417:410-417.

Rödder D, Kielgast J, Bielby J et al. (2009). Global amphibian extinction risk assessment for the panzootic chytrid fungus. Diversity 1:52-66.

Schloegel LM, Ferreira CM, James TY et al.(2010). The North American bullfrog as a reservoir for the spread of *Batrachochytrium dendrobatidis* in Brazil. Anim Conserv 14:53-61.

Smith MA, Green DM (2005). Dispersal and the metapopulation paradigm in amphibian ecology and conservation: are all amphibian populations metapopulations? Ecography 28:110-128.

Smith KF, Behrens M, Schoegel LM et al.(2009). Reducing the risk of wildlife trade. Science 324:594-595.

Smith SA, Seigel RA, Driscoll CP et al. (2014). Detecting the extent of mortality events from ranavirus in amphibians of the Northeastern U.S. http://rcngrants.org/sites/default/files/original_ proposals/RCN% 202012 (1)% 20Ranavirus%20in%20amphibians.pdf. Accessed 21 May 2014.

St-Amour V, Lesbarrères D (2007). Detecting Ranavirus in toe clips: an alternative to lethal sampling methods. Conserv Genet 8:1247-1250.

Storfer A, Alfaro ME, Ridenhour BJ et al.(2007). Phylogenetic concordance analysis shows an emerging pathogen is novel and endemic. Ecol Lett 10:1075-1083.

Sutton WB, Gray MJ, Hoverman JT et al. (2014). Trends in ranavirus prevalence among plethodontid salamanders in the Great Smoky Mountains National Park. EcoHealth DOI:10.1007/ s10393-014-0994-z.

Teacher AGF, Cunningham AA, Garner TWJ (2010). Assessing the long-term impact of *Ranavirus* infection in wild common frog populations. Anim Conserv 13: 514-522.

Thoesen JC (1994). Blue book: suggested procedures for the detection and identification of certain finfish and shellfish pathogens. Version 1. American Fisheries Society, Bethesda.

Thomson DL, Cooch EG, Conroy MJ (2009). Modeling demographic processes in marked populations, Environmental and ecological statistics. Springer, New York.

Todd-Thompson M (2010). Seasonality, variation in species prevalence, and localized disease for ranavirus in cades cove (Great Smoky Mountains National Park) amphibians. M. S., University of Tennessee. http://trace. tennessee. edu/utk_gradthes/665/. Accessed 21 May 2014.

Vose D (2000). Risk analysis—a quantitative guide, 2nd edn. Wiley, Chichester.

Waltzek TB, Miller DL, Gray MJ et al. (2014). New disease records for hatchery-reared sturgeon. I. Expansion of Frog Virus 3 into *Scaphirhynchus albus*. Diseases of Aquatic Organisms 111: 219-227.

Wells KD (2007). The ecology and behavior of amphibians. University of Chicago Press, Chicago.

Wiegand T, Jeltsch F, Hanski I et al. (2003). Using pattern-oriented modeling for revealing hidden information: a key for reconciling ecological theory and application. Oikos 100: 209-222.

Williams BK, Nichols JD, Conroy MJ (2002). Analysis and management of animal populations. Academic, San Diego.

Wilson EB (1927). Probable inference, the law of succession, and statistical inference. J Am Stat Assoc 22: 209-212.

Wobeser GA (2006). Essentials of disease in wild animals. Blackwell, Ames.

Woodhams DC, Alford RA, Briggs CJ et al. (2008). Life-history trade-offs influence disease in changing climates: strategies of an amphibian pathogen. Ecology 89: 1627-1639.

Woodhams DC, Bosch J, Briggs CJ et al. (2011). Mitigating amphibian disease: strategies to maintain wild populations and control chytridiomycosis. Front Zool 8: 8.

Zuur AF, Savaliev AA, Ieno EN (2012). Zero inflated models and generalized linear mixed models with R. Highland Statistics, Newburgh.

索 引

（本索引所标页码为英文版页码，即中译本边码）

A

Accelerated failure time（AFT）models 加速失效时间(AFT)模型 222
Adaptive immune responses 适应性免疫反应
antibody responses 抗体反应 156-157
immunological memory 免疫记忆 158
nonclassical MHC-restricted cells 非经典 MHC 限制细胞 158-159
T cell responses T 细胞反应 157
ADRV. *See* Andrias davidianus ranavirus（ADRV）ADRV,见中国大鲵蛙病毒（ADRV）
Agent-based models（ABMs）基于主体的模型 226
Ambystoma tigrinum virus（ATV）虎纹钝口螈病毒
amphibians 两栖类 25，62，64
coevolutionary history 共进化史 91
fish 鱼 62，64
genomic organization 基因组结构 108
ICTV recognization ICTV 识别 61
interclass transmission 类别间传播 43
KO and KI mutants 基因敲除和基因敲入突变体 124
phylogenetic analysis 系统发生分析 61-63
Andrias davidianus ranavirus（ADRV）中国大鲵蛙病毒
amphibians 两栖类 62，64
phylogenetic analysis 系统发生分析 61-63
Antisense morpholino oligonucleotides（asMO）反义吗啉代低聚核苷 113，115，121-123
Antiviral immune responses 抗病毒免疫反应
IFN,ectothermic vertebrates IFN,变温脊椎动物
cytosolic DNA sensors 胞浆 DNA 传感器 152
Mx proteins Mx 蛋白 153-154
RIG-I-like receptors RIG-I-样受体 152
TLRs Toll-样受体 152
type I IFN I 型 IFN 152-153
type II IFN II 型 IFN 152，153
type III IFN III 型 IFN 152，153
immune parameter 免疫参数 160
ranavirus infections 蛙病毒感染
EHNV 动物流行性造血器官坏死病毒 154
ESV 欧洲六须鲶病毒 154
IFITM1 functions 干扰素诱导跨膜蛋白 1 抗体功能 155
IPNV 传染性胰腺坏死病毒 154
LCDV 淋巴细胞囊肿病毒 154
RGV infections 蛙虹彩病毒感染 154-155
RSIV 真鲷虹彩病毒 154
SAV 鲑鱼甲病毒 154
VHSV 出血性败血症病毒 154

VNNV 神经坏死病毒 154
X. laevis type I IFN 爪蟾 I 型 IFN155-156
ATV. See Ambystoma tigrinum virus（ATV）ATV,见虎纹钝口螈病毒

B

3β-Hydroxysteroid dehydrogenase（βHSD）3β-羟基类固醇脱氢酶 122，124，127-128，162
Bohle iridovirus（BIV）博乐病毒
ICTV recognization ICTV 识别 61
interclass transmission 类别间传播 43
phylogenetic analysis 系统发生分析 61-63
reptiles 爬行动物 62，64

C

Capture-mark reencounter（CMR）model 捕获标记再捕获模型 212，222-223
Chinese giant salamander iridovirus. See Andrias davidianus ranavirus（ADRV）见中国大鲵虹彩病毒
Chi-square tests 卡方检验 217，218
Cod ranavirus（CoIV）鳕鱼蛙病毒 36
Common midwife toad virus（CMTV）普通产婆蟾病毒
amphibians 两栖类 26-27，62，64
genomic organization 基因组结构 108
phylogenetic analysis 系统发生分析 61-63
Cox Proportional Hazard（Cox PH）model Cox 比例风险模型 221
Critical control points（CCP）关键控制点 233，234
Cytopathic effect（CPE）细胞病理效应 191

D

Danish Atlantic cod 丹麦大西洋鳕鱼 36
Data analysis 数据分析
AFT models AFT 模型 222
Chi-square tests 卡方检验 217，218
CMR model CMR 模型 222-223
confidence intervals 置信区间 216-217
contingency table 列联表 217
Cox PH model Cox PH 模型 221-222
Kaplan-Meier（K-M）function K-M 函数 220-221
logistic regression 逻辑回归 218-219
Mantel-Haenszel test Mantel-Haenszel 检验 221
viral titers 病毒滴度 219-220
DNA cytosine methyltransferase（DMTase）DNA 胞嘧啶甲基转移酶 112，162-163
DNA sequence analysis DNA 序列分析 62，66
Dot plot analysis Dot plot 分析 62，64，67，108
dUTP pyrophosphatase（dUTPase）dUTP 焦磷酸酶 117，125,128，130，162
Dynamic models 评价
evaludation 动态模型评价 229-230
IBMs 基于个体建模 226-227
modeling disease intervention strategy 疾病干预策略建模 228-229
parameterize models 参数化模型 229
POM 以模式为导向的建模 227
population matrix models 种群矩阵模型 227-228
SI/SIR models SI/SIR 模型 224-226

E

Electron microscopy（EM）电镜 197
amphibians 两栖类 62，64
epidemics 流行病 79

ICTV recognization ICTV 识别 61
phylogenetic analysis 系统发生分析 61-63
polyclonal antisera 多克隆抗血清 194
typical cytopathic effect 典型细胞病理效应，191
Essential for replication and viability（ERV1）复制和存活的必需要素 125，128
European catfish virus（ECV）欧洲鲶鱼病毒
fish 鱼 29-30
ICTV recognization ICTV 识别 61
polyclonal antisera 多克隆抗血清 194

F

Frog virus 3（FV3）蛙病毒 3
epidemics 流行病 81-82
genomic organization 基因组结构 108
GIV 石斑鱼虹彩病毒 108
ICTV recognization ICTV 识别 61
infecting amphibians 传染性两栖类 23-25
infecting fish 传染性鱼类 31，35
interclass transmission 类别间传播 43
KO and KI mutants KO 和 KI 突变体 124
Molecular and cellular events 分子和细胞事件 2
phylogenetic analysis 系统发生分析 61-63
replication strategy 复制策略 1-2
translational research 翻译研究 2

G

Global Ranavirus Consortium（GRC）全球蛙病毒联盟 4
Granoff, Allan 格拉诺夫，艾伦 1
Grouper iridovirus（GIV）石斑鱼虹彩病毒
fish 鱼 27，36，62，64
genomic sequence 基因组序列 108
phylogenetic analysis 系统发生分析 61-63
whole genome dot plot analysis 全基因组 dot plot 分析 67

H

Herpes simplex virus 1（HSV1）单纯性疱疹病毒 117
Host antiviral immunity 宿主抗病毒免疫
βHSD β-羟基类固醇脱氢酶 162
DMTase,DNA 去甲基化酶 162
dUTPase 脱氧尿苷焦磷酸酶 162
RNAse III-like proteins，RNAse III-样蛋白质 161-162
vCARD 病毒半胱天冬酶激活和募集结构域 162
vIF-2α blocks phosphorylation 病毒翻译启动因子 2α 阻断磷酸化 159，161
vTNFR 病毒肿瘤坏死因子受体 162

I

Immune evasion 免疫逃避
adaptive immune responses 适应性免疫免疫反应
antibody responses 抗体反应 156-157
immunological memory 免疫记忆 158
nonclassical MHC-restricted cells 非经典 MHC 限制细胞 158-159
T cell responses T 细胞反应 157
antiviral immune responses（*see* Antiviral immune responses）抗病毒免疫反应(见抗病毒免疫反应)
innate immune responses 先天性免疫反应
antimicrobial peptide responses 抗菌

肽反应 142
anuran amphibians 无尾两栖类 143-145
ranavirus-induced pathogenicity and mortality 蛙病毒诱导的致病性和死亡 146-147
teleost fish 硬骨鱼类 145-146
urodel amphibians 有尾两栖类 145
macrophage-lineage cells 巨噬细胞系细胞
amphibian vectors 两栖类载体 148-150
FV3 infection FV3 感染 147-148
ranavirus infections 蛙病毒感染 150-152
reservoirs and ranavirus reactivation 贮藏宿主和蛙病毒的再激活 150，151
Immunohistochemistry（IHC）免疫组织化学 194-196
Import risk analysis（IRA）进口风险分析
consequence assessment 后果评价 234
definition 定义 230-231
flow diagrams 流程图 231-233
risk estimation 风险评估 234-235
risk management and communication 风险管理与通报 235
routes of introduction 引入途径 232-234
Individuals-based models（IBMs）基于个体的模型 226-227
Infectious pancreatic necrosis virus（IPNV）传染性胰腺坏死病毒 154
Innate immune responses 先天性免疫反应
antimicrobial peptide responses 抗菌肽反应 142
anuran amphibians 无尾两栖类 143-145
ranavirus-induced pathogenicity and mortality 蛙病毒诱导的致病性和死亡 146-147
teleost fish 硬骨鱼类 145-146
urodel amphibians 有尾两栖类 145
In situ hybridization（ISH）原位杂交 195，197
International Committee on Taxonomy of Viruses（ICTV）国际病毒分类委员会 27，59
Invertebrate iridovirus 1（IIV1）无脊椎动物虹彩病毒 1

K

Kaplan-Meier（K-M）function，Kaplan-Meier（K-M）函数 220-221
Knock down（KD）strategies 基因敲低策略 121-123
Knock in（KI）mutants 基因敲入突变体 124
Knock out（KO）mutants 基因敲除突变体 123-124，146，163

L

Largemouth bass virus（LMBV）大嘴鲈病毒
cutaneous mucus 皮肤黏液，85
die-offs 大规模死亡 75，79
epidemics 流行病 81
fish 鱼 27
genomic sequence 基因组序列 67
mortality 死亡率 146
water transmission 水传播 85
Lymphocystis disease virus（LCDV）淋巴细胞囊肿病毒 1，2，107，154

M

Macrophage-lineage cells 巨噬细胞系细胞
amphibian vectors 两栖类载体 148-150
FV3 infection FV3 传染 147-148
ranavirus infections 蛙病毒传染 150-152

reservoirs and ranavirus reactivation 贮藏宿主和蛙病毒再激活 150, 151
Mantel-Haenszel test 检验 220, 221
Myxovirus resistance (Mx) protein 黏病毒抗性蛋白 153-154

N

Nucleocytoplasmic large DNA viruses 核质大 DNA 病毒 (NCLDV)
Ascoviridae 囊泡病毒科 59-60
Asfarviridae 非洲猪瘟病毒科 59-60
classification 分类 59
comparative analysis 比较分析 60
genome size 基因组大小 59-60
Iridoviridae 虹彩病毒科 60
Mimiviridae 拟菌病毒科 59-60
Phycodnaviridae 藻类 DNA 病毒科 59-60
Poxviridae 痘病毒科 59-60
sequence analysis 序列分析 60-61

O

Open reading frames (ORFs) 开放阅读框 106, 108

P

Pattern-oriented modeling (POM) 以模式为导向的建模 227, 229
Pike-perch iridovirus (PPIV) 梭鲈虹彩病毒 36, 43, 62, 81, 88

Q

Quantitative real-time PCR (qPCR) 定量实时 PCR 188-189

R

Ranaviral disease 蛙病毒病
diagnostic testing 诊断试验
antigen-capture ELISA 抗原捕获 ELISA 190
bioassay/experimental transmission trial 生物测定/实验传播试验 193

Ranaviral disease (*cont.*) 蛙病毒病(续)
conventional PCR 常规 PCR 188
cytology 细胞学 194
detection antibodies 检测抗体 192-193
electron microscopy 电子显微镜 197
IHC 免疫组化 194-196
OIE reference laboratory OIE 参考实验室 180-181
qPCR 定量 PCR 188-189
ranavirus species and strains 蛙病毒种和株 189-190
sample collection protocols 采样方案 184-186
sample types and limitations 样品的类型和局限性 183, 184
in situ hybridization 原位杂交 195, 197
sterile collection techniques 无菌采样技术 184-185
with study goal 研究目标 181-183
test validation and efficiency 试验验证和效率 197-198
virus isolation 病毒分离 190-192
field and clinical findings 野外和临床发现 172-173
gross pathology 大体病理学
amphibians 两栖类 174-175
fish 鱼类 175-176
lesions 病灶 173-174
reptiles 爬行动物 177
histopathology 组织病理学 178-180, 193-194
subclinical infection 亚临床感染 179-180
treatment and vaccine development 治疗

和疫苗开发 198-199
Ranavirus（RV）蛙病毒
anthropogenic stressors 人为应激因子 83-84
design and analysis of（*see* Study design）设计和分析（见研究设计）
diagnosis and pathology（*see* Ranaviral disease）诊断与病理学（见蛙病毒病）
ectothermic vertebrate classes 变温脊椎动物类别 10，12
epidemics 流行病
detection biases 检测偏差 77
seasonal introductions and incidence 季节性引入和发病率 77-78
stage-specific susceptibility 发育阶段特异性的易感性 79，80
temperatures 温度 79-82
epidemiology 流行病学
in amphibians 在两栖类动物中 72-74
in fishes 在鱼类中 74-75
in reptiles 在爬行动物中 75-76
evolution 进化 4，91-93
global distribution 全球分布 10，11
Granoff，Allan，格拉诺夫，艾伦 1
host immunity（*see* Host antiviral immunity）宿主免疫（见宿主抗病毒免疫）
immune evasion（*see* Immune evasion）免疫逃避（见免疫逃避）
infecting amphibians 感染两栖类
ATV 虎纹钝口螈病毒 25
BIV 博乐虹彩病毒 26
CMTV 普通产婆蟾病毒 26-27
distribution cases 分布案例 13，22
ectothermic vertebrate classes 变温脊椎动物类 12，13
epizootic die-offs，流行性大规模死亡 13
FV3 蛙病毒 3 23-25
global distribution 全球分布 11，14-21
infecting fish 鱼类的传染
BIV 博乐虹彩病毒 35-36
CoIV 鳕鱼蛙病毒 36
ECV 欧洲鲶鱼病毒 29-30
EHNV 动物流行性造血器官坏死病毒 28-29
FV3 蛙病毒 3 31，35
GIV 石斑鱼虹彩病毒 32-34，36
LMBV 大嘴鲈病毒 27
PPIV 梭鲈虹彩病毒 36
SCRV 桑堤-库珀蛙病毒 30-34
SERV 短鳍鳗蛙病毒 36
SGIV 新加坡石斑鱼虹彩病毒 32-34，36
infecting reptiles 爬行动物的感染
in chelonians 在龟类 38-41
reptile cases 爬行动物案例 37-40
in squamates 在有鳞动物中 41-42
persistence of 持续存在 88-90
predators and natural stressors 捕食者和自然应激因子 83
replication（*see* Replication）复制（见复制）
risk of extinction 灭绝风险 93-94
selection and coevolution 选择和共进化 90-91
susceptibility 易感性 87-88
taxonomy（*see* Taxonomy）分类学
transmission 传播
direct transmission 直接传播 86
fomites 污染物 84

indirect transmission 间接传播 86
interclass transmission 不同类别间的传播 43-44，86
mosquito transmission 蚊虫传播 85
water-borne transmission 水源性传播 84，85
undisturbed sites 未干扰过的地点 44-46
viral countermeasures 病毒对策 4
Ranaviruses also contain homologs of Tumor Necrosis Factor （TNF） receptor （vTNFR）蛙病毒同样包括肿瘤坏死因子受体同源物 162
Red seabream iridovirus（RSIV）真鲷虹彩病毒 107，154，157
Replication 复制
antiviral immunity 抗病毒免疫 118-119
genomes 基因组
coding capacity 编码容量 107
coding regions 编码区 109
complete genomic sequence information permitted analysis 全基因组信息许可分析 107-108
features 特征 106，107
FV3 蛙病毒 3 106-107
gene products 基因产物 108，109
genetic composition 遗传组成 107
genomic organization 基因组结构 108
hosts 宿主 107
inter-and intragenic variation 基因间和基因内变异 109
microsatellites 微卫星 109
ORFs 开放阅读框 106，107
palindromes 回文 109
percentage of guanine and cytosine （G+C）鸟嘌呤与胞嘧啶百分比 106，107
phylogenetic analysis 系统发生分析 106，107
repeat and variable regions 重复区和可变区 109
size 大小 106-108
host and virus determination 宿主和病毒的确定 117-118
strategy 策略
cytoplasmic events 细胞质事件 112
nuclear events 核事件 111-112
viral entry 病毒的进入 110-111
virus assembly sites 病毒组装地点 112-115
taxonomy 分类学 106，107
viral gene function 病毒基因功能
asMOs 反义吗啉代低聚核苷 121-123
assessment of 评价 125-127
βHSD β 羟基类固醇脱氢酶 127-128
biochemical and genetic approaches 生物化学和遗传学方法 119-121
conditionally lethal mutants 条件性致死突变体 124-125
dUTPase,脱氧尿苷焦磷酸酶 128
ERV1 复制和存活必需蛋白 1 128
knock out mutants 基因敲除突变体 123-124
LITAF 脂多糖诱导的肿瘤坏死因子 129
recombinant SGIV-and I,9 Proteins，重组 SGIV 及-I,9 蛋白 130-131
RGD motif-containing proteins 含 RGD 基序的蛋白质 129-130
RGV50L 蛙虹彩病毒 50L 128-129
RNA interference RNA 干扰

121-123
vIF-2α 病毒翻译启动因子 2α 125-127
virus infection 病毒感染
apoptosis 细胞凋亡 115
host shut-off and selective expression 宿主关闭和选择性表达 116
necrosis 坏死 116
parapoptosis 拟凋亡 115-116
Restriction endonuclease fragment length polymorphism (RFLP) 限制性内切分段长度多态 62, 66
RNA interference (RNAi) RNA 干扰 121-123

S

Salmonid alpha virus (SAV) 鲑鱼甲病毒 154
Santee-Cooper ranavirus (SCRV) 桑堤-库珀蛙病毒
fish 鱼类 30-34
ICTV recognization ICTV 识别 61
Short-finned eel ranavirus (SERV) 短鳍鳗蛙病毒 36, 62,80, 88
Singapore grouper iridovirus (SGIV) 新加坡石斑鱼虹彩病毒 27, 36,61, 67, 75, 108, 116, 195, 198
Study design 研究设计
data analysis 数据分析
AFT models AFT 模型 222
Chi-square tests 卡方检验 217, 218
CMR model CMR 模型 222-223
confidence intervals 置信区间 216-217
contingency table 列联表 217
Cox PH model Cox PH 模型 221-222
Kaplan-Meier (K-M) function Kaplan-Meier (K-M)函数 220-221
logistic regression 逻辑回归 218-219
Mantel-Haenszel test Mantel-Haenszel 检验 221
viral titers 病毒滴度 219-220
dynamic models 动态模型
evaludation 评价 229-230
IBMs 基于个体的模型 226-227
modeling disease intervention strategies 疾病干预策略建模 228-229
parameterize models 参数化模型 229
POM 针对模式的建模 227
population matrix models 种群矩阵模型 227-228
SI/SIR models SI/SIR 模型 224-226
import risk analysis 进口风险分析
consequence assessment 后果评价 234
definition 定义 230-231
flow diagrams 流程图 231-233
risk estimation 风险评估 234-235
risk management and communication 风险管理与通报 235
routes of introduction 引入途径 232-234
laboratory and mesocosm studies 实验室和中试研究 214
ranavirus surveillance 蛙病毒的监测
interpreting infection data 感染数据的解释 211-212
planning 规划 213
random sampling 随机采样 213-214
required sample size 必要样本数 214-216
Susceptible-infected (SI) model 易感传播模型 224-226
Susceptible-infected-recovered (SIR) models, 易感-传染-恢复模型 224-226

T

Transmission electron microscopic (TEM) analysis 透射电子显微镜分析 114

V

Vaccinia virus（VACV）牛痘病毒 117，124，125，127

Viral nervous necrosis virus（VNNV）病毒性神经坏死病毒 154

X

Xenopus laevis type I interferon（XlIFN）爪蟾I型干扰素 119